Risk, Reliability and Safety Engineering

In this era of globalization and competitive scenario there is a conscious effort to ensure that while meeting the reliability targets the potential risk to society is minimal and meet the acceptability criteria towards achieving long term targets, including sustainability of a given technology. The objective of reliability is not only limited to customer satisfaction but also important for design, operating systems, products, and services, while complying risk metrics. Particularly when it comes to complex systems, such as, power generation systems, process systems, transport systems, space systems, large banking and financial systems, pharmaceutical systems, the risk metrics becomes an overriding factor while designing and operating engineering systems to ensure reliability not only for mission phase but also for complete life cycle of the entity to satisfy the criteria of sustainable systems.

This book series in Risk, Reliability and Safety Engineering covers topics that deal with reliability and risk in traditional sense, that is based on probabilistic notion, the science-based approaches like physics-of-failure (PoF), fracture mechanics, prognostics and health management (PHM), dynamic probabilistic risk assessment, risk-informed, risk-based, special considerations for human factor and uncertainty, common cause failure, AI based methods for design and operations, data driven or data mining approaches to the complex systems. Within the scope of the series are monographs, professional books or graduate textbooks and edited volumes on the following topics:

- Physics of Failure approach to Reliability for Electronics
- Mechanics of Failure approach to Mechanical Systems
- Fracture Risk Assessment
- Condition Monitoring
- Risk Based-In-service Inspection
- Common Cause Failure
- Risk-based audit
- Risk-informed operations management
- Reliability Cantered Maintenance
- Human and Institutional Factors in Operations
- Human Reliability
- Reliability Data Analysis
- Prognostics and Health Management
- Risk-informed approach
- Risk-based approach
- Digital System Reliability
- Power Electronics Reliability
- Artificial Intelligence in Operations and Maintenance
- Dynamic Probabilistic Risk Assessment
- Uncertainty
- Aging Assessment & Management
- Risk and Reliability standards and Codes
- Industrial Safety

Potential authors who wish to submit a book proposal should contact:
Neha Sharma, Editor, e-mail: neha.sharma.2@springernature.com

Yiliu Liu · Nicola Paltrinieri · Dimitrios Tzioutzios
Editors

Safety Measures for Hydrogen Fueling Stations

A Multi-disciplinary Risk Management Approach

Springer

Editors
Yiliu Liu
Department of Mechanical and Industrial Engineering
Norwegian University of Science and Technology (NTNU)
Trondheim, Norway

Nicola Paltrinieri
Department of Mechanical and Industrial Engineering
Norwegian University of Science and Technology (NTNU)
Trondheim, Norway

Dimitrios Tzioutzios
Department of Mechanical and Industrial Engineering
Norwegian University of Science and Technology (NTNU)
Trondheim, Norway

ISSN 2731-7811 ISSN 2731-782X (electronic)
Risk, Reliability and Safety Engineering
ISBN 978-981-96-9411-2 ISBN 978-981-96-9412-9 (eBook)
https://doi.org/10.1007/978-981-96-9412-9

This Springer imprint is published by the registered company Springer Nature Singapore Pte Ltd.
The registered company address is: 152 Beach Road, #21-01/04 Gateway East, Singapore 189721, Singapore

Preface

Hydrogen has been recognized as a cornerstone of the global transition to sustainable energy. It has the potential to reshape our energy systems by integrating renewable energy sources and supporting zero-emission technologies. Hydrogen refueling stations (HRSs) are the key infrastructures in enabling the adoption of hydrogen-powered vehicles and advancing green mobility. However, the deployment of these critical infrastructures brings forth a range of multidisciplinary challenges, including safety, reliability, public perception, economic feasibility, and environmental impact.

This book presents a comprehensive exploration of these challenges. It is motivated by an EIG CONCERT-Japan collaborative research project, *SUStainability and cost-reduction of Hydrogen stations through risk-based, multidisciplinary approaches* (SUSHy), in which researchers from Japan, Norway, Poland, Spain, and Turkey contribute their expertise. Most of the authors of this book are project partners.

The book adopts a comprehensive risk management framework, investigating the technical, organizational, societal, and regulatory dimensions of HRS operations. It starts from an introduction to the fundamental properties of hydrogen and the basic considerations of HRSs, and the subsequent chapters offer in-depth analyses of safety concerns, public risk perceptions, advanced methodologies for hazard identification, and state-of-the-art tools such as artificial intelligence for risk assessment. Special attention is also given to historical lessons from hydrogen-related incidents, human and organizational factors influencing safety, and the critical role of safety barriers. In addition, the book addresses public acceptance, economic risk management, and lifecycle environmental assessments to ensure a balanced and sustainable approach to HRS development.

The SUSHy project has demonstrated that multidisciplinary collaboration is key to overcoming the challenges associated with hydrogen infrastructure. By integrating technical innovations with a deep understanding of societal and environmental factors, this book aims to provide a roadmap for policymakers, engineers, researchers, and stakeholders working toward hydrogen-powered future renewable energy applications.

We hope this book serves as a valuable resource and inspires continued innovation in the fields of hydrogen safety and sustainable development.

Trondheim, Norway

Yiliu Liu
Nicola Paltrinieri

Acknowledgements

The works included in this book were undertaken as a part of the research project "Sustainability and Cost-Reduction of Hydrogen Stations through Risk-Based, Multidisciplinary Approaches" (SUSHy). The authors would like to acknowledge the financial support of the EIG CONCERT-Japan platform (Grant number by the Norwegian Research Council is 334340).

Contents

Editors and Contributors

About the Editors

Prof. Yiliu Liu is a professor at the Department of Mechanical and Industrial Engineering, Norwegian University of Science and Technology (NTNU). His main research interests include system reliability and resilience, safety-critical system and safety barrier analysis, and risk-informed decision-making, with the applications in the sectors of renewable energies and sustainability. Currently, he is the co-chair of Sweden-Norway joint section of IEEE Reliability Society, the leader of NTNU thematic group in natural hazard triggered technical accidents (NATECH), the editor-in-chief for *International Journal of Reliability and Safety*, and the director of international RAMS (reliability, availability, maintainability, and safety) master program at NTNU. He is a member of Norwegian Academy of Technical Science (NTVA).

Prof. Nicola Paltrinieri is professor of risk analysis at the Norwegian University of Science and Technology (NTNU), Norway, and adjunct professor in offshore HSE management at the University of Bologna, Italy, since 2016. His research focuses on both the method and the application of risk analysis within sociotechnical systems. Regarding the former, he has investigated the concepts and techniques supporting dynamic risk analysis, from uncertainty to machine learning. Regarding the latter, he has worked on risk analysis for safety-critical emerging technologies (e.g., hydrogen technologies). Prof. Paltrinieri serves as associate editor of the journal *Safety Science*. He is Norwegian delegate of the Working Party on Loss Prevention and Safety Promotion within the European Federation of Chemical Engineering. He is co-chair on Accident and Incident Modelling, European Safety and Reliability Association Technical Committee.

Dr. Dimitrios Tzioutzios is a spatial planning and development engineer with particular research interest in disaster risk reduction, Natech accidents, risk communication and participatory risk management processes. His doctoral research focused on risk communication, information disclosure and citizen engagement issues in the context of chemical accidents caused by natural hazards (Natech). Going beyond academic research, his work also involved the development of a serious game for raising community awareness about Natech risks and fostering stakeholder engagement. Currently, Dr. Tzioutzios' postdoc research focuses on disaster risk reduction, emergency safety systems, community perceptions and technology acceptance for hydrogen fuelling stations and hydrogen technologies, primarily associated with the SUSHy Project.

Contributors

Antonio Javier Nakhal Akel Department of Mechanical and Aerospace Engineering, Sapienza University of Rome, Rome, Italy;
Department of Engineering and Science, Universitas Mercatorum, Rome, Italy

Eulàlia Badia Socio-Technical Research Centre, Energy, Environmental, and Technological Research Centre (CIEMAT), Barcelona, Spain

Marta Bucelli SINTEF Energy Research AS, Trondheim, Norway

Alessandro Campari Department of Mechanical and Industrial Engineering, Norwegian University of Science and Technology (NTNU), Trondheim, Norway

Svetlana Erofeeva Department of Mechanical and Industrial Engineering, Norwegian University of Science and Technology, Trondheim, Norway

Murat Gökçek Department of Mechanical Engineering, Faculty of Engineering, Niğde Ömer Halisdemir University, Niğde, Türkiye

Ning Huan Institute of Material and Systems for Sustainability, Nagoya University, Nagoya, Japan

Raquel Santos Jorge Department of Industrial Economics and Technology Management, Norwegian University of Science and Technology (NTNU), Trondheim, Norway

Anna Kalesnikava Department of Industrial Economics and Technology Management, Norwegian University of Science and Technology (NTNU), Trondheim, Norway

Wojciech Kosman Silesian University of Technology, Gliwice, Poland

Yiliu Liu Department of Mechanical and Industrial Engineering, Norwegian University of Science and Technology (NTNU), Trondheim, Norway

Joaquín Navajas Socio-Technical Research Centre, Energy, Environmental, and Technological Research Centre (CIEMAT), Barcelona, Spain

Nicola Paltrinieri Department of Mechanical and Industrial Engineering, Norwegian University of Science and Technology (NTNU), Trondheim, Norway

Andrzej Rusin Silesian University of Technology, Gliwice, Poland

Krzysztof Rusin Silesian University of Technology, Gliwice, Poland

Roser Sala Socio-Technical Research Centre, Energy, Environmental, and Technological Research Centre (CIEMAT), Barcelona, Spain

Hitomi Sato Institutes of Innovation for Future Society, Nagoya University, Nagoya, Japan

Haiping Shen Department of Industrial Economics and Technology Management, Norwegian University of Science and Technology (NTNU), Trondheim, Norway

Katarzyna Stolecka-Antczak Silesian University of Technology, Gliwice, Poland

Abhishek Subedi Department of Mechanical and Industrial Engineering, Norwegian University of Science and Technology (NTNU), Trondheim, Norway

Farhana Yasmine Tuhi Department of Mechanical and Industrial Engineering, Norwegian University of Science and Technology (NTNU), Trondheim, Norway

Dimitrios Tzioutzios Department of Mechanical and Industrial Engineering, Norwegian University of Science and Technology (NTNU), Trondheim, Norway

Toshiyuki Yamamoto Institute of Material and Systems for Sustainability, Nagoya University, Nagoya, Japan

Chapter 1
Hydrogen Refueling Stations: New Critical Infrastructure for Utilizing Renewable Energy

Yiliu Liu and Nicola Paltrinieri

Abstract Hydrogen is a key enabler in the global transition to sustainable energy systems, offering high energy density, zero emissions, and long-term storage capabilities. Its role in achieving net-zero carbon goals is underscored by its ability to integrate with renewable energy sources like wind and solar, addressing challenges such as energy intermittency and supporting diverse applications. This chapter introduces hydrogen (re)fueling stations (HRSs) as critical infrastructure in the hydrogen economy. It highlights the physical properties of hydrogen that necessitate unique design, safety, and operational considerations and compares hydrogen and natural gas properties. Additionally, the chapter provides an overview of HRS structures and the technical components, and then it discusses the challenges of introducing and managing HRS to set the stage for the risk management strategies detailed in the subsequent chapters.

1.1 Hydrogen in the Green Transition

Hydrogen is emerging as an important solution and critical enabler in the global transition to cleaner, sustainable energy systems to achieve net-zero carbon goals, because it has the following characteristics (IEA 2023):

- Energy density: Hydrogen has a very high gravimetric energy density, meaning that it can provide a significant amount of energy per unit of weight.
- Zero emission: When hydrogen is used as a fuel, either through combustion or in fuel cells, the only product is water vapor, without any pollution or carbon emission.

Y. Liu (✉) · N. Paltrinieri
Department of Mechanical and Industrial Engineering, Norwegian University of Science and Technology (NTNU), Trondheim, Norway
e-mail: yiliu.liu@ntnu.no

Y. Liu et al. (eds.), *Safety Measures for Hydrogen Fueling Stations*, Risk, Reliability and Safety Engineering, https://doi.org/10.1007/978-981-96-9412-9_1

- Longevity: Hydrogen can be stored over a long period without significant energy loss, compared with the other state-of-the-art energy storing approaches, such as Lithium-ion batteries, which can suffer from self-discharge over time. This makes hydrogen ideal for balancing seasonal energy demands (Tian et al. 2025).

Hydrogen is the most abundant element in the universe, but it does not naturally exist in a pure form on the earth. To achieve pure hydrogen (H_2), it needs to be extracted from compounds such as water or hydrocarbons. The extracted hydrogen can then serve as an energy carrier, meaning it can store energy that can be released later for use in various applications (Guduru et al. 2025).

In green transition, hydrogen is expected to be produced through the electrolysis of water using renewable electricity, like wind and solar, so as to minimize carbon footprint. Such kind of hydrogen is called green hydrogen (Ahmad et al. 2025). Unlike the hydrogen produced from fossil fuels (gray hydrogen) or with carbon capture and storage (blue hydrogen), green hydrogen is entirely free of carbon emissions. In the green way, hydrogen production enables the development of renewable energy projects in remote areas with abundant resources, such as offshore wind farms or desert solar installations. Green hydrogen can be produced on site and transported to demand centers.

In addition, the production of green hydrogen can enable the storage and utilization of surplus renewable energy. For example, during periods of high wind or solar generation, excess electricity can be used to produce green hydrogen, otherwise it will be wasted. Therefore, green hydrogen can help address one of the biggest challenges of renewable energy: intermittency. Solar and wind power depend on weather conditions, leading to fluctuations in energy supply. By producing hydrogen through electrolysis, the excess renewable electricity can be stored as hydrogen. This hydrogen can later be converted back into electricity using fuel cells or gas turbines when renewable generation is insufficient or serve as a feedstock for ammonia and methanol production, enabling the creation of sustainable fuels and chemicals.

The integration of hydrogen and renewable energy resources thus can balance energy supply and demand and support long-term energy resilience. More importantly, hydrogen can act as a bridge between renewable electricity generation and versatile end-use applications (IEA 2023), such as transportation, building, and refining. In the transportation sector, hydrogen fuel-cell vehicles (FCEVs) can be regarded as an alternative to traditional internal combustion engines and battery electric vehicles (BEVs). FCEVs can provide extended ranges and faster refueling times, making them promising solutions for heavy-duty applications such as trucks, buses, and trains. In addition, hydrogen also can play a role in the transportation scenarios where electrification is challenging, such as maritime shipping and aviation.

In the coming decades, hydrogen is expected to drive new industries, create jobs, and enable a sustainable energy future. The widespread adoption of hydrogen requires significant investments in infrastructure, including production facilities, storage systems, pipelines, and fueling stations. The focus of this book is on risk management of renewable energy-powered hydrogen fueling/re-fueling stations (HRSs), which are the critical infrastructures for using hydrogen in the transportation sectors.

The purpose of risk management is to minimize the impacts of unexpected events and fluctuations on such facilities and ensure their reliable, safe, cost-effective, and environmentally friendly operations. In the first chapter, we will have a brief overview on the particular physical properties of hydrogen, and the basic structure and operational modes of HRSs.

1.2 Comparison of Hydrogen and Natural Gas

In the current applications and most promising technical routes, hydrogen to be used in the transportation sector exists in the format of compressed gas. Compressed gaseous hydrogen (CH_2 or CGH_2) is kept under pressure, typically at 350 bar (5,000 psi) and 700 bar (10,000 psi).

To realize effective risk management of HRSs, it is necessary to understand the similarities and differences of the physical properties of gaseous hydrogen and the other gas fuels, especially natural gas (mainly methane), which is widely used (Table 1.1).

It can be found that neither hydrogen gas nor natural gas, nor their products is toxic. Due to the above similarities, hydrogen fueling stations can share some similar equipment and layout with the traditional gas stations. On the other hand, hydrogen gas has many different properties with natural gas, such as very small molecule diameters and very wide flammability range, which need more considerations on the design, operation, and safety management of the infrastructures in the hydrogen applications. High pressure for compressing hydrogen gas also means more risks in leakage and more potential failures of containers. We will have more discussions on

Table 1.1 Comparison of physical properties of hydrogen gas and natural gas (Ustolin et al. 2020)

Property	Hydrogen gas	Natural gas (methane)
Gravimetric energy density	119.96 MJ/kg	~ 50 MJ/kg
Combustion products	Water	CO_2 and water
Toxic	No	No
Density at normal temperature and pressure (kg/m^3)	0.0883	0.717
Minimum ignition energy (mJ)	0.017	~ 0.28
Flammability range in air (%vol)	4 to 75	5 to 15
Flame visibility	Almost invisible	Visible
Color and/or odor	None	None (odor added, mercaptans for detection)
Pressure (psi)	5000 or 10,000	2900–3600
Molecule diameter (pm)	120	~ 380

the impacts of these properties in Sect. 1.4. In Chap. 6, the authors provide a more comprehensive comparison between hydrogen and the other fuel gases widely used.

1.3 Hydrogen Refueling Station as a Critical Infrastructure

HRSs are the backbone infrastructures of hydrogen economy in the transportation sector. These stations facilitate the deployment and widespread adoption of hydrogen-powered vehicles by providing accessible refueling options (Fragiacomo et al. 2024).

For an HRS where hydrogen is produced in external plants and delivered to the station by tube trailers or pipelines, it can be called an off-site HRS (Piraino et al. 2024). Such a station at least comprises the following several elements:

- Hydrogen receiving system: This refers to the equipment that receives delivery of compressed gaseous hydrogen from external sources.
- Compression system: Compressors increase the pressure of hydrogen gas to the levels required for storage and dispensing, typically 350 or 700 bar for vehicles.
- Storage system: High-pressure tanks or cryogenic containers store compressed hydrogen gas, ready for dispensing.
- Dispensing unit: This is equipped with precision nozzles and metering systems. Dispensers are expected to provide accurate and safe hydrogen transfer to vehicles. Two dispensing methods can be applied: Cascade process where the fueling of hydrogen is based on the pressure difference between the station tank and the vehicle tank without using a compressor, and direct process, where a compressor is used to enable the fueling process.
- Safety system: This includes diverse devices and facilities, such as various sensors, ventilation, and fire suppression systems. They are integral to an HRS for ensuring equipment reliability and the human, asset, and environmental safety related to the station.

For HRSs equipped with on-site hydrogen production functionality, renewable power generation and water electrolysis are also essential elements. Such kinds of stations can be regarded as on-site HRSs (Piraino et al. 2024). If we consider the renewable energy source, on-site HRSs can be further classified as:

- Wind-powered HRS: Wind-powered electrolyzers convert electricity generated by wind turbines into green hydrogen. This approach is particularly advantageous in regions with abundant wind resources, as it enhances the utilization of locally available energy. In such an HRS, wind turbines, electric cables, and water electrolyzers are also included as technical components.
- Solar-powered HRSs: Solar-powered HRSs utilize photovoltaic panels to transfer solar energy for water electrolysis. Solar integration offers the benefit of daytime hydrogen production, often coinciding with peak energy demand. Solar panels are the particular technical components in such kinds of HRSs.

- Hybrid HRSs: Hybrid stations combine wind, solar, and even energy storage technologies such as batteries maximizing the efficiency and reliability of hydrogen fueling stations. These systems can operate continuously, compensating for the intermittency of individual renewable sources. However, these stations include more components, which may have more reliability and risk issues.

1.4 Challenges of HRSs and Concerns

Hydrogen, especially in its compressed gaseous form, presents unique operation- and safety challenges due to its properties, as well as broader societal and regulatory concerns. Here, we outline some of challenges and concerns related to HRSs (Stolecka-Antczak et al., 2024). Challenges due to physical and chemical properties include:

- Very small molecule diameters: Hydrogen has the smallest molecule, meaning that it can easily permeate through seals, joints, and even some materials that are considered gas-tight for other substances. For example, hydrogen embrittlement can occur when hydrogen diffuses into metal materials like steel, reducing their strength and useful lifetime. This characteristic increases the likelihood of leaks, and thus increases the failure frequency of the containing and sealing technologies in HRSs.
- Very wide flammability range: Hydrogen has a flammability range of 4–75% by volume in air, which is significantly broader than other common fuels like methane or gasoline vapors. This wide range increases the risk of accidental ignition in the event of a leak, even at relatively low concentrations.
- Very light density: Because of this, hydrogen is highly buoyant, meaning that leaks of hydrogen gas tend to rise and disperse quickly. While this property can reduce the risk of ground-level explosions, it also creates challenges in detecting and managing leaks in enclosed spaces, such as the roofs of fueling stations or inside vehicles.
- Very low minimum ignition energy: Hydrogen requires only 0.017 mJ of energy to ignite, which is a very low threshold that even static electricity can be a source of ignition. Thus, it is highly necessary to prevent static discharge during the hydrogen fueling operations with grounding and specialized safety equipment.

At the same time, the technological, economic, and societal factors cannot be skipped:

- Novelty of applications: Although hydrogen production and usage have been common in chemical industries for decades, adapting this technology for widespread use in transportation introduces significant challenges. Since the application is relatively new, the existing data and methods supporting reliability and safety analysis are still lacking. For example, the research on failure analysis and degradation modeling of electrolyzers is very limited (Tuhi et al. 2024).

- Cost-effectiveness: The economic viability of hydrogen as a fuel remains a concern. The high costs of green hydrogen production, transportation, and infrastructure development can make it less competitive compared to fossil fuels.
- Public acceptance: Public acceptance of hydrogen as a safe and reliable fuel is another critical issue (Ning Huan et al. 2024a, b, c). Accidents, such as the explosions of hydrogen refueling stations in Norway in 2019 (International 2019) and in South Korea in 2024 (Collins 2025), and the depot fire accident in French city of Belfort where seven hydrogen fuel-cell buses were destroyed (Martin 2025), have leveraged public fears regarding hydrogen safety. These incidents underscore the need for transparent communication and rigorous safety protocols to build trust.
- Regulatory and standards gaps: As hydrogen technology is still evolving, existing safety regulations and standards may not fully address the specific risks associated with HRSs. Different countries and regions have varying safety standards, which complicates the development of international hydrogen supply chains. Many safety codes, such as those for emergency response or refueling station design, are still being developed or updated to accommodate hydrogen's unique properties (Pique et al. 2017).

These factors also can influence the risk management of producing and utilizing hydrogen. We will discuss these topics in the following chapters.

1.5 Scope of the Book

HRSs, especially those powered by renewable energy, represent a cornerstone of the green transition and a critical infrastructure for the sustainable development of the transportation sector. However, the deployment and operation of HRSs come with unique challenges, particularly concerning safety, reliability, public perception, and environmental impact.

This book provides a comprehensive examination of these challenges and explores solutions through the lens of risk management, incorporating state-of-the-art methodologies for risk identification, analysis, and mitigation. It aims to offer a holistic understanding of the sociotechnical, economic, and environmental dimensions of HRSs, making it an essential resource for researchers, policymakers, and industry stakeholders.

The chapters are structured to guide readers through a systematic exploration of the key aspects of HRS safety and risk management:

Chapter 2 provides a comprehensive review of relevant studies published in scientific journals addressing the safety of hydrogen refueling stations. It includes an analysis of the operational experiences reported globally. Chapter 3 examines public risk perceptions regarding HRSs in Japan, Spain, and Norway. It explores how awareness of accidents influences individual risk assessments, amplifying concerns about both likelihood and severity of potential accidents. These two chapters lay the groundwork by reviewing the safety concerns surrounding HRSs and analyzing public perceptions of hydrogen technologies.

Chapter 4 outlines a general framework for risk management specific to HRSs. It introduces the core principles and methodologies necessary for identifying, analyzing, and mitigating risks in this emerging field, providing a structured approach to enhancing safety and reliability. Then, Chaps. 5–14 discuss the detailed works needed to be conducted under such a framework. More specifically, here is an overview of the contents of the following chapters:

Chapter 5 reviews historical data on accidents and failures in hydrogen production and fuel-cell applications. By analyzing causes, consequences, and lessons learned from 18 incidents in the Hydrogen Incident and Accident Database (HIAD) 2.1, the chapter identifies common failure modes and proposes preventive measures to improve safety practices.

Chapter 6 presents methodologies for hazard identification and consequence analysis in HRSs. It evaluates the severity of hydrogen leakages and their potential to cause fires or explosions. This chapter provides a foundation for subsequent risk assessments by offering detailed models for understanding hazardous events.

Chapter 7 investigates risk identification through the integration of artificial intelligence techniques with traditional methods. It introduces Risk-Based Inspection Strategy for Hydrogen Technologies (HyRIS), a novel AI-driven framework for addressing hydrogen-induced degradation mechanisms, enabling precise risk assessments and tailored inspection strategies.

Chapter 8 illustrates the human and organizational factors influencing safety in HRSs. This chapter emphasizes the role of human behavior, organizational practices, and their interaction with technological systems in fostering a culture of safety and preventing incidents.

Chapter 9 discusses commonly used methods for qualitative and quantitative risk analysis. It introduces approaches for assessing risks based on probabilities and severities, as well as techniques for optimizing HRS layouts to minimize hazards. The chapter provides tools for identifying and mitigating risks across different scenarios.

Chapter 10 addresses safety barrier management in HRSs. Using Acoustic Emission Testing (AET) for composite overwrapped pressure vessels as a case study, it highlights the integration of human reliability analysis and Bayesian Network modeling to enhance the effectiveness of safety barriers. The chapter also emphasizes the role of digital technologies in improving predictive insights and defect detection.

Chapter 11 further examines technical safety barriers by introducing methods for evaluating their performance, reliability, and integrity. Through case studies on pressure relief valve systems, the chapter explores failure modes and provides frameworks for ensuring barrier availability and effectiveness.

Chapter 12 explores emergency preparedness strategies for hydrogen system accidents, focusing on HRSs. It synthesizes research findings and offers practical recommendations for enhancing preparedness through a capabilities-based approach. The chapter highlights the importance of continuous improvement, collaborative planning, and advanced technologies in managing emergency responses effectively.

Chapter 13 examines the economic feasibility and cost-effectiveness of HRSs. Through case studies on renewable energy-powered HRSs, it evaluates hydrogen production costs, retail prices, and infrastructure challenges. The chapter provides insights for policymakers and investors to optimize project costs and promote hydrogen as a cornerstone of sustainable transportation.

Chapter 14 introduces life cycle assessment (LCA) to evaluate the environmental impacts of green hydrogen deployment in the transportation sector. By analyzing scenarios involving green, blue, and baseline hydrogen, the chapter demonstrates the climate mitigation potential of green hydrogen and proposes a future framework for comprehensive environmental risk assessments.

References

Ahmad FF, Rejeb O, Ghenai C (2025) Green hydrogen production and solar to hydrogen ratio using innovative and integrated bifacial solar photovoltaics and cool roof technologies. Int J Hydrogen Energy

Collins L (2025) Safety concerns grow after two explosions at South Korean hydrogen filling stations in same week. Hydrogen Insight

Fragiacomo P, Genovese M, Piraino F, Massari F, Boroomandnia M (2024) Analysis of a distributed green hydrogen infrastructure designed to support the sustainable mobility of a heavy-duty fleet. Int J Hydrogen Energy 51:576–594

Guduru R, Singh R, Patel R, Vij RK (2025) Hydrogen as an energy carrier, chap 2. In: Bera A, Kumar S (eds) Subsurface hydrogen energy storage, pp 31–61

Huan N, Yamamoto T, Sato H, Sala R, Goncalves L (2024a) Perceptions to connections: A multidimensional investigation of hydrogen acceptance. Renew Sustain Energy Rev 200:114612

Huan N, Yamamoto T, Sato H, Sala R, Tzioutzios D, Goncalves L, Paltrinieri N (2024b) Cross-national examination of facilitators and hindrances to the public acceptance of hydrogen refuelling facilities. Sustain Cities Soc 113:105653

Huan N, Yamamoto T, Sato H, Tzioutzios D, Yin H, Sala R (2024c) Does accident awareness affect people's risk perception of hydrogen infrastructure and information-seeking behaviour? Appl Energy 364:123141

IEA (2023) Global hydrogen review 2023. Int Energy Agency

International H (2019) Fire at an H2 station in Norway. H2 International

Martin P (2025) Seven hydrogen buses destroyed or damaged in French depot fire. Hydrogen Insight

Pique S, Weinberger B, De-Dianous V, Debray B (2017) Comparative study of regulations, codes and standards and practices on hydrogen fuelling stations. Int J Hydrogen Energy 42(11):7429–7439

Piraino F, Ustolin F, Campari A, Venetsanos A, Paltrinieri N, Fragiacomo P (2024) Performance analysis and safety aspects of hydrogen refueling infrastructures. Chem Eng Trans 111:403–408

Stolecka-Antczak K, Rusin A, Kosman W, Rusin K, Paltrinieri N (2024) Hazards caused by hydrogen explosion at refueling station due to uncontrolled gas release. In: Kolowrocki D (ed) Advances in reliability, safety and security, part 1. Polish Safety and Reliability Association, pp 231–239

Tian T, Ma Z, Pan T, Wu C, Shu J, Feng R, Ding J (2025) Storage dimensioning and energy management for a grid-connected wind/PV/hydrogen/battery system using MISOCP approach. Int J Hydrogen Energy 100:779–791

Tuhi FY, Bucelli M, Liu Y (2024) Technical failures in green hydrogen production and reliability engineering responses: insights from database analysis and a literature review. Int J Hydrogen Energy 94:608–625

Ustolin F, Paltrinieri N, Berto F (2020) Loss of integrity of hydrogen technologies: a critical review. Int J Hydrogen Energy 45(43):23809–23840

Chapter 2
Safety Concerns on Hydrogen Refueling Stations: A Sociotechnical Panoramic View from Literature and Practice

Joaquín Navajas, Roser Sala, and Eulàlia Badia

Abstract This chapter provides an overview of relevant studies published in scientific journals that address the safety of hydrogen refueling stations as well as an analysis of the operating experience. These studies, presenting multiple perspectives, levels, and approaches, point out the necessity of covering the entire sociotechnical spectrum to ensure the safe operation of these facilities, which are crucial for the deployment of hydrogen vehicles. Importantly, the sociotechnical approach understands safety as a factor resulting from the interaction of technical systems (such as equipment, systems, and processes), human factors (including operators, organizational design, and safety culture), and social factors (such as regulatory and legal considerations).

2.1 Scientific Approaches to the Study of HRS Safety

Hydrogen refueling stations (HRSs) are essential for the development of hydrogen fuel cell vehicles. The expansion of these stations and the uptake of hydrogen vehicles are interlinked: the growth of one drives the growth of the other. This progress depends on several factors, including technological advances, political support, international cooperation, and national incentives (Hassan et al. 2024). A robust national strategy is key to their success, as demonstrated by several pioneering countries (Salimi et al. 2022).

Safety is fundamental to the reliable integration of these installations into everyday life (Landucci et al. 2010; Guo et al. 2024). From this perspective, the following chapter provides an overview of relevant studies published in scientific journals that address the safety of HRS as well as an analysis of the operating experience. These studies comprise a range of topics, including the review of standards and regulations,

J. Navajas (✉) · R. Sala · E. Badia
Sociotechnical Research Centre, Energy, Environmental, and Technological Research Centre (CIEMAT), Barcelona, Spain
e-mail: joaquin.navajas@ciemat.es

Y. Liu et al. (eds.), *Safety Measures for Hydrogen Fueling Stations*, Risk, Reliability and Safety Engineering, https://doi.org/10.1007/978-981-96-9412-9_2

the development of methodologies and techniques for an accurate risk assessment, and the identification of risk scenarios.

2.1.1 Review of Standards and Regulations

Studies based on regulatory reviews significantly contribute to clarifying the regulatory framework available and potentially affecting the entire life cycle of hydrogen facilities, from design and installation to operation and maintenance. In addition, a regulatory assessment facilitates the interoperability of refueling stations, standardizing processes and promotes the development and mobility of fuel cell vehicles (HFCV) internationally. Below are two studies which explore the diversity of applicable standards, identify areas for improvement, and highlight the importance of a robust regulatory framework to develop HRSs.

The study by Salehi et al. (2022) provides a comprehensive review of regulations and standards affecting the economic viability and safety of hydrogen, highlighting the need for regulatory reforms and technological advances needed to overcome barriers within the Australian national context. The review points out the importance of integrating the policy and regulatory dimension into the development of hydrogen technologies. The main study findings are presented in Table 2.1.

A second study by Genovese et al. (2023) reviews norms and standards applicable to HRS construction and maintenance. This review covers both international (ISO, IEC, SAE) and European (CEN/CENELEC) regulations, specifically the Italian ones. The authors highlight the need for further harmonization and regulatory development, with clearer, more consistent regulations on HRS safety and efficiency. They also draw attention to the heterogeneous scope of the different international standards applicable to HRS: (a) ISO standards, which cover HRS operation components, with a focus on quality, safety, manufacturing, and testing; (b) SAE technical guidelines,

Table 2.1 Hydrogen regulation challenges and opportunities according to Salehi et al. (2022)

Challenges/ Opportunities	Description
Relevance of international standards	International standards (ISO, IEC, and SAE, among others) contribute to standardizing hydrogen systems, including infrastructures such as hydrogen refueling stations and fuel cell vehicles. Regulatory focus on safety aspects
Need for a collaborative approach	Collaboration between governments, industries, and safety associations is paramount to develop coherent, consistent regulatory frameworks. Good practices, such as the creation of digital platforms and data-driven tools, are suggested to streamline safety procedures
Proposed regulatory reforms	Need for regulatory reform, especially within the Australian context (subject of this study), to streamline approval processes, reduce costs, and encourage the development of hydrogen infrastructures, such as HRS

Table 2.2 Key elements for a comprehensive HRS regulation according to Genovese et al. (2023)

Relevant regulatory items	Description
Differentiated scope of standards	ISO standards focused on quality and safety; SAE guidelines aimed at hydrogen refueling processes; IEC standards specifically aimed at fuel cells
Station design and operation	Station layout varies depending on the hydrogen supply method, either by tube trailers, by pipeline or by on-site production (methane reforming or electrolysis)
Storage capabilities	HRS-focused standard with daily capacity between 100 and 520 kg of compressed hydrogen (GH_2) and more than 1000 kg of liquid hydrogen (LH2)

which address hydrogen refueling procedures and associated equipment; and (c) IEC standards, specific to fuel cells and their use in electric power systems, although the latter have limited application when it comes to HRS. Table 2.2 shows the relevant regulatory development items based on the review carried out.

2.1.2 *Identification of Risk Scenarios*

The main purpose of the HAZOP methodology, an acronym for Hazard and Operability Study, developed by the chemical industry in the 1960s, is to identify and evaluate risks in industrial processes to prevent accidents and improve safety in chemical plants (Crawley and Tyler 2003). This technique is based on the premise that accidents usually occur due to process variables deviating from their normal operational parameters. HAZOP is a methodology which breaks down a process into micro-processes to identify possible deviations and their consequences. In addition, it is usually undertaken by a multidisciplinary team which evaluates each process step and proposes safety measures. Although it is useful for discovering hidden risks, it has limitations such as subjectivity and high resource consumption (Kletz 2006). Two HAZOP studies applied to a couple of HRS types are presented below.

The first study, by Zhiyong et al. (2010), conducts a HAZOP analysis with the main purpose to establish safety distances by evaluating the consequences of worst-case accident scenarios such as explosions, jet fires, and confined vapor cloud explosions. The study concluded that explosions and confined vapor cloud explosions represent the greatest risks in flash and sustained release scenarios. The analysis also revealed that in the case of flash releases, harmful effect distances increase with the amount of hydrogen released, with nearly no influence of pressure. In the case of sustained releases, the opposite is true: pressure is the key factor, while the amount released has less influence. It was also found that raising the hydrogen equipment and reducing the size of vessels and pipelines would be effective measures to reduce the risk by shortening impact distances. Finally, the study also highlighted the relevance of wind speed in amplifying harmful effect distances, except for jet fire scenarios.

Table 2.3 Focus and findings of the studies by Zhiyong et al. (2010) and Suzuki et al. (2021)

Study	Study focus	Relevant findings
Zhiyong et al. (2010)	Safety distances	Influence of release type, hydrogen volume and pressure
Suzuki et al. (2021)	Explosion risk due to static charge inside MCH tanks	Pre-release ignition, need for electrostatic discharge systems, and monitoring

In the second study, Suzuki et al. (2021) used the HAZOP methodology to identify risk scenarios at stations using methylcyclohexane (MCH) as a hydrogen carrier. These stations have particularities distinguishing them from compressed hydrogen stations, such as the handling of flammable liquids and the management of decomposition by-products, thus introducing specific risks. The study identified 371 scenarios, classifying 105 of them into two groups: (i) those where substances ignite after release to the environment, and (ii) those where ignition occurs during the process, prior to release. The newest findings refer to scenarios specifically derived from MCH (ii), where a risk of explosion of a gaseous toluene and oxygen mixture was identified due to the accumulation of static charge inside the tank. In these scenarios, ignition occurs prior to release, highlighting the risk of explosion of toluene-oxygen mixtures in storage tanks during loading and unloading processes. As safety measures, the authors propose the installation of electrostatic discharge systems to prevent the ignition of explosive mixtures, as well as advanced monitoring systems to detect leaks and pressure increases in MCH and toluene. They also suggest the use of specialized extinguishing systems to control puddle fires and hydrogen jet fires. Table 2.3 summarizes the findings of both studies.

2.1.3 *Quantitative Risk Assessments*

Quantitative Risk Assessments (QRA) aim to numerically quantify the probability and severity of risks previously identified by other techniques, such as HAZOP studies. QRA use mathematical models and statistical data to accurately calculate the risk associated with each scenario. One of the QRA advantages is the possibility to compare risks and prioritize mitigation measures, thus facilitating data-driven decision-making. In addition, they allow the analysis of both expected frequency and potential consequences of any containment event hydrogen loss (Hansen 2020). However, QRA require a large amount of information, which is not always available, and their accuracy largely depends on the quality of the models and data used.

In the context of HRS, QRA have been applied to a wide variety of scenarios, from calculating leak and explosion risks, to assessing the probability of specialized equipment failures, as well as in studies addressing the probability and severity of human error (Rusin et al. 2024). The following is an overview of some relevant QRA-based studies.

Sun et al. (2014), in their analysis of risks associated with hydrogen mobile stations, separately assessed risks to station personnel and the general public. Results revealed that risks to personnel and clients were significantly below acceptable levels according to regulations. On the other hand, risks to third parties were considered acceptable, provided adequate mitigation measures were implemented. The study also identified booster compressor leaks as the main risk factor. Regarding hydrogen supply by road, it was concluded that risks remained within acceptable levels as long as safety measures such as proper route planning and restrictions on transport times were adopted.

In another study, Pan et al. (2016) conducted a QRA on the first renewable hydrogen refueling station in China, concluding that compressors represented the main source of risk, and that flame length in case of fire was a critical factor to establish safety distances. It was recommended to cool tanks with water and to establish minimum distances of 6 m from the compressors. For extreme scenarios, a safety perimeter of up to 50 m was suggested to protect the public, based on the maximum flame length.

Some studies have focused on hydrogen storage. Casamirra et al. (2009) conducted a safety assessment at a refueling station using techniques such as HAZOP among others, with the aim of identifying high-pressure storage equipment failures. Although components with high failure potential were identified, the frequency of critical events such as overpressure in storage vessels was low. Periodic testing of vent lines and frequent compressor maintenance were recommended as preventive measures.

Moradi and Groth (2019) also focused on hydrogen storage, identifying hydrogen overflow and thermal effects as major challenges. In addition, they noted the need to manage failures of thermal Pressure Relief Device (PRD) considered critical to avoid catastrophic tank ruptures.

In the context of hydrogen storage for fuel cell vehicles, Dadashzadeh et al. (2018) conducted a QRA to evaluate the loss-of-life risk in the event of a fire. Their risk distance calculations considered both the blast wave and fireball resulting from a catastrophic tank rupture, with results showing that increased tank fire resistance significantly reduces the risk. The importance of thermal protection in high-pressure tanks to minimize risk was highlighted.

Tsunemi et al. (2019) focused their research on hydrogen leaks, analyzing three HRS scenarios: in external piping near dispensers, in high-pressure accumulator connection piping, and in compressor piping. Using an Event Tree Analysis (ETA), they concluded the leak risk near dispensers was less than 10^{-6} per year at a distance of 6 m. In areas near accumulators and compressors, the risk exceeded 10^{-4} per year within 10 m of the ignition point. The installation of 3 m barriers and a 6 m separation proved effective in keeping risks within acceptable levels.

Other studies have addressed risks of a more complex nature, including those related to human intervention and organizational factors. Zhang et al. (2024a, b) integrated two techniques, System Theoretic Process Analysis (STPA) and Decision-Making Trial and Evaluation Laboratory (DEMATEL), into a QRA. These techniques are typically used independently to analyze complex systems with multiple variables and interactions. This integrated approach allowed them to systematize human tasks and analyze interactions among HRS components. They identified and classified 24 unsafe control actions and provided targeted recommendations to address critical issues.

As for Xing et al. (2022), they analyzed risks in all phases of hydrogen supply, integrating human behavior into their assessment and using the PHARA model to prioritize the consequences of possible accidents and establish safety distances in case of fires or explosions.

Finally, Zhang et al. (2024b) evaluated HRS safety from a resilience perspective. Their approach included expert visits to ten stations in China, the creation of an evaluation index system, and the use of techniques to weight resilience indicators. Results underscored the importance of monitoring, governmental response capacity, and equipment vulnerability.

It is important to note that conducting a QRA is essential to the process of obtaining permits to open a refueling station (West et al. 2022). Thus, these assessments make it possible to justify HRS safety, even in situations where certain parameters do not comply with current regulations. In this regard, LaFleur et al. (2017) proposed for the design of outdoor hydrogen refueling stations, a performance-based approach which evaluates safety based on the ability to prevent and mitigate risks, rather than focusing solely on regulatory compliance.

Table 2.4 shows the diversity of approaches and findings from QRA-based studies.

2.1.4 Studies Based on Numerical Simulations

Numerical simulations applied to the safety of HRS allow detailed modeling of physical processes and phenomena, such as fluid flow and heat transfer. These simulations provide accurate results on system behavior under specific conditions, for instance, it can address the technical feasibility and economic viability of HRS powered by renewable energy sources (Gökçek et al. 2024). These kinds of studies are closely related to QRA. While QRA serves as a general framework for risk assessment, numerical simulation facilitates system modeling and evaluation of different scenarios. Below, several studies based on mathematical models and numerical simulations applied to HRS safety and efficiency are presented.

Al-Shanini et al. (2014) evaluated HRS safety in a land transportation network by accident modeling, integrating preventive barriers related to human factors and organizational failures. They used the "System Hazard Identification, Prediction, and Prevention methodology" (SHIPP), which combines several techniques to identify hazards, with results showing that emergency management and preventive barriers, especially those related to leak prevention and detection, significantly increased the probability of keeping a safe operational station condition. The study concluded that the SHIPP methodology is useful for predicting risks and facilitating risk-based decision-making.

Table 2.4 QRA studies in HRS. Diversity of approaches and results

QRA approach	Main findings
Risk assessment for staff and the public in a mobile HRS (Sun et al. 2014)	Risks significantly below acceptable levels for workers and the public Booster compressor leaks as main risk factor Road risks acceptable with proper planning
HRS renewable hydrogen (Pan et al. 2016)	Compressors as main source of risk Tank cooling, and safety and perimeter distances for the public
Focus on high-pressure storage equipment (Casamirra et al. 2009)	Identification of components with high failure potential Periodic testing of vent lines and frequent compressor maintenance
Hydrogen storage (Moradi and Groth 2019)	Hydrogen overflow and thermal effects Need to manage potential PRD failures
Human losses in case of fire (Dadashzadeh et al. 2018)	Calculation of risk distances Importance of increasing tank fire resistance
Hydrogen leaks in pipelines (Tsunemi et al. 2019)	Leakage scenario analysis, risk calculation, and safety distances Installation of barriers
Human and organizational factor (Zhang et al. 2024a, b)	Identification of 24 unsafe control actions
Complete supply process (Xing et al. 2022)	Integration of human behavior, calculation of safety distances
Responsiveness and vulnerability of 10 HRS (Zhang et al. 2024b)	Creation of resilience indicators
Performance-based design (LaFleur et al. 2017)	Flexibility in regulatory compliance, use of QRA to justify unconventional designs

In the design domain, Serdaroglu et al. (2015) developed empirical equations to model the impact of environmental conditions on hydrogen station operation. Their simulations made it possible to estimate the influence of factors such as average annual surface temperature, location, size, and surface color of storage tanks, concluding that careful design and implementation of operational strategies could optimize HRS efficiency.

Regarding the risk of damage caused by jet fires in critical equipment and fire spread, Tian et al. (2024) determined through numerical simulations that the firewall should be at least 2 m higher than the leak point to prevent fire from spreading. They also established that the relationship between safe distance behind the firewall and the hydrogen storage pressure was linear, providing practical guidance for installation of safety measures in HRS.

Li et al. (2023) focused their research on hydrogen dispersion in mobile stations. Using simulations and experimental data, they analyzed a number of ventilation scenarios and environmental conditions, concluding that the opening of doors and windows, together with low wind speeds, increases the risk of leakage. In severe cases, they calculated that the average airborne hydrogen volume was 87.9 m^3, while the minimum safe distance in case of an explosion was 313.2 m.

Searle et al. (2024) proposed a flexible methodological approach combining three techniques: analysis of potential hydrogen demand, optimal infrastructure design, and agent-based simulation to study system operation. This model was applied to the planning of a strategic hydrogen refueling network for heavy-duty vehicles in northern England and showed that system optimization depends essentially on the likely number of daily deliveries of hydrogen trailers.

Another innovative approach was that of Yang et al. (2024), who used a hybrid model based on deep learning (WD-KNN-CNN) and Computational Fluid Dynamics (CFD) to predict the location and intensity of hydrogen leaks. Although the model focused on incident response, the fact that it was based on training data not including actual leaks caused it to have limitations as it could predict only one leak source.

For fire risk at urban stations, Shi et al. (2020) developed a procedure integrating computational methods with statistical techniques to reduce uncertainty and speed up calculations. The study concluded that this technique can reduce uncertainty by 96% and computational costs by 97%, which is a significant risk assessment improvement.

Some simulations introduced user behavior as the focus of computation. For example, Isaac and Saha (2021) developed a mathematical simulation to understand hydrogen vehicle refueling patterns in South Africa. In essence, the study designed an algorithm which, using a theoretical model (Markov chain), attempted to represent and analyze the refueling process. The study found that a person spent approximately 3.8 min at a refueling station during weekdays, spending about 2 min refueling, a time reduced to 1.3 min with a higher refueling rate. Daily and weekly refueling patterns showed temporary peaks between 11 a.m. and 4 p.m. on weekdays, and between 10 a.m. and 9 p.m. on weekends.

The choice of software used for simulations also influences the reliability of results. Kwon et al. (2022) compared two specific programs (Hy-KoRAM and Phast/Safeti) and found notable differences in terms of hydrogen leakage risk quantification. Hy-KoRAM tended to give more conservative estimates, while Phast/Safeti offered more accurate assessments, based on specific parameters and actual operating conditions.

Finally, the irruption of machine learning processes allowed the development of hybrid studies combining elements of physical modeling and machine learning, with the aim of optimizing HRS operation (Wang et al. 2022). In this study, a machine learning model was used to predict key refueling process parameters. The final process temperature and pressure (below 85 °C and 70 MPa, respectively, in the vehicle tank), as well as the vehicle State-Of-Charge (SOC) performance, which should be between 95 and 100%, were identified as critical parameters. Results suggested that the robust regression model outperformed other approaches by providing effective real-time predictions.

Table 2.5 shows the diverse approaches and findings of numerical simulation-based studies.

Table 2.5 Studies based on numerical simulations

Simulation purpose	Main findings
Ground transportation network, including human and organizational factors (Al-Shanini et al. 2014)	Leak prevention and detection, as well as emergency management, significantly impacted safe HRS operation
Impact of annual storage tank temperature (Serdaroglu et al. 2015)	Location, size, surface color, and shade in storage tanks
Firewall design and fire intervention strategies (Tian et al. 2024)	Optimal firewall height, linear relationship between safe distance and storage pressure
Hydrogen dispersion behavior in mobile stations (Li et al. 2023)	Impact of ventilation and environmental conditions on explosion risk
Optimal refueling infrastructure design (Searle et al. 2024)	Importance of hydrogen delivery frequency on system efficiency
Hydrogen leak prediction (Yang et al. 2024)	Hybrid model to predict leak location and intensity
Fire risk assessment at urban stations (Shi et al. 2020)	Reducing uncertainty and computational costs by integrating simulation with statistical techniques
Identification of user refueling patterns (Isaac and Saha 2021)	Establishment of average dwell times, distinguishing between weekday and weekend time peaks
Comparison of simulation software (Kwon et al. 2022)	Software discrepancies in risk quantification (Hy-KoRAM and Phast/Safeti)
Prediction of key refueling process parameters (Wang et al. 2022)	Identification of critical process temperature and final pressure parameters

Diversity of approaches and results

2.1.5 *Reviews of Published Scientific Literature*

Literature reviews provide a contrasting overview of validated HRS safety knowledge and allow the identification of gaps and areas requiring further research. Below, some results of studies focusing on the review of existing literature are presented. These studies underline the importance of adopting a holistic approach to conceiving HRS safety, incorporating technical, organizational, and human aspects, as well as revealing the scientific and collaborative networks around hydrogen research.

Apostolou and Xydis (2019), through a comprehensive review of scientific literature, identify three essential technical components for HRS safety and efficiency. Table 2.6 shows the main results obtained in their review.

The review also identified differentiated risks according to the two basic HRS types (with or without hydrogen (H_2) production):

- H_2 on site: These stations must primarily manage risks related to the delivery and handling of large hydrogen volumes. Road transport methods, using trucks or pipelines, require specific certifications and the use of highly safe equipment due to the volatile properties of hydrogen.

Table 2.6 Safety components arising from the literature review by Apostolou and Xydis (2019)

Technical components	Safety impact
Pressure and temperature control equipment	The studies published all agree on the importance of controlling the refueling process, which must be regulated to ensure it does not exceed 700 bar Hydrogen gas temperature should be reduced to – 40 °C to prevent vehicle tanks from overheating during fast refueling (maximum 85 °C) With these parameters, the reviewed literature identifies compressors and high-pressure storage tanks as critical equipment
Hydrogen purity factor	Hydrogen must meet strict purity standards (greater than 99.97%). The presence of impurities can damage fuel cell systems. To ensure fuel quality, HRS must feature advanced purification systems, which are critical infrastructure components
Integrated safety systems	HRS must be equipped with leak detection sensors, pressure relief valves, and waterless fire suppression systems. These elements ensure that, in case of emergency, control mechanisms are quickly activated to prevent major accidents

- H_2 offsite: Hydrogen production processes, such as methane reforming or water electrolysis, involve inherent risks. In these stations, the integration of low-pressure hydrogen purification and storage systems is crucial to ensuring safe operations.

Genovese et al. (2023) conducted a comprehensive literature review of hydrogen refueling procedures, covering aspects such as storage, distribution lines, pressure regulators, pre-cooling, and vehicle filling procedures. Table 2.7 summarizes the main technical factors.

In a subsequent review, Genovese and Fragiacomo (2023) complement this essentially technical view by identifying four relevant complementary aspects which impact HRS safety. Review results are shown in Table 2.8.

Through a bibliometric study Wei et al. (2022) provide a structural, comprehensive view of scientific stakeholders and the field of hydrogen safety research.

Table 2.7 Safety components derived from the review of Genovese et al. (2023)

Components	Safety impact
Design of storage systems	Optimal design of cascade storage systems is key to minimizing energy losses
Dispensers	In hydrogen dispensing lines, it is critical to avoid both pulsations and pressure losses
Joule–Thomson effect	The effect produced when gas changes temperature after expanding through a valve or nozzle without exchanging heat with its surroundings, greatly influences the temperature of hydrogen after passing through the pressure regulator
Pre-cooling systems	They are critical to ensuring fast refueling without hydrogen overheating

Table 2.8 Safety components derived from the review by Genovese and Fragiacomo (2023)

Components	Safety impact
Human factor	Need to systematically investigate the role of operators in HRS safety, including aspects such as training, fatigue, and ergonomics
Corrosion	Corrosion of HRS components may compromise structural and functional facility integrity, representing a significant risk
Operational standards	Lack of studies thoroughly analyzing operational standards and their safety impact
Advanced methodologies	Need to develop and apply more sophisticated methodologies to assess and mitigate HRS risks, such as event-based analysis (EBA)

For this purpose and through the use of visualization tools and bibliographic data analysis, scientific publications on hydrogen safety between 1957 and 2021 were examined, allowing the identification of key research areas, emerging trends, and future challenges in this field. Table 2.9 shows the main findings of that study.

In a similar study, Suzuki et al. (2022) comprehensively reviewed international studies on HRS safety, with a particular focus on work conducted in Japan. Bibliometric tools (such as VOSviewer and KH Coder) were used to identify the institutions,

Table 2.9 Bibliometric insights on hydrogen safety (1957–2021) by Wei et al. (2022)

Review results	Relevance
Distribution of publications	Importance of certain regions and journals in the accumulation of knowledge on the topic
Emergent research topics	Three main themes are identified in hydrogen safety research over time: – Storage and detection (1957–2013): approach of production, storage, and detection of hydrogen leaks – Combustion and explosion (2013–2017): studies on thermal hydrogen behavior, including combustion, explosion, and deflagration – Ignition and spread (2017–2021): recent research on the mechanisms of hydrogen ignition and the propagation of flames and blast waves
Collaborative networks and lead authors	The most active countries in hydrogen safety research are China, the United States, and Germany Ulster University and Hefei University have the highest publication production rate Research shows that the United States leads in number of hydrogen safety publications The "*International Journal of Hydrogen Energy*" is considered as the leading journal in the field of hydrogen safety
Future trends and challenges	Need for large-scale studies on hydrogen safety (from a hydrogen economy perspective and new technologies to be developed)

Table 2.10 Safety components derived from the review by Suzuki et al. (2022)

Review results	Relevance
Studies focusing on HRS safety	189 publications were identified highlighting Japan, with institutions such as Yokohama National University and the National Institute of Advanced Industrial Science and Technology (AIST) among the main contributors
Qualitative and quantitative risk assessment	Applied mainly to compressed hydrogen stations using advanced methodologies, including the simulation of hydrogen leak scenarios and accidents with "domino effects" A need is identified to extend these assessments to other technologies (hydrogen production by electrolysis and multi-energy stations)

authors, and countries involved in relevant HRS safety research in recent years. The main review findings are listed in Table 2.10.

2.2 Analysis of HRS Events and Accidents

The review and analysis of unwanted events (or abnormal processes) is paramount in high-risk industries, as it allows learning from mistakes and continuously improving safety. This practice is firmly established in high-risk industrial sectors, such as nuclear, aeronautics, and chemistry, where safety is an overriding priority and a social and regulatory requirement. This approach, known as "learning from events," is a key pillar in preventing future accidents. As Paltrinieri et al. (2012) point out, many incidents are "known unknowns": events that could have been predicted if previous data had been rigorously and thoroughly analyzed. This section presents research focused on the collection and analysis of operational data in industries such as chemical and hydrogen refueling stations, with the aim to identify patterns and trends potentially predicting and preventing adverse events.

2.2.1 Gathering Event Information in Databases

A systematic compilation of incidents allows this information to be shared with the rest of the industry. In parallel to the development of green hydrogen technologies, repositories are being developed to capture, analyze, and share operating experience. The studies presented below show pioneering processes in the creation of hydrogen technology databases, as well as a current comparative between different available databases.

The first study, conducted by Weiner et al. (2011), developed a repository of good practices in hydrogen use, called Hydrogen Safety Best Practices. This database compiled information based on operating experience. The public dissemination of this database sought to facilitate the exchange of knowledge and experience within

Table 2.11 Main results of comparison by West et al. (2022)

Database	Strengths	Weaknesses
H2Tools Lessons Learned	Provides qualitative descriptions of failures and errors based on narratives, identifying lessons learned	Provides insufficient information to perform rigorous risk assessments
Hydrogen incidents and accidents database (HIAD)	Designed to collect both qualitative and quantitative data	Most data are qualitative and incomplete due to the voluntary nature of reporting
National renewable energy lab's (NREL) Composite data products (CDPs)	Collects system-level data	Does not provide detailed information on failure modes and mechanisms
Center for hydrogen safety (CHS) Equipment and component failure rate data submission form	Designed to collect information at an equipment component level	Lacks data on component lifetimes, as well as on their failure modes and mechanisms

Own development

the field of hydrogen safety. The authors stressed the need for international collaboration to improve and expand the use of safe practices within the framework of the International Association for Hydrogen Safety (IA HySafe). This tool is still available online today as a manual providing recommendations on safe hydrogen handling.

Second, Galassi et al. (2012) describe the creation and fundamentals of the Hydrogen Incident and Accident Database (HIAD), conceived as a systematic repository of information on hydrogen-related accidents and incidents. In fact, HIAD originates from the effort to develop hydrogen technologies within a context of transition to more sustainable energy systems. This database was designed as a multidisciplinary tool to store and communicate quality information on unwanted events in order to improve safety and contribute to risk analysis. Initially developed under the HySafe Network of Excellence (2004–2009), HIAD continues to evolve under the coordination of the HySafe IA and is currently available in version 2.1 online.

Lastly, a recent comparative analysis of current databases by West et al. (2022) highlights the strengths and weaknesses of a number of databases considering their value to improve HRS safety. Table 2.11 summarizes the results of such analysis.

2.2.2 *Analysis of Hydrogen Industry Events*

A core aspect in the analysis of operating experience information was the search for event root causes within the chemical industry. Over time, studies became more specialized, delving deeper into the different stages of the hydrogen value chain. Below, a range of studies based on the systematic analysis of events collected in different databases are shown.

In a study focused on hydrogen processing, Mirza et al. (2011) analyzed 32 incidents extracted from the HIRD database (Hydrogen Incident Reporting Database, now H2Tools), with the aim of obtaining a typology of events and understanding the underlying causes. Event causes were categorized as "Technical," "Operator error," and "Design error." Analysis results showed that 43.7% of events resulted in fires, 31.3% in explosions, 15.6% in fires and explosions, and 9.4% in "near misses," which are basically events which could have caused a damaging effect. In terms of impact on the facility, 59.4% of events resulted in plant damage, and 25% of them adversely impacted both personnel and the plant. As for event causes, "Maintenance" and "Management" were identified as primary and secondary causes, respectively.

In an analysis meant to establish risk factors in hydrogen logistics, Lam et al. (2019) used a network modeling approach including both factors and their effects in an interrelated manner. They selected 100 incidents that occurred between 1983 and 2013, taking into account their severity, presence of multiple risk factors, availability of detailed data, and relevance within the hydrogen logistics context. For data collection, the following databases were used: HIAD, H2Tools, KHK, and eMARS. The selection of events was based on criteria of completeness and representativeness of hydrogen logistics-related risks. Events of the following typology were included: (a) operational errors in hydrogen refueling, (b) collisions of trucks carrying liquid hydrogen, (c) connection failures between truck and facility during hydrogen supply, (d) hydrogen leaks in storage tanks, and (e) failures in refueling hoses. Events related to hydrogen production and events derived from natural gas reforming, coal and biomass gasification, or failures during the electrolysis process, were excluded. The study identified improper operation as the main event contributor, with hydrogen leaks being the main effect. It was also noted that a combination of control measures on both elements (factors and effects) could reduce the occurrence of hydrogen logistics events by up to 76%.

In an investigation covering the entire hydrogen value chain, Alfasfos et al. (2024) analyzed 82 events documented in various databases, such as HIAD and eMARS, among others. The analysis provided a broad view of the hydrogen value chain, pointing out some processes with the highest risk. Table 2.12 shows some of the most relevant results.

Wen et al. (2022) reviewed 706 documented events in HIAD 2.0 (Hydrogen Incident and Accident Database), with the goal of identifying recurring patterns and generating recommendations based on lessons learned. The authors categorized these events into four main areas: (a) System design; (b) System manufacture, installation, and modification; (c) Human factors; and (d) Emergency response. For each of these areas, a set of lessons learned was extracted from the analysis. In terms of design, one of the main causes of adverse events was the lack of intrinsic safety measures, including problems such as corrosion not detected on time due to a lack of regular inspections or materials being incompatible with hydrogen. In the manufacturing, installation, and modification phases, factors such as the absence of adequate ventilation systems, component fatigue, and lack of redundancy in critical systems were major event contributors. Regarding human and organizational factors, the lack of regular maintenance and personnel training accounted for a large number of adverse

Table 2.12 Main findings of the study by Alfasfos et al. (2024)

Stage/Process/Hydrogen type	Main findings
Transportation and distribution	Most events occurred during hydrogen distribution (40%) and storage (24%)
Compression and electrolysis	Compression, although energy and pressure intensive, accounted for only 11% of events Electrolysis-related events were 9%
Type of hydrogen	Most events involved gaseous hydrogen (gaseous hydrogen pipelines and tube trailers) rather than liquid hydrogen
Human and organizational factors	The causes in 87% of events were related to human and organizational factors, mainly maintenance failures, lack of training, or lack of situational awareness

events. Finally, with regard to safety management, deficiencies in inspection plans and a lack of clear protocols for rapid hydrogen leak isolation were also identified.

A systematic analysis by Campari et al. (2023) of 628 events recorded in the HIAD 2.0 database using Business Analytics (BA) attempted to understand hydrogen-induced material failures. Table 2.13 shows the main causes linked to their analysis.

The study by Campari et al. also emphasized the need for a multidisciplinary approach, underlining the need for collaboration between materials science and

Table 2.13 Main findings of the study by Campari et al. (2023)

Identified factor	Main findings
Inspection and maintenance deficiencies	Inadequate planning of inspection and maintenance activities was a common cause of major accidents Most events could have been prevented with regular audits and proper attention to critical points, such as welds and joints
Hydrogen embrittlement	A relevant factor the lack of specific inspection plans and clear indications on component service life in hydrogen environments
Corrosion and high-temperature hydrogen attack	The corrosion by hydrogen-containing acid compounds, as well as the attack of high temperature hydrogen were identified as relevant causes for containment losses and unwanted hazardous substance releases
Inspection standards and practices	Inspection and maintenance practices and standards were assessed as inadequate for hydrogen technologies because they do not consider hydrogen embrittlement The risk-based inspection (RBI) methodology could be beneficial if adapted to consider this type of material degradation

Table 2.14 Main findings of the study by Badia et al. (2024)

Stage/Process/ Systems	Main findings
Electrolyzers and storage tanks	They were identified as critical equipment due to risks of overheating or corrosion The need to implement rigorous inspection and maintenance protocols to prevent leaks or explosions was pointed out
Management of storage activities	Hydrogen storage processes had the greatest potential to cause major events, especially in activities involving work inside tanks or during cylinder filling
Distribution and transport	The few events occurred during hydrogen distribution through large pipelines did not cause significant personal injuries Road transport was identified as a critical and dangerous activity due to the risk of collision
Human factors	Connections between pipelines and compressors, along with welding failures, revealed that human factors (errors in operation, maintenance, or installation) play a key role in safety events

Reliability, Availability, Maintainability and Safety (RAMS) engineering, thereby improving safety against hydrogen-induced material failures.

Finally, an analysis of events recorded in the HIAD 2.0 database (Badia et al., 2024) selected 224 of them that occurred in the different hydrogen value chain stages (production, storage, transport, and industrial applications) which could be extrapolated to refueling stations. The analysis combined quantitative and qualitative data. On the one hand, yearly specific data on people were examined, such as the number of injured and deceased, making it possible to calculate a mortality rate and assess the risk associated with each type of event. On the other hand, event descriptions were examined in detail, identifying the main causes and types of failures originating them. This combination of quantitative and qualitative approaches provided a detailed characterization of what happened, thus enabling the quantification of event impacts as well as the generation of knowledge about their causes. Table 2.14 shows the main analysis results.

2.2.3 Review of HRS Events

The analysis of HRS-related events provides crucial information on safety and risk aspects in these facilities. Studies focused on the use of these events with a special interest on risk prevention and improved system safety are shown below.

Kodoth et al. (2018) analyzed 26 events recorded in the Japanese KHK database and estimated the accident rate. Regarding the typology of analyses, 23 of 26 recorded accidents involved hydrogen leaks and 3 involved explosions. The authors considered that a critical element of HRS safety is to monitor and control leak causes. In addition, researchers applied a time correlation model showing that uncertainty increased as stations operated for longer periods of time.

In a comparative analysis, Sakamoto et al. (2016) found differences and similarities in the 21 events and accidents that occurred in Japan and the USA in relation to Hydrogen Refueling Stations (HRS) between 2005 and 2014. Events were grouped into six categories: leaks (types I, II, and III), explosions, ruptures, and others. Analyses were also focused on events in different types of refueling stations, including those producing hydrogen on-site from natural gas and those receiving hydrogen in liquid or gaseous form from other locations. Table 2.15 presents a summary of some key study findings.

To prevent type-1 leaks (damage to main system bodies), Sakamoto et al. (2016) pointed out the need to carefully consider the operating environment during design. To reduce type-2 leaks (joints, valves, and seals), it is suggested to use sufficiently strong welded joints and to reduce piping thickness. Finally, to prevent type-3 leaks (human error), the development of safety measures aimed at preventing errors by users of fuel cell vehicles was suggested.

In a more recent study, within the framework of the international project SUSHy (SUStainability and cost-reduction of Hydrogen stations through risk-based, multidisciplinary approaches), Navajas et al. (2023) reviewed HRS events by selecting records from four international databases, namely ARIA, H2Tools, HIAD 2.0, and KHK. This compilation made it possible to identify a total of 211 events, distributed by year as shown in Fig. 2.1.

The graph presents the distribution of the total number of events at Hydrogen Refueling Stations (HRS), as identified in the databases analyzed. Each bar corresponds to a specific year, and its height reflects the proportion of events recorded that year. The percentages above each bar provide a more detailed view of the relative frequency of events in each period. Event frequency shows an upward trend over time, with a significant peak in 2016. However, the number of events decreases in some years, thus suggesting a lack of a clear variability pattern.

Table 2.15 Main findings of the study by Sakamoto et al. (2016)

Factor	Main findings
Leaks	The most common cause of events was gasket, valve and seal leaks, mainly due to design errors, such as an underestimation of component-applied loads and a failure to consider the operating environment during design
Human error	They played an important role in some events, particularly in the case of self-staffed refueling stations
Material fatigue	Essentially in components subjected to cyclic loading over long periods
Comparison between USA and Japan	In Japan, threaded joints are prevalent, whereas welded joints are more common in the USA The difference is attributed to factors such as ease of maintenance, complexity of installations, and lack of data on the strength of welded joints in hydrogen environments

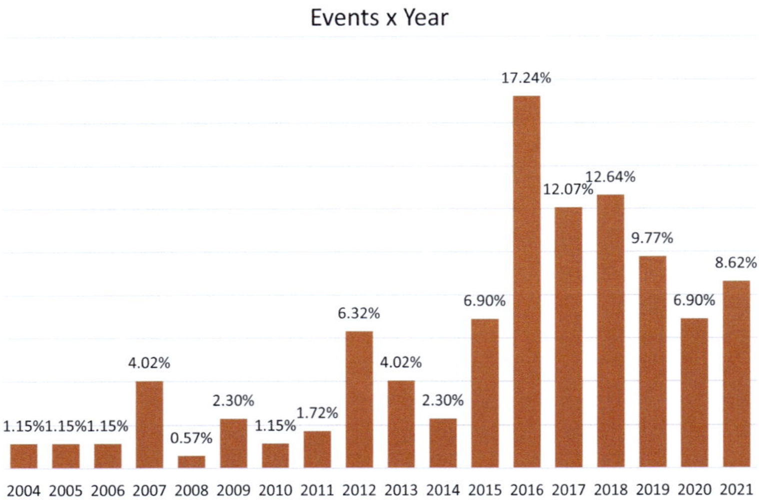

Fig. 2.1 Distribution of HRS events per year

In order to analyze events, a specific protocol called Incident Contributing Factor Analysis (ICFA) was developed, in accordance with proven methodologies used in high safety industries such as nuclear (MORT) and offshore drilling (SACA). The protocol was validated by means of a pilot analysis of 5 events. Table 2.16 shows the main findings of systematic ICFA protocol application to all events identified.

Finally, it should be noted that some studies focused on the analysis of HRS operation. In this regard, Samuelsen et al. (2020) evaluated HRS performance at the University of California, Irvine (UCI), during the first five years of commercial operation of Fuel Cell Electric Vehicles (FCEVs). The main technical problems identified had to do with compressor and dispenser leaks, as well as with instrument

Table 2.16 Main findings of the study by Navajas et al. (2023)

Factor	Main findings
Design and equipment	Design flaws were present in 60% of events High number of failure of components and equipment not specifically designed for HRS Lack of "service life" considerations in available parts, especially dispensers and compressors
Human and organizational factors	Human failure present in almost 20% of events Human error associated with inspection, maintenance, installation, and refueling operations Shortcomings in risk identification and prevention in almost 5% of events
Operating experience-based learning	Recurring hydrogen refueling events Operating experience on the system not widely applied at all hydrogen sites

air compressor failures. Although these technical problems decreased over time, the average refueling rate remained significantly below the target set by the U.S. Department of Energy (DOE). According to the authors, this highlighted the need to enhance equipment design so as to adapt to actual daily demand.

Within the 3Emotion project, Caponi et al. (2023) analyzed the performance of five hydrogen refueling stations and 34 Fuel Cell Buses (FCBs) over a four-year period. Although the amount of hydrogen supplied per refueling was uniform, refueling times varied considerably, thus affecting operational safety conditions. In addition, the low usage rate of stations, averaging less than 30%, suggested an underutilized capacity which could be optimized without increasing safety risks.

2.3 Conclusions

The diversity of methodological approaches discussed in this chapter reflects the inherent complexity of hydrogen refueling station safety and underlines the need to combine various perspectives for effective safety management of these facilities. Research on HRS safety has progressed through multiple and complementary approaches.

First, several methodologies have contributed valuable insights: regulatory studies that clarify and harmonize regulatory frameworks to support international standardization; risk scenario analyses, primarily using HAZOPs, which offer a systematic understanding of potential hazards; and quantitative risk assessments that enable objective prioritization of mitigation measures. Additionally, numerical simulation studies have provided detailed models of physical processes, while literature reviews offer integrative perspectives, highlighting both advancements and areas for further investigation.

Second, incident reviews are crucial for deriving lessons from accumulated operational experience to improve HRS safety. Reliable, systematic databases are essential for identifying patterns and common causes of incidents. These data platforms facilitate knowledge-sharing among facilities, reinforcing the development of safe practices. Operational experience has highlighted common contributing factors to incidents, such as design flaws, technical failures, and operational shortcomings. Research underscores the importance of multidisciplinary approaches that integrate safety, engineering, and materials science. This integrated approach enhances safety across each stage of the value chain, considering technical, organizational, and human factors that contribute to risk reduction and safety management.

References

Alfasfos R, Sillman J, Soukka R (2024) Lessons learned and recommendations from analysis of hydrogen incidents and accidents to support risk assessment for the hydrogen economy. Int J Hydrogen Energy 60:1203–1214. https://doi.org/10.1016/j.ijhydene.2024.02.226

Al-Shanini A, Ahmad A, Khan F (2014) Accident modelling and safety measure design of a hydrogen station. Int J Hydrogen Energy 39(35):20362–20370. https://doi.org/10.1016/j.ijhydene.2014.05.044

Apostolou D, Xydis G (2019) A literature review on hydrogen refuelling stations and infrastructure. Current status and future prospects. In: Renewable and sustainable energy reviews, vol 113. Elsevier Ltd. https://doi.org/10.1016/j.rser.2019.109292

Badia E, Navajas J, Sala R, Paltrinieri N, Sato H (2024) Analysis of hydrogen value chain events: implications for hydrogen refueling stations' safety. Safety, 10(2). https://doi.org/10.3390/safety10020044

Campari A, Nakhal Akel AJ, Ustolin F, Alvaro A, Ledda A, Agnello P, Moretto P, Patriarca R, Paltrinieri N (2023) Lessons learned from HIAD 2.0: inspection and maintenance to avoid hydrogen-induced material failures. Comput Chem Eng 173. https://doi.org/10.1016/j.compchemeng.2023.108199

Caponi R, Monforti Ferrario A, Del Zotto L, Bocci E (2023) Hydrogen refueling stations and fuel cell buses four year operational analysis under real-world conditions. Int J Hydrogen Energy 48(54):20957–20970. https://doi.org/10.1016/j.ijhydene.2022.10.093

Casamirra M, Castiglia F, Giardina M, Lombardo C (2009) Safety studies of a hydrogen refuelling station: determination of the occurrence frequency of the accidental scenarios. Int J Hydrogen Energy 34(14):5846–5854. https://doi.org/10.1016/j.ijhydene.2009.01.096

Crawley FK, Tyler BJ (2003) Hazard identification methods Rugby. EPSC/IChemE, UK

Galassi MC, Papanikolaou E, Baraldi D, Funnemark E, Håland E, Engebo A, Haugom GP, Jordan T, Tchouvelev AV (2012) HIAD—hydrogen incident and accident database. Int J Hydrogen Energy 37(22):17351–17357. https://doi.org/10.1016/j.ijhydene.2012.06.018

Dadashzadeh M, Kashkarov S, Makarov D, Molkov V (2018) Risk assessment methodology for onboard hydrogen storage. Int J Hydrogen Energy 43(12):6462–6475. https://doi.org/10.1016/j.ijhydene.2018.01.195

Genovese M, Cigolotti V, Jannelli E, Fragiacomo P (2023a) Current standards and configurations for the permitting and operation of hydrogen refueling stations. Int J Hydrogen Energy 48(51):19357–19371. https://doi.org/10.1016/j.ijhydene.2023.01.324

Genovese M, Fragiacomo P (2023) Hydrogen refueling station: overview of the technological status and research enhancement. J Energy Storage 61:106758. https://doi.org/10.1016/j.est.2023.106758

Gökçek M, Paltrinieri N, Liu Y, Badia E, Dokuz AŞ, Erdoğmuş A, Urhan BB, Yoldaş Ö (2024) Optimum sizing of hybrid renewable power systems for on-site hydrogen refuelling stations: case studies from Türkiye and Spain. Int J Hydrogen Energy 59(2023):715–729. https://doi.org/10.1016/j.ijhydene.2024.02.068

Guo L, Su J, Wang Z, Shi J, Guan X, Cao W, Ou Z (2024) Hydrogen safety: an obstacle that must be overcome on the road towards future hydrogen economy. Int J Hydrogen Energy 51:1055–1078. https://doi.org/10.1016/j.ijhydene.2023.08.248

Hansen OR (2020) Hydrogen infrastructure—efficient risk assessment and design optimization approach to ensure safe and practical solutions. Process Saf Environ Prot 143:164–176. https://doi.org/10.1016/j.psep.2020.06.028

Hassan Q, Algburi S, Jaszczur M, Al-Jiboory AK, Al Musawi TJ, Ali BM, Viktor P, Fodor M, Ahsan M, Salman HM, Sameen AZ (2024) Hydrogen role in energy transition: a comparative review. In: Process safety and environmental protection, vol 184. Institution of Chemical Engineers, pp 1069–1093. https://doi.org/10.1016/j.psep.2024.02.030

Isaac N, Saha AK (2021) Analysis of refueling behavior of hydrogen fuel vehicles through a stochastic model using Markov chain process. Renew Sustain Energy Rev 141. https://doi.org/10.1016/j.rser.2021.110761

Kletz T (2006) HAZOP and HAZAN, 4th edn. IChemE, Rugby, UK

Kodoth M, Aoyama S, Sakamoto J, Kasai N, Shibutani T, Miyake A (2018) Evaluating uncertainty in accident rate estimation at hydrogen refueling station using time correlation model. Int J Hydrogen Energy 43(52):23409–23417. https://doi.org/10.1016/j.ijhydene.2018.10.175

Kwon D, Choi SK, Yu C (2022) Improved safety by crossanalyzing quantitative risk assessment of hydrogen refueling stations. Int J Hydrogen Energy 47(19):10788–10798. https://doi.org/10.1016/j.ijhydene.2021.12.211

LaFleur AC, Muna AB, Groth KM (2017) Application of quantitative risk assessment for performance-based permitting of hydrogen fueling stations. Int J Hydrogen Energy 42(11):7529–7535. https://doi.org/10.1016/j.ijhydene.2016.06.167

Lam CY, Fuse M, Shimizu T (2019) Assessment of risk factors and effects in hydrogen logistics incidents from a network modeling perspective. Int J Hydrogen Energy 44(36):20572–20586. https://doi.org/10.1016/j.ijhydene.2019.05.187

Landucci G, Tugnoli A, Cozzani V (2010) Safety assessment of envisaged systems for automotive hydrogen supply and utilization. Int J Hydrogen Energy 35(3):1493–1505. https://doi.org/10.1016/j.ijhydene.2009.11.097

Li Y, Wang Z, Shi X, Fan R (2023) Safety analysis of hydrogen leakage accident with a mobile hydrogen refueling station. Process Saf Environ Prot 171:619–629. https://doi.org/10.1016/j.psep.2023.01.051

Mirza NR, Degenkolbe S, Witt W (2011) Analysis of hydrogen incidents to support risk assessment. Int J Hydrogen Energy 36(18):12068–12077. https://doi.org/10.1016/j.ijhydene.2011.06.080

Moradi R, Groth KM (2019) Hydrogen storage and delivery: review of the state of the art technologies and risk and reliability analysis. Int J Hydrogen Energy 44(23):12254–12269). https://doi.org/10.1016/j.ijhydene.2019.03.041

Navajas J, Badia E, Sala R, Sato H, Paltrinieri N (2023) Development of a protocol for the systematic analysis of events at hydrogen refuelling stations. Chem Eng Trans 105:127–132. https://doi.org/10.3303/CET23105022

Paltrinieri N, Dechy N, Salzano E, Wardman M, Cozzani V (2012) Lessons learned from Toulouse and Buncefield disasters: from risk analysis failures to the identification of atypical scenarios through a better knowledge management. Risk Anal 32(8):1404–1419. https://doi.org/10.1111/j.1539-6924.2011.01749.x

Pan X, Li Z, Zhang C, Lv H, Liu S, Ma J (2016) Safety study of a wind–solar hybrid renewable hydrogen refuelling station in China. Int J Hydrogen Energy 41(30):13315–13321. https://doi.org/10.1016/j.ijhydene.2016.05.180

Rusin A, Stolecka-Antczak K, Kosman W, Rusin K (2024) Influence of driver error on the level of hydrogen refuelling station risk. Int J Hydrogen Energy 73:73–85. https://doi.org/10.1016/j.ijhydene.2024.05.436

Sakamoto J, Sato R, Nakayama J, Kasai N, Shibutani T, Miyake A (2016) Leakage-type-based analysis of accidents involving hydrogen fueling stations in Japan and USA. Int J Hydrogen Energy 41(46):21564–21570. https://doi.org/10.1016/j.ijhydene.2016.08.060

Salehi F, Abbassi R, Asadnia M, Chan B, Chen L (2022) Overview of safety practices in sustainable hydrogen economy—an Australian perspective. Int J Hydrogen Energy 47(81):34689–34703). https://doi.org/10.1016/j.ijhydene.2022.08.041

Salimi M, Hosseinpour M, Borhani NT (2022) The role of clean hydrogen value chain in a successful energy transition of Japan. Energies 15(16). https://doi.org/10.3390/en15166064

Samuelsen S, Shaffer B, Grigg J, Lane B, Reed J (2020) Performance of a hydrogen refueling station in the early years of commercial fuel cell vehicle deployment. Int J Hydrogen Energy 45(56):31341–31352. https://doi.org/10.1016/j.ijhydene.2020.08.251

Searle C, Searle KD, Wang S, Greening P (2024) The ATHENA framework: analysis and design of a strategic hydrogen refuelling infrastructure. Int J Hydrogen Energy 52:767–781. https://doi.org/10.1016/j.ijhydene.2023.04.190

Serdaroglu G, Khzouz M, Gillott M, Shields A, Walker GS (2015) The effect of environmental conditions on the operation of a hydrogen refuelling station. Int J Hydrogen Energy 40(47):17153–17162. https://doi.org/10.1016/j.ijhydene.2015.05.003

Shi J, Chang Y, Khan F, Zhu Y, Chen G (2020) Methodological improvements in the risk analysis of an urban hydrogen fueling station. J Cleaner Prod, 257. https://doi.org/10.1016/j.jclepro.2020.120545

Sun K, Pan X, Li Z, Ma J (2014) Risk analysis on mobile hydrogen refueling stations in Shanghai. Int J Hydrogen Energy 39(35):20411–20419. https://doi.org/10.1016/j.ijhydene.2014.07.098

Suzuki T, Izato YI, Miyake A (2021) Identification of accident scenarios caused by internal factors using HAZOP to assess an organic hydride hydrogen refueling station involving methylcyclohexane. J Loss Prev Process Ind 71. https://doi.org/10.1016/j.jlp.2021.104479

Suzuki T, Shiota K, Nakayama J, Izatoa YI, Miyake A (2022) Overview of the risk assessments for hydrogen refuelling stations in Japan. Chem Eng Trans 90:307–312. https://doi.org/10.3303/CET2290052

Tian Y, Zhang XY, Shan MM, Qi M, Shu CM, Li B, Liu Y (2024) Methodology for optimally designing firewalls in hydrogen refueling stations. Int J Hydrogen Energy 49:1196–1209. https://doi.org/10.1016/j.ijhydene.2023.08.051

Tsunemi K, Kihara T, Kato E, Kawamoto A, Saburi T (2019) Quantitative risk assessment of the interior of a hydrogen refueling station considering safety barrier systems. Int J Hydrogen Energy 44(41):23522–23531. https://doi.org/10.1016/j.ijhydene.2019.07.027

Wang Y, Wang S, Decès-Petit C (2022) Less is more: Robust prediction for fueling processes on hydrogen refueling stations. Int J Hydrogen Energy 47(67):28993–29005. https://doi.org/10.1016/j.ijhydene.2022.06.239

Wei R, Lan J, Lian L, Huang S, Zhao C, Dong Z, Weng J (2022) A bibliometric study on research trends in hydrogen safety. Process Saf Environ Prot 159:1064–1081. https://doi.org/10.1016/j.psep.2022.01.078

Weiner SC, Fassbender LL, Quick KA (2011) Using hydrogen safety best practices and learning from safety events. Int J Hydrogen Energy 36(3):2729–2735. https://doi.org/10.1016/j.ijhydene.2010.03.037

Wen, J. X., Marono, M., Moretto, P., Reinecke, E. A., Sathiah, P., Studer, E., Vyazmina, E., & Melideo, D. (2022). Statistics, lessons learned and recommendations from analysis of HIAD 2.0 database. Int J Hydrogen Energy 47(38):17082–17096. https://doi.org/10.1016/j.ijhydene.2022.03.170

West M, Al-Douri A, Hartmann K, Buttner W, Groth KM (2022) Critical review and analysis of hydrogen safety data collection tools. Int J Hydrogen Energy 47(40):17845–17858. https://doi.org/10.1016/j.ijhydene.2022.03.244

Xing Y, Wu J, Bai Y, Cai J, Zhu X (2022) All-process risk modelling of typical accidents in urban hydrogen refueling stations. Process Saf Environ Prot 166:414–429. https://doi.org/10.1016/j.psep.2022.08.047

Yang G, Kong D, He X, Yu X, Jiang K (2024) Prediction of hydrogen leakage location and intensity in hydrogen refueling stations based on deep learning. Int J Hydrogen Energy 68:209–220. https://doi.org/10.1016/j.ijhydene.2024.04.234

Zhang J, Zhang S, Liang Z, Lang X, Shi M, Qiao J, Wei J, Dai H, Kang J (2024a) A risk assessment method based on DEMATEL-STPA and its application in safety risk evaluation of hydrogen refueling stations. Int J Hydrogen Energy 50:889–902. https://doi.org/10.1016/j.ijhydene.2023.07.250

Zhang J, Zhang S, Qiao J, Wei J, Wang L, Li Z, Zhuo J (2024b) Safety resilience evaluation of hydrogen refueling stations based on improved TOPSIS approach. Int J Hydrogen Energy 66:396–405. https://doi.org/10.1016/j.ijhydene.2024.04.129

Zhiyong L, Xiangmin P, Jianxin M (2010) Harm effect distances evaluation of severe accidents for gaseous hydrogen refueling station. Int J Hydrogen Energy 35(3):1515–1521. https://doi.org/10.1016/j.ijhydene.2009.11.081

Chapter 3
Public Perceptions of Hydrogen Refuelling Stations

Ning Huan, Toshiyuki Yamamoto, and Hitomi Sato

Abstract Assessing the societal impact of hydrogen energy and its supporting refuelling infrastructure offers critical insights for promoting broader hydrogen energy applications in a socially accepted manner. This chapter provides an introductory overview of a case study examining public risk perceptions of hydrogen refuelling stations and the corresponding information-seeking preferences, which inform the development of communication strategies to foster public engagement and support.

Public perceptions of hydrogen refuelling stations (HRSs) are a critical aspect of the societal impact of hydrogen refuelling infrastructure and have garnered increasing attention with the global expansion of HRS networks. Hydrogen fuel derived from renewable energy sources is widely regarded as a promising pathway for mobility decarbonisation, offering high energy efficiency and substantial emission reductions in both heavy-duty and passenger transport. Since the launch of the Toyota Mirai in Japan in December 2014, automobile manufacturers in the United States, Canada, Germany, and South Korea have introduced commercial hydrogen fuel cell vehicles to the market, making them feasible options for the public. At the same time, this also underscores the need for a more extensive network of hydrogen refuelling facilities in urban areas.

HRSs have undergone rapid global expansion over the past decade. The number of operational stations worldwide increased from 320 in 2017 to 540 in 2020, surpassing 1000 by the end of 2024 (Genovese and Fragiacomo 2023; Ludwig-Bölkow-Systemtechnik GmbH 2025), with Asia, Europe, and North America leading the deployment efforts. Our analysis draws on questionnaire survey data from Japan

N. Huan (✉) · T. Yamamoto
Institute of Material and Systems for Sustainability, Nagoya University, Nagoya, Japan
e-mail: huan@bjtu.edu.cn

T. Yamamoto
e-mail: yamamoto@civil.nagoya-u.ac.jp

H. Sato
Institutes of Innovation for Future Society, Nagoya University, Nagoya, Japan
e-mail: sato@trans.civil.nagoya-u.ac.jp

Y. Liu et al. (eds.), *Safety Measures for Hydrogen Fueling Stations*, Risk, Reliability and Safety Engineering, https://doi.org/10.1007/978-981-96-9412-9_3

and two European countries, Spain and Norway, to assess public perceptions of HRSs. In Japan, GIS-based analysis has indicated that, as of July 2023, approximately 2.83% of the total population lived within an 800 m radius of an HRS (Huan et al. 2024). This proportion is expected to rise substantially as the hydrogen refuelling infrastructure network continues to expand in the forthcoming future. Hence, examining public perceptions of HRSs is becoming increasingly important for researchers and practitioners to understand public opinions during this unprecedented expansion, which plays a decisive role in facilitating the socially acceptable advancement of hydrogen energy applications.

Another noteworthy phenomenon is that as operational experience with HRSs accumulates, hydrogen-related accidents—such as hydrogen explosion incidents at HRSs in South Korea, Norway, and the USA (Campari et al. 2023)—have also gained public attention and raised safety apprehension to some extent. Accordingly, addressing these safety concerns represents a primary focus of recent research efforts aimed at assessing public perceptions. A nuanced understanding of the perceived effects of HRSs from the public's perspective is constructive for policymakers, industry stakeholders, and researchers in maintaining public trust in hydrogen safety, ultimately facilitating the successful implementation and acceptance of hydrogen energy technologies.

This chapter presents preliminary results from the parallel questionnaire survey data collected in 2023, providing a cross-country comparison focusing on three aspects: risk perceptions, risk information-seeking preferences, and the acceptance of nearby HRSs. Furthermore, the relationships between hydrogen knowledge, awareness of hydrogen-related accidents, and individual attitudes were analysed.

3.1 Data Description

A web-based survey on public perceptions and acceptance of HRSs was designed, piloted, and refined between August and October 2022. Three parallel electronic surveys, each conducted in the local language, were administered simultaneously in Japan (JP), Spain (ES), and Norway (NO) between December 2022 and January 2023, with target sample sizes of 1000 for Japan, 1000 for Spain, and 500 for Norway. To enhance the representativeness of the sample, gender and age distributions were controlled to align with the demographic composition of each country's population (Table 3.1). Given the limitations in digital literacy and accessibility among minors and individuals over 70 years old, the sampling age range was restricted to 18 to 69 years.

Table 3.1 Gender and age distribution of the sample sets

Screening variable	Category	Frequency		
		JP sample	ES sample	NO sample
Sex	Male	506	500	257
	Female	494	500	243
Age	≤ 25	95	106	88
	26 ~ 40	282	297	142
	41 ~ 50	240	237	100
	51 ~ 60	204	218	96
	≥ 61	179	142	74
Total sample size		1000	1000	500

3.2 Risk Perception Assessment

We defined risk as accidents occurring at HRSs that may result in physical injury and/or property damage to nearby residents. Following the general dimensions of risk perception analysis (Wilson et al. 2019), respondents' risk perceptions of HRSs were measured in terms of perceived accident probability (PAP) and the perceived severity of accident consequence (PAC), which were defined as two attitudinal constructs in our analysis. Furthermore, considering that literature indicates that past experiences with activities involving risk contribute to the variability in individual risk perceptions (Barnett and Breakwell 2001), we employed a bi-group approach to measuring risk perceptions based on individuals' awareness of hydrogen-related accidents. Accordingly, respondents were classified based on whether they had personally experienced or been informed about hydrogen-related accidents in any of the following contexts: chemical plants, HRSs, or other facilities. In the JP, ES, and NO sample sets, the number of accident-aware respondents was 206, 283, and 231, respectively, accounting for 20.6%, 28.3%, and 46.2% of their total sample sizes. The accident-aware rate in the Norwegian sample was significantly higher than that in the other two countries.

3.2.1 Perceived Accident Probability and Consequence

Despite extensive past surveys measuring individuals' perceived risk or safety in various regions, most have primarily focused on overall perceptions of HRSs. Our analysis aims to present a refined and experimental measurement method that examines individuals' perceptions of specific operational processes within an HRS. Specifically, four technical units—hydrogen production, transportation, storage, and dispensing—were identified as potential sources of risk and serve as the focus of the risk assessment. For each process, the constructs of PAP and PAC correspond to four measurement items assessed on a five-point Likert scale, reflecting two dimensions of risk levels that range from very low to very high.

The distribution of items for PAP in the ES and NO sample sets is shown in Fig. 3.1; for the results pertaining to the JP data, refer to Huan et al. (2024). Generally, Japanese respondents exhibited a higher proportion of moderate perceived risk likelihood compared to respondents in Spain and Norway. Accident-aware groups demonstrated higher proportions of 'high' and 'very high' risk perceptions across all four items in the three countries. Among items PAP1 to PAP4, hydrogen transportation was perceived as the process with the highest risk probability. For the accident-aware group, the proportions perceiving risk at a high level or above were 29.1%, 30.4%, and 21.2% in the JP, ES, and NO sample sets, respectively. In contrast, the highest proportion of respondents who perceived low risk probability was for hydrogen production, with proportions of 23.3%, 36.7%, and 49.8% in the accident-aware group across the three sample sets. The results indicate that the proportion of accident-aware individuals perceiving 'low' and 'very low' risk levels was higher than that of the accident-unaware group in Japan and Norway, but lower in Spain. This suggests that accident awareness does not exert a uniform effect on perceived accident probability. Instead, it shows a polarising effect in Japan and Norway while consistently increasing perceived risk likelihood in Spain.

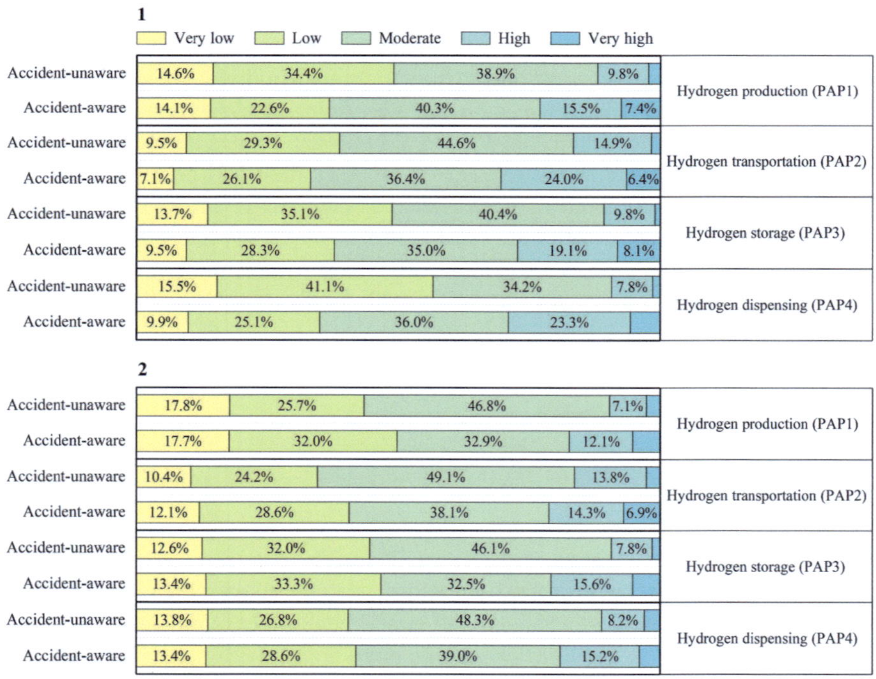

Fig. 3.1 Distribution of perceived accident probability

The most notable difference between the two groups was observed for hydrogen dispensing (PAP4) in the Spanish sample set, where the proportions of 'low' and 'very low' risk perceptions were 56.6% and 35% for the accident-unaware and accident-aware groups, respectively. Overall, Norwegian respondents displayed the most optimistic perceptions of risk likelihood among the three countries. Even within the accident-aware group, the proportions perceiving the lowest level of risk for PAP1 to PAP consistently exceeded 10%. Although public perceptions of risk associated with the four operational processes may not always align with actual risks, these findings suggest genuine concerns about safety regarding HRSs.

The second construct, PAC, was measured using items PAC1 to PAC4, with their distribution in the ES and NO sample sets presented in Fig. 3.2. Overall, the proportion of the public perceiving accident consequences as 'high' or 'very high' is greater than that for PAP (Fig. 3.1). The distribution patterns of these two dimensions of risk perception suggest that the potential severity of accidents may be the public's main concern and warrants additional efforts in risk communication practices.

In the JP sample set, respondents exhibited higher perceived consequences than those from the other two countries, particularly among individuals aware of hydrogen-related accidents. Notably, the proportions considering the consequences of accidents during hydrogen production and storage as highly severe exceeded 40%. In the two European sample sets, particularly in the NO data, the differences

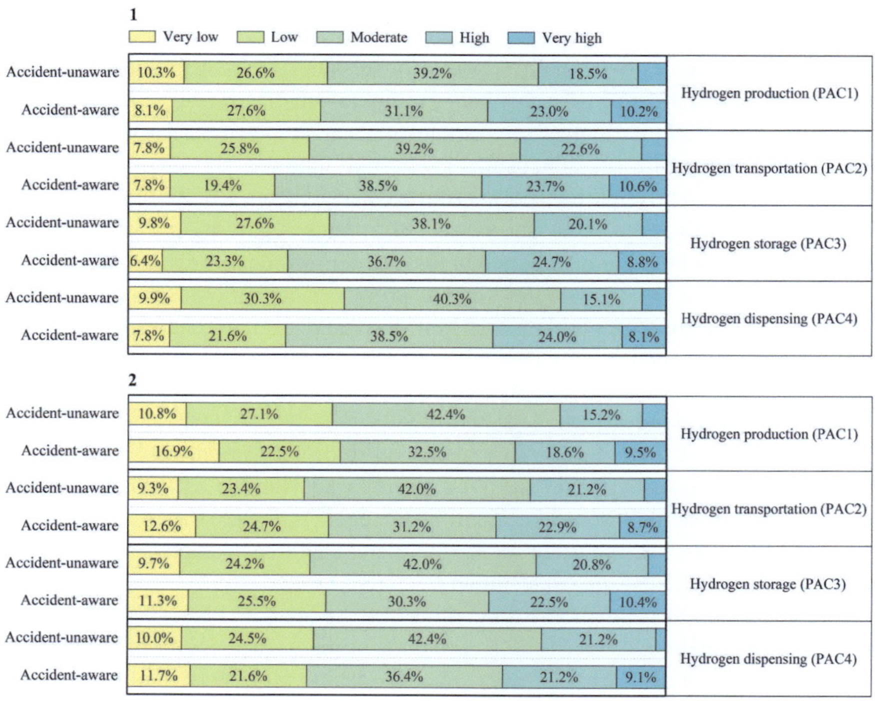

Fig. 3.2 Distribution of perceived accident consequence

between accident-aware and accident-unaware groups were much less discernible. While accident awareness appeared to heighten risk perception regarding the likelihood of accidents, its impact on the perceived severity of accident consequences was relatively limited.

Another notable phenomenon was that in the ES and NO data, a higher proportion of individuals indicated the highest level of perceived risk consequences compared to the JP sample set, demonstrating a tendency towards more extreme perceptions of risk among the European respondents. In contrast to PAP, the distribution of PAC, especially regarding high-risk perceptions, did not show significant differences across the four items. This indicates that, despite a larger group believing that the risk probability of hydrogen transportation is high, there is no corresponding perception of high severity for potential accidents.

3.2.2 Risk Information Seeking Preference

In addition to the two risk perception-related constructs, the surveys further explored relevant behavioural tendencies related to risk perceptions by defining a third construct: risk information seeking (RIS). This construct is grounded in the observation that risk perceptions motivate information-seeking behaviour to actively acquire knowledge for risk preparedness (Kim and Grunig 2011; Rogers et al. 1983). We assume that when individuals perceive a higher risk probability and/or consequence related to a particular object, their appetite for relevant risk preparedness information may increase, driven by the instinct to mitigate negative feelings. Accordingly, three items were designed to measure respondents' subjective importance regarding a series of risk-related information, using the question, 'How important do you think it is to receive the following information or training on hydrogen fuelling station accidents?' Specifically, item RIS1 addresses information regarding the geographical areas that could potentially be affected in the event of accidents at HRSs. Item RIS2 provides instructions on preparedness and response actions for potential accidents at HRSs, serving as guidelines for locals on how to respond to emergencies, such as fires or explosions. The final item, RIS3, aims to assess respondents' interest in understanding information about existing plans and preparedness measures established by the government and industries.

The distribution of items for RIS in the ES and NO sample sets is shown in Fig. 3.3; for the distribution in the JP data, refer to Huan et al. (2024). The JP samples showed the highest proportion of respondents rating the importance of such information as 'slightly important' or lower among the three countries, with approximately 25% of accident-unaware individuals and around 20% of accident-aware individuals across the three items. In contrast, the Spanish and Norwegian samples placed greater importance on this risk communication, with over half of the respondents in Spain and nearly half in Norway considering the relevant information as very important or higher.

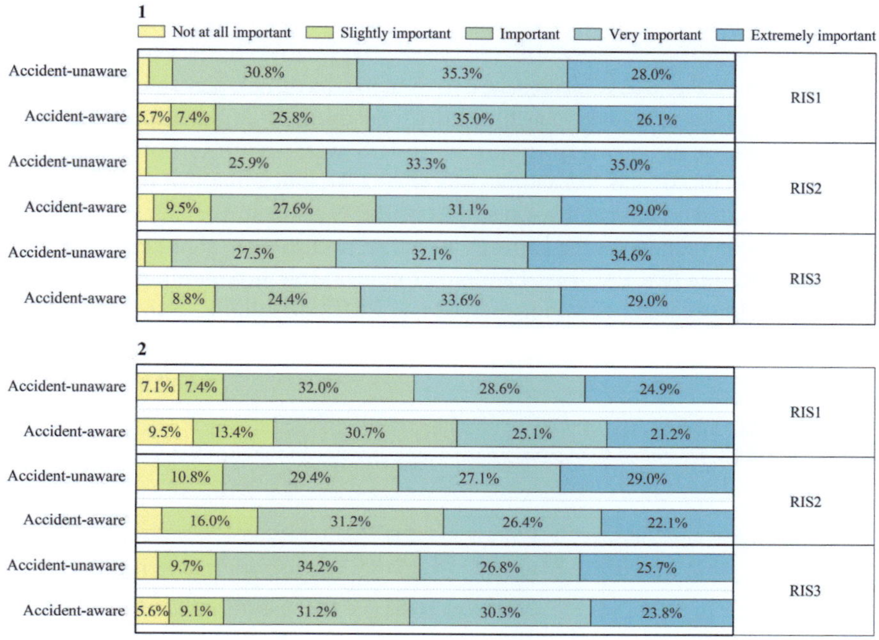

Fig. 3.3 Distribution of risk information seeking

A tendency of positive correlations between accident awareness and RIS was observed in the JP sample set, where the proportion of individuals in the accident-aware group who regarded risk information as very important was higher than that in the accident-unaware group. This trend was not as evident in the ES and NO data, despite a similarly low proportion of accident-aware individuals in Europe who rated such information as 'slightly important' or lower. This also underscores the need for further investigation into the mechanisms by which accident awareness affects risk perception and individuals' behavioural responses, based on the observed phenomena and the exploration of other relevant psychological factors.

3.2.3 Relationship Analysis

Based on the hypotheses concerning the formation of RIS, we further examined the relationships between risk perception and the preference for risk information-seeking, specifically linking PAP to RIS, and PAC to RIS. A simple conceptual framework with three constructs and two hypothesised pathways was developed, and a multigroup factor analysis was performed, with the factor loadings reported in Table 3.2. All estimates are significant and exceed the recommended threshold value of 0.7 (Nunnally 1978), indicating satisfactory internal consistency for each construct.

Given that PAP and PAC represent two dimensions of risk perception, we introduced a covariance link between these two constructs to mitigate potential bias. The estimates for the pathway from PAP to RIS were significant only in the ES data, but not in the JP and NO sample sets (Fig. 3.4), indicating that the hypothesis positing a positive relationship between PAP and RIS is not universally supported. In contrast, the coefficients for the pathway from PAC to RIS were significant across all groups, suggesting that PAC plays a more prominent role in triggering individuals' risk information-seeking behaviour relative to PAP, particularly in the JP and NO sample sets.

Table 3.2 Factor loadings across groups

Constructs	Items	Factor loadings		
		JP sample	ES sample	NO sample
Perceived accident probability	PAP1	0.809***	0.756***	0.763***
	PAP2	0.875***	0.768***	0.813***
	PAP3	0.888***	0.840***	0.842***
	PAP4	0.855***	0.809***	0.833***
Perceived accident consequence	PAC1	0.877***	0.787***	0.792***
	PAC2	0.923***	0.806***	0.838***
	PAC3	0.933***	0.888***	0.879***
	PAC4	0.907***	0.846***	0.838***
Risk information seeking	RIS1	0.911***	0.825***	0.845***
	RIS2	0.937***	0.903***	0.881***
	RIS3	0.943***	0.855***	0.860***

Note *** $p < 0.001$

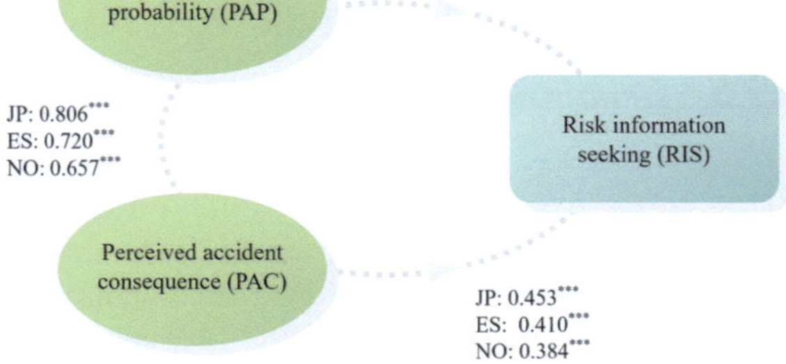

Fig. 3.4 Distribution of risk information seeking

3.3 Awareness and Acceptance

Public acceptance of HRSs was assessed with an emphasis on capturing individual subjective perceptions regarding support for or opposition to the placement of hypothetical HRSs within their communities. HRSs are typically categorised as onsite or offsite stations, depending on whether they include a hydrogen production unit. Additionally, HRSs can be classified as either newly constructed or integrated into existing facilities, such as petrol stations or publicly accessible parking lots. Given that onsite newly constructed HRSs are the most typical type, we focused on this category to gauge respondents' opinions. The acceptance was measured on a five-point Likert scale, ranging from 'strongly oppose' to 'strongly support', as illustrated in Fig. 3.5.

Notably, the criteria for grouping the knowledgeable sample are based on respondents' self-reported subjective knowledge level regarding HRSs, with individuals rating their knowledge as moderate or higher included in this classification. This resulted in 134, 389, and 59 participants in the JP, ES, and NO sample sets, respectively, accounting for 13.4%, 38.9%, and 11.8% in the corresponding sample sets. Additionally, accident awareness used in the risk perception analysis was employed here as the third grouping criterion.

A notable phenomenon is the exceptionally high proportion of Japanese respondents adopting a neutral stance, reaching 69.9%. However, this proportion decreased to 51.5% and 55.3% in the knowledgeable and accident-aware groups, respectively. Among the knowledgeable group, the support rate increased to 37.3%, while in the accident-aware group, although the opposition rate also rose compared to the benchmark, the support rate increased from 17.9% to 26.2%. In contrast to the JP data, the ES and NO samples exhibited a more pronounced distribution of support and opposition, with 37.3% and 41.6% holding neutral positions, respectively. Across all three countries, knowledge level consistently showed a positive correlation with acceptance, highlighting its role as a key factor in fostering broader societal support. Furthermore, accident awareness was identified as a significant driver of opposition, while also exerting a limited positive influence on support. This highlights the need for further investigation into how accident awareness shapes individual perceptions. It is speculated that severe accidents may amplify negative perceptions, whereas well-managed minor incidents may foster public confidence.

3.4 Implications

Measuring public perceptions of HRSs serves as a foundation for effective risk communication strategies, which are essential for integrating hydrogen refuelling infrastructure into residential areas and advancing future mobility. Cross-country comparisons highlight the necessity of distinguishing between risk probability and potential consequences to establish a more refined framework for assessing risk

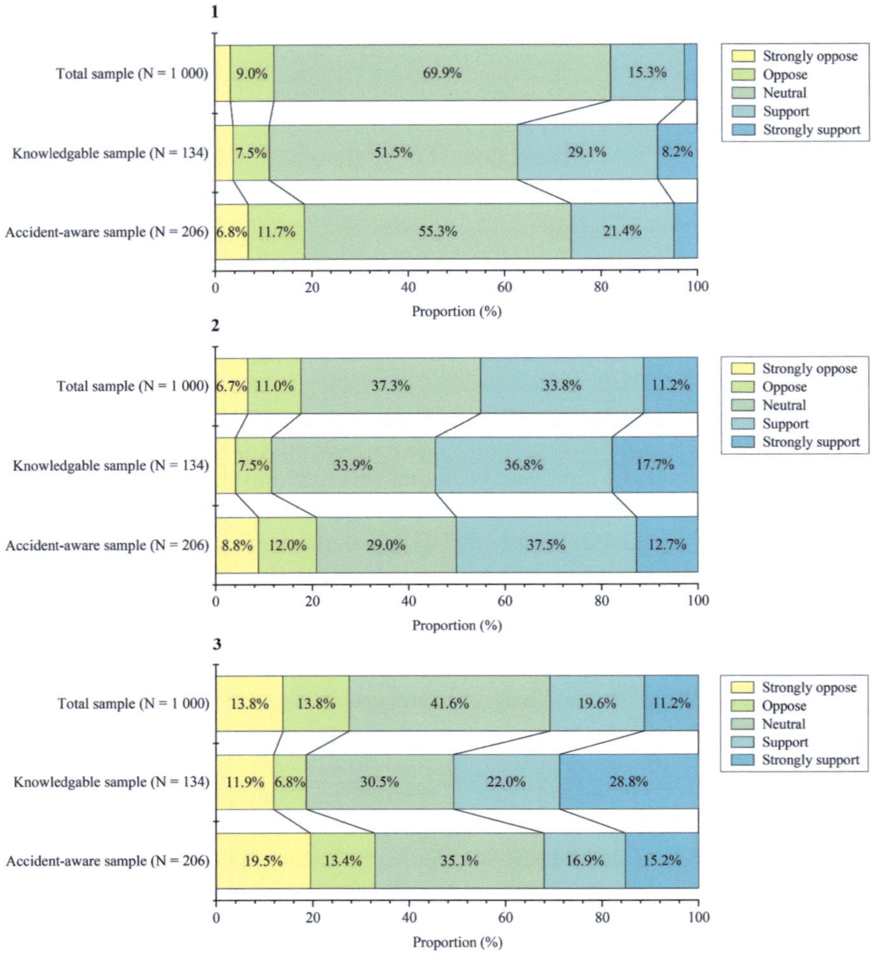

Fig. 3.5 Distribution of local acceptance of HRSs

perception, particularly in the aftermath of hydrogen-related accidents. Accordingly, addressing the public's demand for risk-related information constitutes an effective means of alleviating concerns and reinforcing public support for hydrogen infrastructure deployment. Given the observed influence of knowledge levels and accident awareness on acceptance, communication efforts should be designed in line with public psychological orientations, ensuring that educational initiatives achieve their intended effectiveness in fostering acceptance.

Acknowledgements The authors wish to express their gratitude to Dr. Roser Sala and Dr. Lila Goncalves from the Socio-Technical Research Centre at the Centro de Investigaciones Energéticas,

Medioambientales y Tecnológicas (CIEMAT), along with Dr. Dimitrios Tzioutzios from the Norwegian University of Science and Technology, for their invaluable contributions to the survey design, data collection, and analysis throughout this research.

References

Barnett J, Breakwell GM (2001) Risk perception and experience: hazard personality profiles and individual differences. Risk Anal 21:171–178. https://doi.org/10.1111/0272-4332.211099

Campari A, Nakhal Akel AJ, Ustolin F, Alvaro A, Ledda A, Agnello P, Moretto P, Patriarca R, Paltrinieri N (2023). Lessons learned from HIAD 2.0: inspection and maintenance to avoid hydrogen-induced material failures. Comput Chem Eng 173. https://doi.org/10.1016/j.compchemeng.2023.108199

Genovese M, Fragiacomo P (2023) Hydrogen refueling station: overview of the technological status and research enhancement. J Energy Storage. https://doi.org/10.1016/j.est.2023.106758

Huan N, Yamamoto T, Sato H, Tzioutzios D, Yin H, Sala R (2024) Does accident awareness affect people's risk perception of hydrogen infrastructure and information-seeking behaviour? Appl Energy 364:123141. https://doi.org/10.1016/j.apenergy.2024.123141

Kim JN, Grunig JE (2011) Problem solving and communicative action: a situational theory of problem solving. J Commun 61:120–149. https://doi.org/10.1111/j.1460-2466.2010.01529.x

Ludwig-Bölkow-Systemtechnik GmbH (2025) Statistics: Hydrogen infrastructure (WWW Document). URL https://www.h2stations.org/statistics/. Accessed 24 Feb 25

Nunnally JC (1978) Psychometric theory, 2nd edn. McGraw-Hill, New York

Rogers RW, Cacioppo JT, Petty RE (1983) Cognitive and physiological processes in fear appeals and attitude change: a revised theory of protection motivation. Social psychophysiology: a sourcebook. The Guilford Press, New York, USA, pp 153–177

Wilson RS, Zwickle A, Walpole H (2019) Developing a broadly applicable measure of risk perception. Risk Anal 39:777–791. https://doi.org/10.1111/risa.13207

Chapter 4
Framework for Multidisciplinary Risk Management

Marta Bucelli, Yiliu Liu, and Nicola Paltrinieri

Abstract Hydrogen, with its unique physiochemical properties, presents distinct challenges for safety and risk management across its value chain. This chapter outlines a multidisciplinary framework for risk management tailored to hydrogen applications, emphasizing its use as a fuel in hydrogen refueling stations. A thorough examination of the social, economic, and environmental sustainability pillars forms the foundation for a proposed risk management framework. The chapter also addresses the limitations of current quantitative risk assessments due to a lack of hydrogen-specific data and emphasizes the need for advanced leakage detection systems and robust safety measures. The proposed framework provides a comprehensive approach to managing hydrogen-related risks, as the foundation of the following chapters.

4.1 Particular Requirements of Hydrogen Applications

Hydrogen is a colorless, odorless, non-toxic substance. The flammability range of hydrogen is included in the range 4–75% vol in air, and its Minimum Ignition Energy (MIE) is calculated as 0.017 mJ, which is 10 times lower than the one of methane. Due to its wide flammability range and the low MIE, hydrogen is classified as a dangerous substance. Furthermore, physiochemical properties of hydrogen are also peculiar compared with other conventional fuels. When burning, the hydrogen flame is scarcely visible and difficult to detect without technical equipment, such as thermal or infrared cameras. The boiling point of hydrogen is 20.39 K. Therefore, at atmospheric pressure, hydrogen is in gaseous states, and its density (0.089 kg/m^3) is around 14 times lower than the one of air. In these conditions, in open and non-confined spaces, hydrogen tends to dissipate relatively fast.

M. Bucelli (✉)
SINTEF Energy Research AS, Trondheim, Norway
e-mail: marta.bucelli@sintef.no

Y. Liu · N. Paltrinieri
Norwegian University of Science and Technology (NTNU), Trondheim, Norway

Y. Liu et al. (eds.), *Safety Measures for Hydrogen Fueling Stations*, Risk, Reliability and Safety Engineering, https://doi.org/10.1007/978-981-96-9412-9_4

Due to the physiochemical properties, hydrogen is normally stored and transported as a compressed gas or liquefied cryogenic liquid. Releases of hydrogen gas stored at high pressure result in clouds driven by momentum before the dispersion process by buoyancy takes over. This behavior can potentially lead to jet fires upon immediate ignition or hazardous flammable clouds near ground level.

Handling hydrogen as a cryogenic liquid represents an operational hazard for frostbite and cryogenic burns. NASA provides a guide for hydrogen hazards analysis on components and systems (NASA 2003).

Due to its small size, hydrogen can permeate and embrittle commonly used industrial materials, such as a wide range of metals (Abohamzeh et al. 2021), leading to severe mechanical property degradation and potentially damage to the equipment integrity.

In 2023, a comprehensive assessment study of the climate effects of hydrogen was performed by (Sand et al. 2023). The study considered different atmospheric chemistry models to investigate the effects of hydrogen on different biogeochemical processes (Sand et al. 2023). When burning coal and methane gas, CO_2 is directly emitted into the atmosphere in the exhausts; when burning hydrogen, water vapor only is generated. Anyway, it was investigated that, due to the high reactivity of hydrogen with methane, ozone, and other components in the atmosphere, it influences climate change. To quantify this impact, the Green House Gas (GHG) potential (GWP100) of hydrogen was assessed and calculated as 11.6 (Sand et al. 2023). Any release of hydrogen during production, transportation, and usage represents therefore a hazard from both safety and environmental points of view. The potential benefits of switching to a hydrogen as a fuel will finally depend on the magnitude and frequency of these leaks and fugitive releases of hydrogen (Sand et al. 2023).

The peculiar characteristics of hydrogen, as described above, result in specific needs for risk management for its application, use, distribution, storage, transport, and production. In the following, a multidisciplinary framework for safety risk management is proposed as a generalized methodology suitable for the whole hydrogen value chain, with special considerations for the use of hydrogen as a fuel in hydrogen refueling stations. It should be noted that certain applications and uses of hydrogen might require specific assessments and analyses, which are beyond the scope of the work presented further in this chapter.

4.2 The Classical Risk Management Framework

Figure 4.1 presents the conventional risk management framework as defined in the ISO 31000 guidelines (International Organization for Standardization 2018).

The risk management framework proposed by (International Organization for Standardization 2018) is a iterative step procedure that aims at analyzing, evaluating, and controlling risks (Nyborg et al. 2011) and can be applied to risks of various nature. It was developed based on generic models dedicated to industry handling dangerous

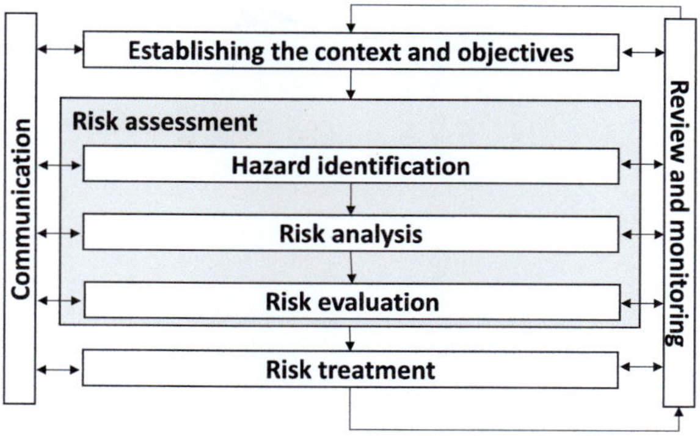

Fig. 4.1 Risk management framework according to the ISO 31000:2018 guidelines (International Organization for Standardization 2018)

substances at risk of major accidents. The framework can include both qualitative and quantitative methods for risk analysis and assessment.

Azadnia et al. (2023) identifies 43 risk factors for green hydrogen supply in the European region. The risk factors are presented in seven categories: economic, supply chain processes and governance, technology and infrastructures, environmental, market, social and policy and regulations. Table 4.1 presents more detailed considerations of such categorization.

In practice, each of these factor categories may bring challenges to the stakeholders in the hydrogen value chain. For example, as illustrated in Table 4.1, the impact of economic factors can come from the high initial investment required for hydrogen production and delivery technologies, which means high economic risk to the owners of hydrogen projects. Azadnia et al. (2023) state that the high investment are the highest ranked risk factors to promote the development of a hydrogen-based economy. These two factors are followed by the lack of sufficient capacity for green hydrogen production (via electrolyzer) and by the fragmented and incomprehensive policy and regulation frameworks. The supply chain processes and governance category underscore issues like insufficient electrolyzer capacity, improper facility locations, and inadequate storage and transportation infrastructure, which can hinder the integration of hydrogen into existing energy systems. Similarly, the market risk factors reflect the concerns about demand fluctuations, consumer awareness, and acceptance of hydrogen technologies, which are pivotal for the profitability of hydrogen projects and then their long-term survivability.

In addition, environmental risks can be from the implications of land, water, and potential pollutions, as well as the broader effects of climate change on renewable energy availability for hydrogen production. Meanwhile, social risk factors emphasize the safety, health, and welfare needs for skilled personnel, and public acceptance of hydrogen refueling stations and storages. All these factors underline the complexity of transitioning to a hydrogen-based economy and also highlight that the classical risk management focusing on safety risk needs to transform to the comprehensive and multidisciplinary risk management.

Table 4.1 Risk factors for green hydrogen supply in Europe adapted from Azadnia et al. (2023)

Economic	Supply chain processes and governance	Technology and infrastructures	Environmental	Market	Social	Policy and regulations
High investment capital for hydrogen production	Improper location of facilities	Lack of sufficient electrolyzer capacity	Natural disasters	Fluctuation of demand of green hydrogen	Skilled and trained personnel	Government incentives
Fluctuations in unit cost of electricity	Power grid disruptions	Lack of storage, transportation, and delivery capacity	Land, water, and air pollution	Customer failure	Societal acceptance of hydrogen technologies	Policy development
High operation and maintenance costs	Information and data exchange among stakeholders	Quality issues with produced hydrogen	Climate change and availability of renewables sources for green hydrogen production	Substitute product	Health and safety	Regulation development
Inflation and currency exchange rates	Supply chain failure	Integrations of green hydrogen in the supply chain	Terrorism and war	Consumer demand		Political stability
Taxes and tariffs	Inventory management and forecasting	Machine and equipment failure	Development of cleaner technologies	Consumer acceptance of green hydrogen		
Long-term financial viability and profit	Availability of power from renewable energy suppliers	Rapid technological development and uncertainty in the hydrogen supply chain	Managing of accidental hydrogen releases	Consumer awareness		

4.3 Multidisciplinary Risk Management Framework

In this chapter, a multidisciplinary or multi-criteria risk management framework is therefore proposed. Three governance pillars are considered as follows: managing risks related to social sustainability (expressed as safety only in this framework), risks related to economic sustainability, and risks related to environmental sustainability.

An overview of the governance pillars of the multidisciplinary risk management framework proposed in this work is provided in Fig. 4.2.

Such a multidisciplinary sustainability risk management model can be a *generic* model used for process industry as a whole, and not a specific framework for the management of the specific properties of hydrogen or its applications.

In the following of this chapter, the three pillars will be briefly discussed.

4.3.1 Social Sustainability Risk Management

The United Nations (UN) define *social sustainability* as the identification and management of business impacts, both positive and negative, on people (United Nations n.d.). Therefore, HSE defines safety risk management as the process to control health and safety risks caused by different types of hazards (Health and Safety Executive n.d.).

Social sustainability risk management is similar to the classical risk management approach. The first step consists in establishing the context for the identification, quantification, and management of the safety risks. Firstly, the dangerous properties of a substance (or mixture), such as flammability and toxicity, are assessed. The storage and transport of the substance (or mixture) might also represent a safety hazard, for example in the case of cryogenic liquids or high-pressure gases. Furthermore the technical components and sub-components of the system of interest as well as their functions and interactions (NASA 2003) should be defined. It is crucial to

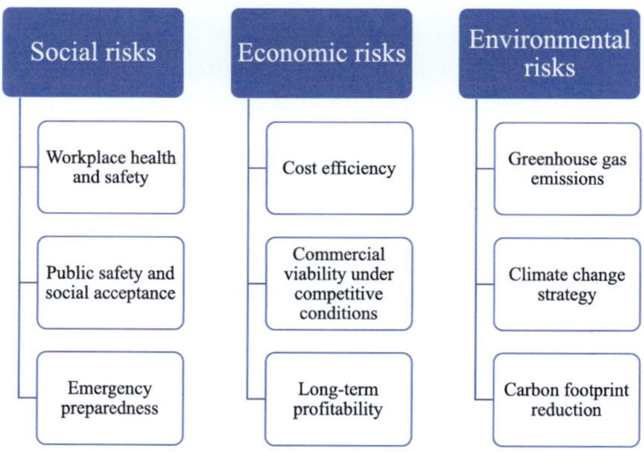

Fig. 4.2 Governance pillars for the multidisciplinary risk management framework proposed

assessing and understanding the battery limit of the analyzed systems and its relative stakeholders.

Acceptance criteria are also a critical input to the assessment of safety risks. It is often debated that acceptance criteria for safety risk management are unclear or missing (Haland and Paaske 2008). The techniques for hazard identification used commonly in a wide range of industrial sectors include Hazard Identification (HAZID), Hazard and Operability Analysis (HAZOP), and Failure Mode and Effect Analysis (FMEA).

There is a requirement to perform a hazard assessment for systems and processes that involve dangerous substances, but no specific requirement is set on the use of one technique rather than another.

Hydrogen has been used for several years on a laboratory scale or as raw material in the chemical and process industry. In many chemical and industrial processes, hydrogen is a by-product that needs storage and handling. Therefore, its use and handling are not novel to certain industrial sectors, and the intrinsic hazards associated with hydrogen are generally known and highlighted in dedicated safety material data sheets. Procedures for handling these hazards are in place and monitored continuously.

Nevertheless, the use of hydrogen on a large scale and its application in different industrial sectors (for example, as a fuel) entail some hazards. These can be summarized as

- Hazards due to the amount of hydrogen stored on site and in equipment.
- Hazards due to the complexity of the operations or systems or their combination.
- Hazards due to simultaneous operations.

For hydrogen refueling stations, the main hazard factor is represented by the release of hydrogen to the environment (Najjar 2013). This might lead to either a fire and/or explosion, immediate or delayed, in confined spaces or in the open air.

Another risk factor to consider is the impact of human and operational factors to accidents. Among the hydrogen-related events reported in accident databases (HIAD, SAFEN), a large proportion is due to causes directly related to human intervention or improper installation and monitoring. (Perelli and Genna 2022) highlights how the role of human-induced errors in the system should not be underestimated and whether proper awareness and training should be provided to ensure that the risks related to storing, using, and handling hydrogen are understood and safety managed.

Hydrogen is classified as a dangerous substance due to its extreme flammability. Therefore, the risks to people, the environment, and the assets related to the use of hydrogen as a fuel should be analyzed.

According to the MIMAH methodology, when released to the environment, depending on its storage conditions, hydrogen can result in one of three consequence categories: dispersion, fire, and explosion.

Ustolin et al. (2023) performed an investigation of the possible consequences for each consequence category defined above. The results are presented in Table 4.2.

The quantification of the frequency of occurrence of loss of containment and the evaluation of their consequences are two challenging steps in hydrogen-specific risk

Table 4.2 Consequences of hydrogen releases (Ustolin et al. 2023)

Consequence category	Consequence	Hydrogen storage state
Dispersion	Pool formation	Liquid
	Pool dispersion	Liquid
	Jet	Two phase, gas
	Dispersion	Gas
Fire	Fire	Liquid, two phase, gas
	Pool fire	Liquid
	Boil over	Liquid
	Jet fire	Two phase, gas
	Fireball	Gas
	Flash fire	Gas
Explosion	Vapor cloud explosion	Gas
	Overpressure	Liquid, two phase, gas
	Missiles	Liquid, two phase, gas
	Pressure peaking phenomenon	Gas
	Deflagration	Liquid, two phase, gas
	Detonation	Two phase, gas
	Deflagration to detonation transition	Two phase, gas
	Boiling Liquid Expanding Vapor Explosion (BLEVE)	Liquid, two phase
	Condensed phase explosion	Liquid

analysis. The quantitative risk analysis of hydrogen refueling stations is carried out using both frequency data and consequence tools that are firstly developed for other industries, such as chemical, oil, and gas (Huser et al. 2021).

The lack of hydrogen-specific data to perform a quantitative analysis of risks is a significant challenge and limitation. Often data from the oil and gas industry are used as a basis for the assessment of leakage frequencies. (LaChance 2009) for example used the generic leak frequencies developed for pipelines carrying hydrocarbons and applied a Bayesian approach to infer hydrogen-specific pipeline leakage frequencies.

Consequence models can be categorized in three types: integral, empirical, and numerical (Ustolin et al. 2023). The first two classes of models focus on specific phenomena and scale. They are generally easier to apply and faster to run. The main drawbacks of integral and empirical models consist of the challenge in implementing in the modeling other phenomena or complex domain, such as industrial plants (Ustolin et al. 2023). Therefore, numerical models are developed. These are based on accurate calculations of different phenomena, and they can represent complex geometries. They require highly skilled and trained personnel to obtain any results. Furthermore, the time for running simulations is significantly higher than for empirical and integral models (Ustolin et al. 2023).

The first and foremost measure to reduce the risks related to hydrogen storage, use, and handling is to develop and implement a reliable and efficient hydrogen leakage detection system (Najjar 2013). The detection system shall be able to promptly identify loss of containment and initiate the safe depressurization and blow down of the equipment.

Another important aspect is to define technical and organizational measures to reduce the risks of fire and explosion events. The effect of operational conditions on the likelihood of ignition is identified as an uncertainty (Correa-Jullian and Groth 2022).

In the following parts of this book, Chaps. 5–9 will introduce different methods for social sustainability risk management or safety risk management, including historical incident review, model-based risk analysis and consequence analysis, AI-based risk inspection, and human and organizational factors in risk management. Chapters 10–12 will discuss the risk mitigation measures, including diverse barrier management approaches and emergency management strategies.

4.3.2 Economic Sustainability Risk Management

Economic sustainability is crucial for the successful deployment of hydrogen refueling stations as part of a broader hydrogen energy ecosystem. Economic sustainability risk management for hydrogen refueling stations (HRSs) focuses on ensuring their long-term financial viability and economic efficiency (Gökçek et al. 2024). As the new energy infrastructures, HRSs face significant economic challenges, for example:

- High capital investment: Establishing hydrogen refueling stations requires substantial upfront investments in equipment such as compressors, high-pressure storage tanks, and dispensers.
- High operational and maintenance costs: The operation of HRSs is with high energy costs for compression and precooling, as well as ongoing maintenance expenses considering the technical complexity.
- Uncertain market demand: Low demand can make HRSs economically unviable in the early stages of market development.
- Policy and regulatory risks: Inconsistent or insufficient policies regarding hydrogen infrastructure may impede the stable profitability of HRSs. In addition, future policies on carbon pricing, environmental taxes, or societal resistance may affect the economic sustainability of HRSs.

Chapter 13 in this book will present the assessment of economic feasibility and cost-effectiveness of HRSs. It will consider hydrogen production costs, retail prices, and infrastructure challenges.

4.3.3 Environmental Sustainability Risk Management

Environmental sustainability risk management for hydrogen applications focuses on minimizing the environmental impacts across its production, storage, distribution, and utilization lifecycle. A structured approach is needed to evaluate and manage environmental risks.

In the environment management context, the lifecycle assessment (LCA) methodologies are widely used. LCA offers a holistic framework to evaluate the environmental impacts of hydrogen technologies. Generally, LCA encompasses four phases (Wulf et al. 2018): goal and scope definition, inventory analysis, impact assessment, and interpretation. These steps allow for a systematic evaluation of emissions, resource depletion, and other environmental impacts across all stages of the hydrogen value chain.

In short, LCA for hydrogen applications, especially for hydrogen refueling stations, can include the following tasks:

- Emission analysis: Evaluating greenhouse gas (GHG) emissions during hydrogen production and HRS operation, especially in comparing traditional fuels and other renewable energy approaches.
- Resource use assessment: Assessing the consumption of critical resources such as water, rare earth metals, and energy during hydrogen production and infrastructure development. This is particularly relevant for electrolysis, which demands substantial water input and energy.
- Non-climate impact analysis: Evaluating the potential issues like ecotoxicity, human toxicity, and land use changes associated with hydrogen infrastructure. These factors often stem from raw material extraction and the disposal of spent components.

From the risk management perspective, environmental sustainability risks associated with hydrogen technologies include

- Unexpected emission of GHG due to equipment failures and human errors.
- Potential high volumes of water in electrolysis, which could exacerbate water scarcity in arid regions.
- Waste materials during the operations of HRSs and at the end of their lifecycle HRSs, such as failed or discarded storage tanks and pipelines.
- Possible disruption of ecosystems and biodiversity due to land use changes when large-scale hydrogen projects are deployed.

To effectively address the above environmental risks, our proposed approach is to integrate LCA into the multidisciplinary risk management frameworks. Under such an integrated framework, it is expected to

- Identify and quantify environmental risks at each stage of the lifecycle of an HRS.
- Compare alternative technologies and scenarios.
- Develop strategies to mitigate environmental risks while keeping the HRS profitable and operating smoothly.

Chapter 14 in this book will present the applications of LCA in the hydrogen applications.

Acknowledgements This work is conducted with the support from the HYDROGENi Research Centre (Grant No. 333118), and funded by industrial partners and the Research Council of Norway under the Norwegian research program FMETEKN.

References

Abohamzeh E, Salehi F, Sheikholeslami M, Abbassi R, Khan F (2021) Review of hydrogen safety during storage, transmission, and applications processes. J Loss Prev Process Ind 72

Azadnia AH, McDaid C, Andwari AM, Hosseini SE (2023) Green hydrogen supply chain risk analysis: a European hard-to-abate sectors perspective. Renew Sustain Energy Rev 182

Correa-Jullian C, Groth KM (2022) Data requirements for improving the quantitative risk assessment of liquid hydrogen storage systems 47(6). https://doi.org/10.1016/j.ijhydene.2021.10.266

Gökçek M, Paltrinieri N, Liu Y, Badia E, Dokuz AŞ, Erdoğmuş A, Urhan BB, Yoldaş Ö (2024) Optimum sizing of hybrid renewable power systems for on-site hydrogen refuelling stations: case studies from Türkiye and Spain. Int J Hydrogen Energy 59(2023):715–729. https://doi.org/10.1016/j.ijhydene.2024.02.068

Haland E, Paaske B (2008) Main report-survey of hydrogen risk assessment methods

Health and Safety Executive (n.d) Managing risks and risk assessment at work. https://www.hse.gov.uk/simple-health-safety/risk/steps-needed-to-manage-risk.htm

Huser A, Bucelli M, Zakariyya K, Skinnemoen MM, Nadim P (2021) Quantitative risk analysis of scaled-up hydrogen facilities. https://hysafe.info/uploads/papers/2021/27.pdf

International Organization for Standardization (2018) ISO31000:2018 risk management-guidelines

LaChance (2009) Analyses to support development of risk-informed separation distances for hydrogen codes and standards

Najjar YS (2013) Hydrogen safety: the road toward green technology. Int J Hydrogen Energy 38:13653–14521

NASA (2003) Guide for hydrogen Hazards analysis on components and systems

Nyborg M, Arvidsson K, Johansson J, Liljemark S, Olsson L (2011) Risk analysis methodology for CO_2 transport including quantified risk calculation. Energy Procedia 4:2816–2823. https://doi.org/10.1016/j.egypro.2011.02.186

Perelli S, Genna G (2022) Hazards identification and risk management of hydrogen production and storage installations. Chem Eng Trans 96:193–198

Sand M, Skeie RB, Sandstad M, Krishnan S, Myhre G, Bryant H, Derwent R et al (2023) A multi-model assessment of the global warming potential of hydrogen. Commun Earth Environ 4(203). https://doi.org/10.1038/s43247-023-00857-8

United Nations (n.d.) Social sustainability. https://Unglobalcompact.Org/What-Is-Gc/Our-Work/Social

Ustolin F, Campari A, Giannini L, Baboi E, Paltrinieri N (2023) Identification of consequences of failure for hydrogen equipment. Chem Eng Trans 98. https://doi.org/10.3303/CET2398032

Wulf C, Reuß M, Grube T, Zapp P, Robinius M, Hake J-F, Stolten D (2018) Life cycle assessment of hydrogen transport and distribution options. J Clean Prod 199:431–443. https://doi.org/10.1016/j.jclepro.2018.07.180

Chapter 5
Historical Failure Analysis of Hydrogen Refueling Stations

Farhana Yasmine Tuhi, Marta Bucelli, and Yiliu Liu

Abstract Hydrogen-based mobility and other applications highly rely on the development of a robust hydrogen refueling infrastructure. However, some hazardous properties of hydrogen, including its wide flammability range, low ignition energy, and explosiveness, raise significant safety and reliability concerns for the deployment of refueling stations. This chapter reviews past accidents in hydrogen refueling stations (HRSs) and water electrolysis, using data from the Hydrogen Incident and Accident Database (HIAD) 2.1. Eighteen incidents were analyzed, with focus on their causes, consequences, injuries, fatalities, and lessons learned. It was found that major accidents, including explosions, were mostly due to corrosion, electrolyzer failures, and incompatible materials, while fires and hydrogen releases were linked to defective parts and leaks. Although HRS accidents were more frequent, water electrolysis plant incidents had more severe outcomes. The analysis identifies common failure modes and suggests preventive measures, emphasizing continuous monitoring, proper material testing, strict protocols, comprehensive risk assessments, and training programs.

5.1 Introduction

As a part of decarbonizing the transportation industry, hydrogen has emerged as a low-carbon and sustainable alternative to the conventional fossil fuels. Due to its ability to produce electricity through fuel cells, hydrogen has attracted a lot of attention. Car manufacturers have recently begun introducing fuel cell electric vehicles (FCEVs) to the market and are now entering a precommercial phase. They are transitioning from prototype demonstration vehicles to small-scale production (Ball and Weeda 2015). Globally, there are around 600 FECVs on the road: ~ 300 in USA, ~ 160

F. Y. Tuhi · Y. Liu (✉)
Department of Mechanical and Industrial Engineering, Norwegian University of Science and Technology (NTNU), Trondheim, Norway
e-mail: yiliu.liu@ntnu.no

M. Bucelli
SINTEF Energy Research AS, Trondheim, Norway

Y. Liu et al. (eds.), *Safety Measures for Hydrogen Fueling Stations*, Risk, Reliability and Safety Engineering, https://doi.org/10.1007/978-981-96-9412-9_5

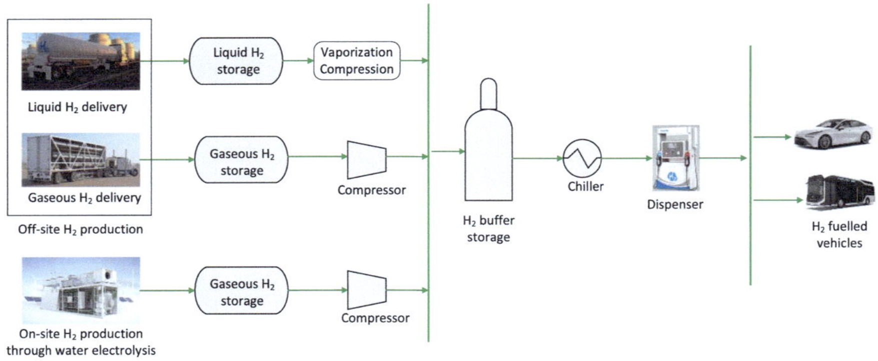

Fig. 5.1 Basic elements in hydrogen refueling stations (adapted from Apostolou and Xydis 2019; Pique et al. 2017))

in Europe, ~ 50 in Japan, and ~ 130 in South Korea (Körner 2014). In addition, there are a little over one hundred hydrogen buses used in various cities around the world (Ball and Weeda 2015). Hydrogen-based mobility depends not only on the widespread deployment of FCEVs in the market but also on the establishment of a sufficient hydrogen refueling infrastructure to meet the expectations and needs of FCEV owners. Station designs can be categorized based on the type of hydrogen production technology used and whether the fuel is generated onsite or delivered from a central production unit (Apostolou and Xydis 2019; Pique et al. 2017). The layout of a typical hydrogen refueling station (HRS) is illustrated in Fig. 5.1. Currently, there are around 300 public HRSs worldwide, located in North America, Europe, and Asia. In contrast, there are about 400,000 conventional service stations globally, highlighting that hydrogen refueling infrastructure is still in its early stages (Ball and Weeda 2015).

However, due to hydrogen's unique properties, such as its wide flammability range (4–75% in air by volume (Ono et al. 2007)), low ignition energy (0.017 mJ (McCarty 1981)), and explosiveness, maintaining the safe and reliable operation of a refueling station requires significant effort. Several accidents have raised public concerns about the safe use of hydrogen. Additionally, these accidents cause environmental and financial losses, hindering the deployment of hydrogen as an alternative fuel. Therefore, this chapter presents previous accidents in HRS to uncover the causes of failures, their consequences, and other significant information. Analyzing accident databases has become a crucial practice for enhancing reliability and safety across various industries, such as chemical, petrochemical, and nuclear sectors (Bowonder and Miyake 1988; Kosai and Yamasue 2019; Paltrinieri et al. 2012; Tamascelli et al. 2022). Historical accidents offer valuable insights into failure modes, root causes, and operational anomalies that may precede incidents (Khan and Abbasi 1999; Pasman 2009). Currently, in the hydrogen sector, there are two significant databases: the Hydrogen Incidents and Accidents Database (HIAD 2.1), created by the European Commission Joint Research Center (JRC) (HIAD 2.1 2024), and Hydrogen Tools

(H2TOOLS), developed by the Pacific Northwest National Laboratories with funding from the U.S. Department of Energy (H2Tools 2024). These databases typically include detailed variables that describe the causes, progression, and outcomes of past events, providing essential data for conducting reliability and safety analyses. Also, they gather comprehensive information on past accidents involving hydrogen and its derivatives, which can be utilized to enhance operational efficiency, implement safety measures, and prevent future incidents.

This study is based on data from the HIAD 2.1 database, which aims to provide detailed accident data, including the location of the accident, a description of the facility, operational conditions, consequences in terms of human and property losses, and lessons learned to support reliability and risk assessment. Currently, HIAD 2.1 contains well documented 713 hydrogen-related accidents, making it a credible source for valuable information on undesired hydrogen-related events. The collected records are exported into an Excel workbook with six spreadsheets labeled: events, facility, consequences, lessons learned, event nature, and references. Each accident is given a unique event ID. Notably, HIAD 2.1 provides traceable links to source information such as the French ARIA, European eMARS, British IChemE, Japanese RISCAD, several American databases, scientific articles, newspapers, and industrial reports (Wen et al. 2022). The quality of event descriptions depends on the information provided by these primary sources. Additionally, the reports in the HIAD 2.1 database are intended for public use, so confidential information is not included. This workbook is available to users upon request.

5.2 Database Research Results

In this section, the accidents retrieved from the HIAD 2.1 are presented. Each accident is identified with its unique event ID and event title as provided in the database.

#778—Explosion at the electrolysis unit of a chemical factory: In 1975, an explosion at the Laporte Industries Limited factory in Ilford, UK, caused significant damage to an alkaline electrolyzer plant and tragically resulted in the death of a plant operator. The investigation found that heavy sludge deposition in some cells of the electrolyzer led to the corrosion of electrodes and gas separators. This corrosion allowed a mixture of hydrogen and oxygen to form, which subsequently ignited.

#243—Explosion on an electrolyzer: In 1998, an explosion took place in the oxygen storage tank of an electrolyzer plant in France. While it was believed that hydrogen had entered the oxygen gas holder, the exact cause remains unknown. Fortunately, there were no injuries. The explosion was so powerful that it propelled the gas storage dome 135 m away, the collar 33 m away, and concrete elements 80 m away.

#1002—Water electrolyzer explosion: In 2005, an accident occurred in a high-pressure hydrogen water electrolysis unit at a Japanese university campus. The incident caused the electrolysis cell to catch fire, damaged pipes, and scattered peripheral equipment throughout the area. The initiating factor was identified as the exposure of the titanium electrode to oxygen under high pressure and temperature. This allowed oxygen to infiltrate the hydrogen stream, leading to an explosion.

#953—Hydrogen release at a public HRS: In 2009 in the USA, hydrogen leaked in a public HRS storing 6000 L of liquid hydrogen. After refueling a vehicle, the compressor was turned on to refill the high-pressure storage tank by compressing 4 bar hydrogen from a liquid hydrogen tank up to the 380 bars. After running for about 2 h, a crankshaft bearing of the compressor failed allowing greater movement of the shaft, which led to a shaft seal leaking hydrogen. The compressor was shut down on low suction pressure and the whole system was shut down by the emergency responders. The station's activity was interrupted, and the rescue team stopped the leak by closing the valves.

#401—Detection of hydrogen release: In 2012, a hydrogen refueling station in Norway was shut down during the filling process due to gas detection. The leak was extremely small, therefore, difficult to locate, but the gas was directed upwards toward the sensor, which triggered the station's shutdown. The safety equipment performed as intended, ensuring a safe shutdown. During inspection, the tiny leak was found about 8 cm from the end fitting, in the middle of the hose. It was suspected that the hose failure was due to extreme temperature swings, as it was the first time experiencing temperatures between − 15 and − 18 °C. The hose was sent back to the manufacturer for testing, but interestingly no leak was found then. The cause was never officially determined, but the issue was resolved.

#1038—Dispenser nozzle freezing during hydrogen refueling: In 2015 in Japan, the nozzle/receptacle connection froze while refueling the on-board hydrogen storage system at a precooling temperature of − 40 °C. Despite this, the freezing did not cause any damage to the apparatus or result in hydrogen leakage. The nozzle/receptacle connection can still fuel safely even if it becomes stuck due to ice. Under certain environmental conditions, air humidity can precipitate as ice around the nozzle/receptacle connection.

Although the connection could not be decoupled, no damage occurred to the components, and waiting for de-icing was sufficient. The standards for nozzle-receptacle designs were subsequently adapted. Research presented at the International Conference on Hydrogen Safety (ICHS) 2015 confirmed that the likelihood of hydrogen leakage or damage during nozzle/receptacle connection freezing is low, ensuring safety. Freezing at the hydrogen station will only prolong refueling time.

#463—Release at a dispenser of an HRS: In 2016 in the UK, a minor leak was discovered in the breakaway coupling of a 35 MPa dispenser at a bus refueling station during a controlled test refuel. The leak developed as the temperature was decreasing to cool the hydrogen to between − 33 and − 40 °C before entering the dispenser. The test was attended by the station operator's staff. The dispenser was taken out of service until the manufacturer installed a replacement breakaway coupling. Interestingly, the dispenser had passed the leak test during the installation and commissioning phase, but this test was conducted at ambient temperature. It is believed that the leak only occurred at low temperatures.

#884—Explosion in the storage of an HRS: In 2012, a significant hydrogen leak occurred at a refueling station on the outskirts of Oslo, Norway. The leak, estimated between 1.5 and 3 kg of hydrogen, happened shortly after refueling when the compressor began refilling the high-pressure tank. The pressure increase caused an uncompressed gasket to fail after 2.5 years of operation. Hydrogen then breached both the high-pressure and low-pressure gaskets, rapidly leaking into the atmosphere and forming an explosive cloud, which resulted in a detonation. The explosion broke windows in buildings 65 m away, and the loud blast alarmed the neighborhood. A dog panicked and jumped from a fourth-floor balcony. Parts of the fence were projected into a nearby roundabout, but no vehicles were damaged. Two car drivers were hospitalized with minor injuries caused by airbag deployment due to the shockwave. The Fire Department estimated that about 1400–2000 people were evacuated from the adjacent business district and nearby residential area. Equipment damages were estimated at $175,000.

The accident was attributed to two untightened bolts securing the gasket between the high-pressure tank and the pipeline during assembly. Although the system passed inspection initially, as pressurization did not cause an immediate failure, the gasket withstood the 900-bar pressure for over two years before eventually failing due to fatigue. Additionally, the monitoring system at the hydrogen refueling station (HRS) was not programmed to detect abnormal pressure loss in the high-pressure tank, which could have indicated a leak. The manufacturing company, NEL, took responsibility for the accident, identifying the faulty assembly as the root cause. NEL promptly shut down all similar plants worldwide for bolt inspections. The station operator, Uno-X, has not reopened their hydrogen stations even after 10 months and may have been permanently deterred from the technology.

#970—Hydrogen explosion at renewable hydrogen production facility: In 2019, a hydrogen buffer tank exploded at a renewable hydrogen production facility in South Korea, known as the Gangneung accident. Investigators believe the explosion was caused by a static spark, with the oxygen concentration exceeding the explosion threshold. The presence of oxygen in the hydrogen stream was attributed to improper operation of the water electrolyzer, which had been running below the minimum required power. However, the accident is still under investigation, and a final report is not published yet.

#889—Fire and explosion in a hydrogen production and storage facility: In 2020, an explosion occurred at a hydrogen fuel plant in the United States where hydrogen was produced through electrolysis. The blast damaged several nearby homes and was felt up to ten miles away. In some cases, cabinet doors inside homes blew open, scattering their contents. Fortunately, the on-site storage tanks did not detonate, and no employees working at the site were injured. The cause of the explosion remains unknown.

#1084—Explosion at an electrolyzer at an HRS: In 2020, an explosion occurred at a hydrogen refueling station's electrolyzer during a restart after maintenance. A loud "bang" was heard, and CCTV footage showed a cloud of water vapor being emitted from the oxygen vent line of the electrolyzer container. The gas generation system was safely shut down, and an investigation was initiated to determine the cause and potential impacts.

The on-site investigation found no visible damage to the electrolyzer equipment, including the water purification system and the oxygen exhaust pipe. However, damage was observed in the stack, where a crack had developed in the first foil on the bottom end-plate, likely causing a leak and an increase in pressure drop. This drop was not significant enough to be detected during pressure tests, allowing the stack to continue generating pressure as the generation rate exceeded the leak rate. Analysis revealed that set-points designed to detect this type of failure were adjusted from their original settings, preventing the plant from automatically alarming and stopping operations at the appropriate time. A key part of the study was to determine if the cloud was water vapor or smoke. Since no evidence of a flame or heat damage to the pipework was found, it was concluded that the cloud was water vapor, not a result of a fire. The water vapor was released due to the combination of small amounts of hydrogen and oxygen gas within the stack system.

#1037—Leakage of potassium hydroxide from a water electrolyzer: In 2021, a renewable hydrogen production plant in Germany experienced a KOH leakage from an alkaline electrolyzer. The fire alarm system was triggered by smoke, causing the electrolyzer to automatically shut down into safety mode, thanks to the installed safety system. Fortunately, there were no injuries or environmental damage. The investigation found that deposits had formed in the flow channels of the stack, leading to a series of internal events within the stack that ultimately caused the leakage.

#1018—Hydrogen car small fire during fueling: In 2022, a hydrogen car caught fire at a hydrogen fueling station in South Korea. During refueling, the owner heard an unusual noise and opened the fuel filler door to investigate. A small fire ignited when the owner discharged static electricity upon touching the fueling cap. Fortunately, the station's fire suppression system quickly extinguished the blaze.

An analysis revealed that material from the fuel inlet receptacle filter detached shortly after refueling, causing hydrogen fuel to leak into the fuel filler compartment. The manufacturer identified this as an isolated incident and concluded that a supplier had used a prototype epoxy-bonded filter instead of the original laser-welded filter. As a precaution, the manufacturer recalled several 2019 models sold in the United States, totaling 54 cars. Affected owners have been notified and are required to bring their vehicles to a local dealership for a free replacement of the hydrogen fuel inlet receptacle. Additionally, the manufacturer will reimburse any owners who previously paid for repairs related to this issue.

#1063—Leakage at a hydrogen refilling station in an airport: In 2022 in France, a hydrogen leak was detected early in the morning by hearing whistles near a pipe at an airport's hydrogen refueling station. The operators alerted the airport firefighters ten minutes later. Upon arrival, the firefighters established a 150-m security perimeter around the installation. The station, located on the border between public and private spaces, was surrounded by a fence with a gate accessible only by a code unknown to the emergency services, complicating their access. Additionally, the proximity of the airport fuel depot raised concerns about a possible domino effect in the event of an explosion of the trailer. Airport activity was impacted for 30 min, though air traffic was not affected. The emergency services had difficulty contacting the airport operator, who sent a technician only at the end of the day. The technician locked the shut-off valve of the trailer acting as hydrogen storage. Two days later, a technician precisely located and fixed the leak. The operator estimated that 5.7 kg of hydrogen was released.

The immediate cause of the leak was the failure of the joint connecting the hydrogen dispenser to the refueling hose. A lack of communication between the operator and emergency services hindered the implementation of preventive and mitigating measures. Additionally, the access logistics were not designed to optimize a quick emergency response.

#1064—Leakage at an HRS in an airport: Approximately one month after incident #1063, another hydrogen leak occurred at the same hydrogen refueling station. Around 4:30 a.m., the station's hydrogen detector was triggered, activating a sound and light alarm. The station entered safe mode, isolating the hydrogen storage system. The connection pipes were purged via the vent, and the power supply was cut off. An on-call technician from the company managing the station arrived at 6:15 a.m., confirmed that the installation was safe, and stopped the alarm. The leak was found on the compressor's filter, which was located in a restricted and confined space monitored by a gas detector. The detector triggered the alarm because the hydrogen concentration in the atmosphere had reached its safety limit (25% of the lower explosive limit). Despite this, the amount of dispersed hydrogen was very low. The defective filter was replaced.

Hydrogen refueling stations can also have onsite hydrogen production unit powered by renewable energy, as shown in Fig. 5.1. In that case, accidents that occurred in water electrolysis plants used for producing hydrogen must be analyzed as well. Very recently, (Tuhi et al. 2024) reviewed all the past accidents in a water electrolysis plant using HIAD 2.1 database. It is also important to look at the work to have a full understanding on the failure scenarios of a water electrolysis plant used for producing hydrogen. The event IDs for these accidents are as follows: HIAD #778, HIAD #243, HIAD #1002, HIAD #970, HIAD #889, HIAD #1084, HIAD #1037, and HIAD #1065. For further information on these accidents, it is suggested to look at the review article.

#1065—Fire on an alkaline electrolyzer: In 2022, leakage of KOH on an alkaline electrolyzer caused an electric arc. This ignition led to a temperature rise and subsequently fire. In response, the KOH was drained into a tank, and the electrolyzer was shut down. The safety technician quickly controlled the fire, and the unit was restarted approximately six hours later.

#1072—Fire on a hydrogen bus during refueling: In 2023, a fire erupted in a recently deployed hydrogen bus during refueling at a public HRS. Explosions were heard and seen from the bus's tanks that had just been filled. Although the investigation has not yet confirmed the cause, the most plausible explanation based on limited news and images is that the "explosion-like" events were triggered by the activation of thermally activated pressure relief devices, and the high-pressure tanks on board did not fail under pressure.

The Fire Department responded to reports of a fire at the site, closing a section of the highway as firefighters worked. According to a release from the bus company, one bus was destroyed, and the hydrogen station's dispenser was damaged. The damaged bus was valued at $1.1 million. Due to the halt of the entire fleet of 10 fuel cell buses, passengers may experience delays and temporary service disruptions. The Fire Department is conducting an investigation, and no conclusions can be drawn until the results are made public.

#1062—Leakage at a hydrogen bus refueling station: In 2023 a leakage occurred at a depot where a local bus company operates 20 fuel cell buses. A hissing sound was heard coming from a hydrogen tank at the refueling facility, prompting the local fire department to be alerted. However, the causes behind such a leakage are not reported in the HIAD 2.1 database. After approximately two hours, the fire brigade successfully closed the main hydrogen supply, temporarily resolving the issue. Local buses were able to resume service, but they could not refill their hydrogen cylinders.

Table 5.1 presents a summary of the reviewed accidents, including the information of event ID, event title, year and place of accident, cause of failure, consequences, and lessons learnt retrieved from the HIAD 2.1 database.

5.3 Discussion on the Results

From the HIAD 2.1 database, in total 18 accidents were found related to HRS and water electrolysis hydrogen production. Among these accidents, 8 were related to water electrolysis plant producing green hydrogen and rest of them were for HRS.

Table 5.1 Available information of accidents extracted from HIAD 2.1 database

Event ID	Event title	Year/Place	Failure cause	Consequence	Lessons learnt
#778	Explosion at the electrolysis unit of a chemical factory	1975/UK	Corrosion inside electrolyzer cell	Explosion Injury—1 Fatality—1	The importance of continuous monitoring of H2 and O2 concentration is highlighted. Besides, internal conditions of an electrolyzer must be analyzed to detect anomalies like sludge formation
#243	Explosion on an electrolyzer	1998/ France	Formation of H_2–O_2 mixture	Explosion Injury—0 Fatality—0	Unknown
#1002	Water electrolyzer explosion	2005/Japan	Exposure of Ti electrode to oxygen under high pressure and temperature	Explosion Injury—0 Fatality—0	Given the property of metals to ignite in the presence of high-pressure oxygen, materials should be tested under the anticipated operating conditions before constructing any equipment. Besides, risk assessment needs to be performed in both testing and temporary operation phase
#953	Hydrogen release at a public hydrogen refueling station	2009/USA	Deterioration of a component of a hydrogen compressor	Unignited hydrogen release Injury—0 Fatality—0	A gas detector was added in close proximity to the compressor shaft as a corrective action to detect any leakage. As preventing measures, predicting bearing failure were also considered. In addition, the manufacturer was contacted, and the bearing design was analyzed for possible improvements

(continued)

Table 5.1 (continued)

Event ID	Event title	Year/Place	Failure cause	Consequence	Lessons learnt
#401	Detection of hydrogen release	2012/ Norway	Leakage in the middle of the hose	Unignited hydrogen release Injury—0 Fatality—0	The failure was due to a defective part, and it was demonstrated that all safety equipment operated as designed and shut the system down
#1038	Dispenser nozzle freezing during hydrogen refueling	2015/Japan	The low temperature of the hydrogen caused the water vapor in the atmosphere to freeze around the nozzle-receptacle connection	No hydrogen release Injury—0 Fatality—0	The results of the research presented at the ICHS2015 confirmed that the possibility of hydrogen leakage or damage to the apparatus during nozzles/receptacle connection freezing is low and that it is safe
#463	Release at a dispenser of a HRS	2016/UK	Leak in the break-away coupling for a 35 MPa dispenser	Unignited hydrogen release Injury—0 Fatality—0	It is recommended to conduct a leak test during commissioning at the lowest temperature expected during operation. Additionally, a more advanced automated leak detection system should be developed to identify such leaks when trained personnel are not available

(continued)

Table 5.1 (continued)

Event ID	Event title	Year/Place	Failure cause	Consequence	Lessons learnt
#884	Explosion in the storage of a HRS	2019/ Norway	The two bolts securing the gasket were not tightened properly eventually leading to the failure	Explosion Injury—2 Fatality—0	High-pressure components need to follow precise procedures for assembly, with double witnesses. Manufacturer of the storage system enacted strict protocols for assembly with double witness (as in aviation industry). A more robust design of the outer perimeter of the refueling station was recommended to shield the nearby structures from projectiles
#970	Hydrogen explosion at renewable hydrogen production facility	2019/South Korea	O_2 entering into the hydrogen buffer tanks	Explosion Injury—6 Fatality—3	To perform emergency shutdown of hydrogen production and automatic isolation of hydrogen storage, the installation of an on-site diagnostic system is proposed
#889	Fire and explosion in a hydrogen production and storage facility	2020/USA	Unknown	Explosion Injury—0 Fatality—0	Investigation is still ongoing

(continued)

Table 5.1 (continued)

Event ID	Event title	Year/Place	Failure cause	Consequence	Lessons learnt
#1084	Explosion at an electrolyzer at a refueling station	2020/UK	Formation of hydrogen–oxygen mixture gas due to a crack in the bottom end-plate of the electrolyzer stack	Explosion Injury—0 Fatality—0	A dedicated team was set up by the electrolyzer manufacturer to review and report on the stacks optimal operating conditions for avoiding stack damage leading to cross-over situations. Also, the management of the alarms, their set-points, and escalations routes were reviewed in all active plants and improved data analysis function within the control room. A significant improvement has been the earlier escalation of alarms to the technology experts
#1037	Leakage of potassium hydroxide from a water electrolyzer	2021/ Germany	Sludge deposition in the electrolyzer stack	KOH leakage Injury—0 Fatality—0	The electrolyzer manufacturer carried out an inspection campaign to strengthen the procedures for operation, inspection, and maintenance of electrolyzers

(continued)

Table 5.1 (continued)

Event ID	Event title	Year/Place	Failure cause	Consequence	Lessons learnt
#1018	Hydrogen car small fire during fueling	2022/South Korea	Inappropriate material of the fuel inlet receptacle filter	Fire Injury—0 Fatality—0	The automatic fire fighting system of the station was able to extinguish the fire. Also, the car manufacturer performed a recall and replacement action and identified the cause. The corrective measures taken by the manufacturer to avoid repetition of such accidents are unknown
#1063	Leakage at a hydrogen refilling station in an airport	2022/France	The immediate cause of the leak is the failure of the joint connecting the hydrogen dispenser to the refueling hose	Unignited hydrogen release Injury—0 Fatality—0	The design of the access to the installation has to be reviewed to ensure prompt emergency services. The possibility of a domino effect resulting by the vicinity of airport fuel storage must be considered in the risk assessment performed during the design phase
#1064	Leakage at a hydrogen refilling station in a airport	2022/France	The initiating cause of the event was a loss of confinement of a gasket inside the filter of the hydrogen compressor	Unignited hydrogen release Injury—0 Fatality—0	Following the event, the airport operator requested the hydrogen station operator to inform local companies about risk management at hydrogen production facilities and to adjust the on-site intervention of technicians based on the needs of the local companies

(continued)

Table 5.1 (continued)

Event ID	Event title	Year/Place	Failure cause	Consequence	Lessons learnt
#1065	Fire on an alkaline electrolyzer	2022/France	KOH leakage	Fire Injury—0 Fatality—0	The frequency of predictive and preventive maintenance has been raised, and operator awareness has been heightened
#1072	Fire on a hydrogen bus during refueling	2023/USA	The cause assumed to be the opening of the thermally activated pressure relieved valve. Investigation is ongoing	Fire Injury—0 Fatality—0	Unknown
#1062	Leakage at a hydrogen buses refilling station	2023/ Netherlands	Unknown	Unignited hydrogen release Injury—0 Fatality—0	Unknown

Only a single accident (#1084) involved a HRS with on-site hydrogen production facility. This is presented as a Venn diagram in Fig. 5.2.

The accidents occurred due to a variety of factors. The most dangerous consequence, explosions, was often caused by corrosion of electrolyzers or component failures that led to the formation of flammable hydrogen–oxygen mixtures. Additionally, using materials without testing their compatibility under all operating conditions resulted in explosions in some cases. The database reports on accidents show

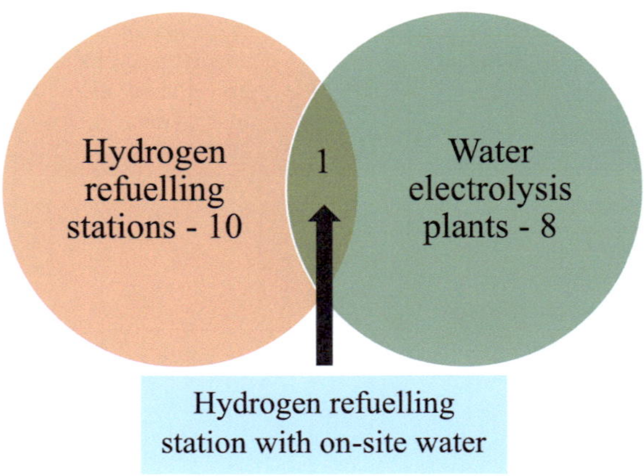

Fig. 5.2 Number of accidents extracted from the HIAD 2.1 database

that sudden or unexpected changes in operating conditions, like cell voltage, cell temperature, or gas concentrations, can indicate a failure. On the other hand, in HRS most accidents were caused by leaks in components such as gaskets, break-away couplings, and refueling hoses.

Although the number of accidents at HRSs is higher than at water electrolysis plants, the consequences are more severe for the latter. Electrolyzer failures led to serious outcome such as explosion causing injuries, and even fatalities. In contrast, most accidents at HRS involved hydrogen leaks, which were managed with prompt emergency services to prevent explosions, as seen in incidents #401, #463, #1063, and #1062. However, when HRSs have on-site hydrogen production, the risks increase, and any hydrogen leak can result in an explosion, as seen in #884. Additionally, some accidents have involved fires and the release of toxic KOH. The consequences are illustrated in Fig. 5.3 showing the number of accidents and their outcomes.

Common themes and patterns emerge from these accidents, highlighting the importance of continuous monitoring, proper material testing, strict assembly, risk assessment, and maintenance strategies. Accidents such as #778 and #1084 emphasize the necessity of real-time monitoring of hydrogen and oxygen gas concentrations to prevent the formation of flammable mixtures. Additionally, real-time monitoring helps analyzing electrolyzer operating conditions to avoid stack damage and cross-over situations. Furthermore, accidents #953 and #463 highlight the need for installing a leak detection system to promptly identify hydrogen gas leaks. Advanced automatic leak detection systems should also be installed to detect leaks in the absence of human operators.

Proper material testing is another critical aspect of ensuring the safety and reliability of hydrogen refueling stations. Materials used in the construction of electrolyzers, storage tanks, and other components must be tested under the actual operating conditions they will face. Accidents #1002 and #1018 highlight the importance

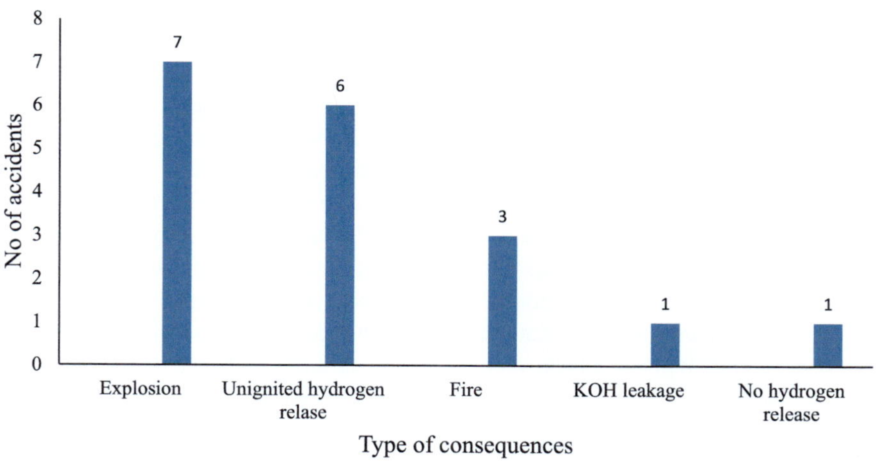

Fig. 5.3 Number of accidents and their types of consequences

of testing material compatibility under all operating conditions, including exposure to high pressure, high temperature, and the lowest temperature environments. Testing materials under these conditions can reveal potential weaknesses and allow for the selection of more suitable materials that can withstand the demands of hydrogen production and storage.

Strict assembly and maintenance protocols are essential for preventing accidents caused by human errors. Many incidents have been attributed to improper assembly, such as untightened bolts (#884) or defective parts (#401). To prevent such errors, implementing double witness procedures—where two individuals verify the correct assembly of critical components—can be effective, as highlighted in the lessons learned from incident #884. Regular maintenance and inspection of equipment are also crucial for identifying and addressing wear and tear before it results in failures, as mentioned in #1037 and #1065.

Comprehensive risk assessments are necessary to identify and mitigate potential hazards during the design and operation phases of hydrogen refueling stations. Risk assessments should consider a wide range of factors, including the possibility of domino effects, where an incident in one part of the facility could trigger additional failures in other areas, as mentioned in #1063. According to #1002, risk assessment should be performed in both testing and temporary operation phase. After accident #1064, the hydrogen station operator was requested to inform the local companies about the risk management of a HRS. By conducting thorough risk assessments, operators can implement measures to mitigate identified risks and enhance the overall safety of the facility.

Training and awareness programs for operators and technicians are vital for ensuring that safety protocols and emergency procedures are followed correctly. After accident #1065, the manufacturer raised the operator awareness. Human error during the assembly of the gasket was the primary cause of one of the major accidents in HRS (#884). It is crucial to emphasize the proper assembly of components, adherence to standard procedures, the importance of maintenance actions, and, importantly, educating operators about the relatively newer hydrogen gas.

Another crucial aspect learned from the accidents is the design of the HRS installation. The incident in #884 underscored the necessity for a robust outer perimeter to protect nearby structures from projectiles in the event of an accident. Additionally, the incident in #1063 highlighted the need for easier access to ensure prompt emergency responses, as the code system entrance complicated access.

In short, the HIAD 2.1 database provides valuable insights into the causes of component failures under actual operating conditions. It includes concise descriptions of accident scenarios in water electrolysis plants, emphasizing key components and offering a clearer understanding of failure causes and consequences. The lessons learnt, as detailed in the database, recommend precautions and necessary design and operational modifications to ensure the plant operates safely and reliably. The event descriptions help illustrate how failures and their consequences propagate throughout the plant, making the database a crucial resource for ensuring safe operation. Furthermore, analyzing these accidents aids in planning proper maintenance activities and improvement in system design to prevent failures and minimize consequences.

However, though the HIAD 2.1 database provides valuable insight regarding the failure causes, consequences, and lessons learnt from each past accident, it lacks quantitative data for performing reliability analysis or risk assessment. Additionally, the information available in the HIAD 2.1 database can only be synthesized into narratives and safety advice, offering general best practices for improving safety. Therefore, the absence of hydrogen specific database remains as a critical gap for implementing quantitative reliability and risk assessments.

5.4 Conclusions

In this chapter, an overview of the past accidents that occurred in HRS, and water electrolysis plants is provided. Hydrogen accident database HIAD 2.1 was reviewed, and the past accidents were analyzed. Information such as failure cause, consequence, injury and fatality, and lessons learnt was discussed in detail for each accident. In total 18 accidents were reported to the database related to HRS and water electrolysis plants. The most severe outcome, such as explosion, often stems from corrosion, electrolyzer failure, and the use of incompatible materials. In case of fire and hydrogen release, failure of components such as defective part, leakage in refueling hose, and inappropriate material are responsible. While HRS accidents are more frequent, the consequences are generally more severe in water electrolysis plants, where failures can lead to explosions and fatalities. Effective emergency response and prompt leak detection are crucial in managing hydrogen leaks and preventing catastrophic events.

By analyzing these accidents, common failure modes can be identified, the consequences of these failures can be understood, and effective measures to prevent future accidents can be implemented. Overall, the lessons learnt from these accidents highlight the necessity of continuous monitoring, proper material testing, strict assembly and maintenance protocols, comprehensive risk assessments, and training and awareness programs to ensure safe and reliable operation of hydrogen systems. By adopting these measures and fostering a culture of safety, the hydrogen industry can ensure the safe and sustainable deployment of hydrogen technologies, paving the way for a cleaner and more sustainable energy future.

Acknowledgements In addition to SUSHy project, this work is conducted with the support from the HYDROGENi Research Centre (Grant No. 333118), funded by industrial partners and the Research Council of Norway under the Norwegian research program FMETEKN.

References

Apostolou D, Xydis G (2019) A literature review on hydrogen refuelling stations and infrastructure. Current status and future prospects. Renew Sustain Energy Rev 113:109292. https://doi.org/10.1016/j.rser.2019.109292

Ball M, Weeda M (2015) The hydrogen economy—vision or reality? 1 1This paper is also published as Chapter 11 'The hydrogen economy—vision or reality?' In: Ball M, Basile A, Nejat Veziroglu T (eds) Compendium of hydrogen energy volume 4: hydrogen use, safety and the hydrogen economy. Elsevier. ISBN: 978-1-78242-364-5

Bowonder B, Miyake T (1988) Managing hazardous facilities: lessons from the Bhopal accident

H2Tools (2024) HydrogenTools. https://h2tools.org/

HIAD 2.1 (2024) Hydrogen incident and accident database. https://hysafe.info/hiad-2-0-free-access-to-the-renewed-hydrogen-incident-and-accident-database/

Khan FI, Abbasi SA (1999) Major accidents in process industries and an analysis of causes and consequences. J Loss Prev Process Ind 12(5):361–378. https://doi.org/10.1016/S0950-4230(98)00062-X

Körner A (2014) Technology roadmap hydrogen and fuel cells

Kosai S, Yamasue E (2019) Recommendation to ASEAN nuclear development based on lessons learnt from the Fukushima nuclear accident. Energy Policy 129:628–635. https://doi.org/10.1016/j.enpol.2019.02.058

McCarty RD (1981) Selected properties of hydrogen (engineering design data)

Ono R, Nifuku M, Fujiwara S, Horiguchi S, Oda T (2007) Minimum ignition energy of hydrogen–air mixture: effects of humidity and spark duration. J Electrostat 65(2):87–93. https://doi.org/10.1016/j.elstat.2006.07.004

Paltrinieri N, Dechy N, Salzano E, Wardman M, Cozzani V (2012) Lessons learned from Toulouse and Buncefield disasters: from risk analysis failures to the identification of atypical scenarios through a better knowledge management. Risk Anal 32(8):1404–1419. https://doi.org/10.1111/j.1539-6924.2011.01749.x

Pasman HJ (2009) Learning from the past and knowledge management: are we making progress? J Loss Prev Process Ind 22(6):672–679. https://doi.org/10.1016/j.jlp.2008.07.010

Pique S, Weinberger B, De-Dianous V, Debray B (2017) Comparative study of regulations, codes and standards and practices on hydrogen fuelling stations. Int J Hydrogen Energy 42(11):7429–7439. https://doi.org/10.1016/j.ijhydene.2016.02.158

Tamascelli N, Solini R, Paltrinieri N, Cozzani V (2022) Learning from major accidents: a machine learning approach. Comput Chem Eng 162:107786. https://doi.org/10.1016/j.compchemeng.2022.107786

Tuhi FY, Bucelli M, Liu Y (2024) Technical failures in green hydrogen production and reliability engineering responses: Insights from database analysis and a literature review. Int J Hydrogen Energy 94:608–625. https://doi.org/10.1016/j.ijhydene.2024.11.129

Wen JX, Marono M, Moretto P, Reinecke E-A, Sathiah P, Studer E, Vyazmina E, Melideo D (2022) Statistics, lessons learned and recommendations from analysis of HIAD 2.0 database. Int J Hydrogen Energy 47(38):7082–17096. https://doi.org/10.1016/j.ijhydene.2022.03.170

Chapter 6
Hazard Identification and Analysis

Andrzej Rusin, Katarzyna Stolecka-Antczak, Wojciech Kosman, and Krzysztof Rusin

Abstract The chapter presents potential safety problems in hydrogen refueling stations, which are mainly caused by uncontrolled hydrogen leakages, as well as how these hazards can be analyzed. Firstly, the severity of the leakage is determined. Basing on the mass and the pressure of the leaking hydrogen, one may evaluate the range of the cloud formed by the leakage. This cloud is the space for a potential fire or explosion. Then, the size of a fire or an explosion is evaluated for chosen locations of the source of ignition. This analysis allows to understand the severity of hazardous events.

6.1 Hydrogen as a Fuel

Nowadays hydrogen is used mostly in the refinery industry, the fertilizer production, and industrial processes. Over time, it has gained the status of an important and fast-developing pillar of the transformation in the power generation branch. The applications that utilize its potential are sought after in the transportation sector (Moriarty and Honnery 2022; DNV 2022).

This colorless and odorless gas has a wide flammability range and low minimal ignition energy, which may affect the safety of the processes that include hydrogen. Commonly known hazards related to the hydrogen properties, which are the possibilities of a fire, or an explosion may occur at any stage involving vehicles fueled with hydrogen. They may also occur in the processes of hydrogen generation, storage, or transportation. In the case of hydrogen-fueled vehicles, a potential risk is related to both: the refueling at the station and driving with a possibility of a road collision (Wu 2008; Momirlan and Veziroglu 2005).

A hydrogen fire or an explosion becomes possible whenever a gas leak appears. However, not every leak results in hazardous events. Occurrence of a fire depends among others on the presence of an ignition source and an oxidizer as well as on

A. Rusin · K. Stolecka-Antczak · W. Kosman (✉) · K. Rusin
Silesian University of Technology, Gliwice, Poland
e-mail: wojciech.kosman@polsl.pl

Y. Liu et al. (eds.), *Safety Measures for Hydrogen Fueling Stations*, Risk, Reliability and Safety Engineering, https://doi.org/10.1007/978-981-96-9412-9_6

the hydrogen concentration in the air within certain limits that pertain to hydrogen flammability (Blazquez and Thorn 2010).

Compared to other gases hydrogen has a wide range of flammability. The limits of flammability are 4 and 75% of hydrogen concentration in the air. For comparison, methane is flammable in the limits of 5–15% and propane 2.1–10.1%.

Hydrogen minimal ignition energy of 0.02 mJ is approximately 10 times smaller than the energy required to ignite methane. Lower and upper heating values of 120 MJ/kg and 142 MJ/kg respectively are over two times higher for hydrogen than for other gaseous fuels. The lower and upper heating values for methane are 50 and 55.5 MJ/kg, and for propane 46.3 and 50.2 MJ/kg. Self-ignition temperature of 585 °C is also high for hydrogen. Corresponding value for methane is 540 °C while for propane 490 °C (Suleman et al. 2015; Spiegel 2018).

An important issue that widens the range of hydrogen applications and increases the development of hydrogen industry, including the usage of hydrogen as a fuel, is the possibility to acquire hydrogen from different sources. These include electrolysis, steam reforming of methane, and photocatalysis. Additionally, if hydrogen generation processes are conducted using renewable energy sources, then they add to the clean energy transformation and create the basis for zero-emission energy generation.

While adding to the development of hydrogen industry, one must not forget that actions involving hydrogen processing, storage, and transportation must be analyzed in terms of potential hazards to people and the surroundings that result from hydrogen physicochemical properties. Analyses of this kind allow to minimize the risk of events that are dangerous for people and the environment.

6.2 Hydrogen Refueling Station

Refueling stations store and deliver hydrogen to passenger cars, trucks, and buses. Common delivery pressure levels are 35 and 70 MPa. In order to give a glimpse of hydrogen services, Fig. 6.1 presents total number of refueling stations in Europe for cars and trucks. Statistical data show that the number of stations constantly increases. For example in Poland in 2023, there were no hydrogen refilling stations at all. Currently, in 2025 there are seven stations available. In 2023 there were 170 stations in Europe, and this number increased to 200 in the first months of 2025. However, it should be noted that some of the stations are not public but remain for industry internal use only. The countries with the highest number of stations are Germany with over 80 facilities and France with 30 (H2-Map.eu).

An example of a station layout is shown in Fig. 6.2. The main sections of a typical station are a storage area, a dispenser section, and a shop/cashier. Not all sections are available for drivers that refuel their vehicles—only trained employees are allowed to enter the storage area and so are the drivers that deliver the trailers with compressed hydrogen to the station. The hydrogen may also be delivered to

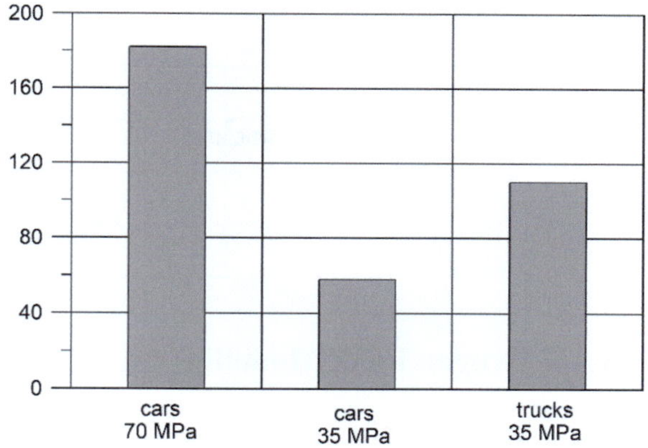

Fig. 6.1 Number of hydrogen refueling stations in Europe

the station from a pipeline or a hydrogen generation system that may include for example an electrolyzer (Rusin et al. 2024a).

The pressure of the hydrogen supplied by any of the three sources mentioned above is lower than the storage pressure. For this reason, the storage section includes a compressor, which delivers the gas to the high-pressure storage tanks. The values of the pressure in characteristic locations of the station are given in Fig. 6.3. The low-pressure storage tanks located downstream of the compressor create a buffer in the flow required to stabilize the compression process.

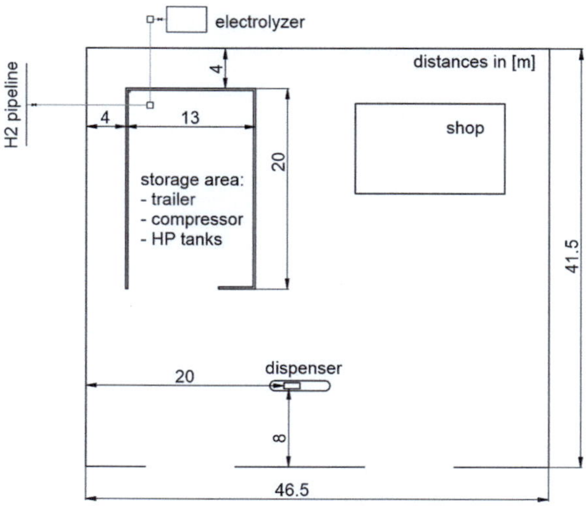

Fig. 6.2 Hydrogen station layout

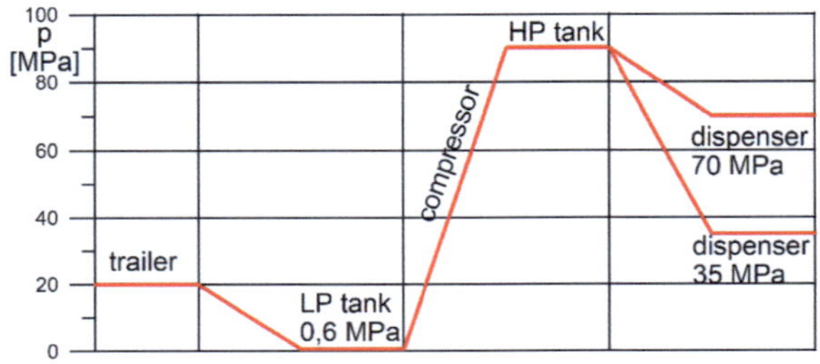

Fig. 6.3 Pressure in the hydrogen station

6.3 Hydrogen Leak Modeling

Uncontrolled hydrogen release from the structures located in a refueling station may lead to consequences like fires or explosions that are dangerous for people and the environment. The consequences start with a hydrogen leak due to a damage of a structure. This may be a result of—for example—a human error that caused a direct hit of a vehicle into structures in a station. The leaks may also result from an incorrect operation of the machines or devices in a hydrogen installation (Rusin et al. 2024a, b; Sun et al. 2014).

A hydrogen flow through a hole in a punctured structure may be determined using different analytical models. These models may be of various complexity and may include the influence of meteorological conditions, topological conditions or the properties specific to the location of the hydrogen release. The topological conditions include the shape of the area, its slope, and undulation as well as the arrangement of buildings and structures. The wind velocity and direction, and atmospheric stability belong to the meteorological conditions.

Hydrogen release and movement of the released gas are best simulated in three-dimensional models.

6.3.1 Unified Dispersion Model

The moment a leak occurs, hydrogen starts to create an invisible cloud in the air. Its shape and dimensions depend on the pressure of the hydrogen and the size of the puncture that caused the leak. There are different analytical models available to evaluate the process of the cloud creation. The one used in this work is Unified Dispersion Model, which is applied in the PHAST[1] software. The model covers different types of gas releases, for example releases that are fed constantly with new portions of the gas flowing out of the damaged structure or releases that are cut off almost immediately after detection by a safety system.

Figure 6.4 presents the limits of the volume, where hydrogen concentration may be ignited—see the beginning of this chapter. The clouds are shown in a side view

[1] PHAST is a trademark registered by DNV.

above the ground level and were obtained using PHAST software. The figure shows three clouds released from three structures in a refueling station: a trailer, a high-pressure storage tank, and a dispenser. The released hydrogen is additionally pushed by a wind blowing at 2 m/s. The puncture that caused the leak has a form of a circular hole with the diameter of 10 mm. The clouds are shown at the very moment they separate from the source of the leak and start flowing in the direction of the blowing wind, continuing to disperse.

Figure 6.5 shows the period between the beginning of the hydrogen leak and the moment of separation of the cloud from the source of the release. These are shown for a range of wind velocities from 2 to 9 m/s and for three main structures. Hydrogen pressure and the stored mass are different in each structure. If the ignition occurs before the time of the separation, the center of the fire or the explosion is close to the point of the release. Delayed ignition might potentially take place at a point, which is distant from the source of the release.

The obtained graphs show that the hydrogen clouds released from three main refueling station structures have the ranges of 15–19 m measured on a ground level from the point of the release. Above the ground the dimensions of the clouds are different. In a trailer leakage, the maximal distance of a cloud from the source is 26 m. For a dispenser this distance is 36 m, and for an HP tank it is 40 m.

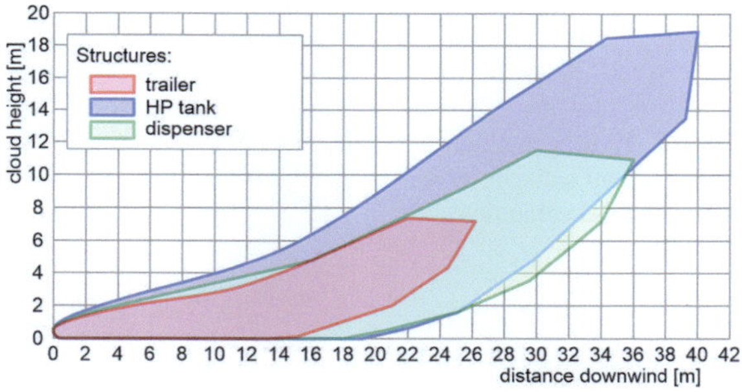

Fig. 6.4 Hydrogen released from main structures in a refueling station

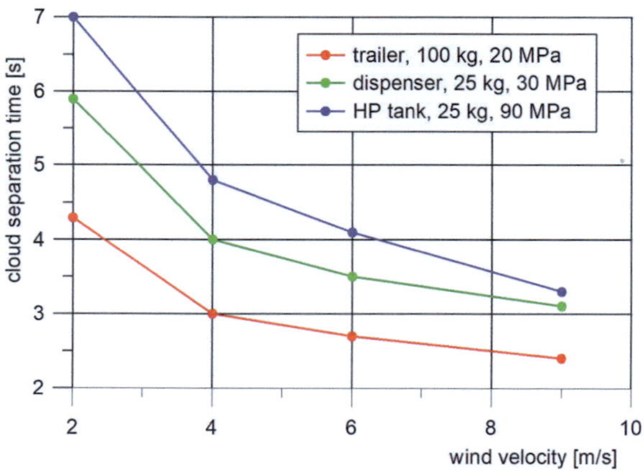

Fig. 6.5 Cloud separation time for hydrogen released from main structures in refueling station

An important factor that affects the cloud dimensions is wind velocity (Lin et al. 2024). A comparison of clouds with hydrogen concentration within the flammability limits is shown in Figs. 6.6, 6.7 and 6.8 for different wind velocities. The figures depict gas releases from the main structures of the station. The clouds are shown for wind velocities between 2 and 9 m/s at the moment of separation from the leak source. The graphs show that increased wind velocity results in longer range of a cloud along the ground level. However, the height of the cloud is smaller for faster wind. This is due to the dispersion of the hydrogen in the air under a strong wind.

Hydrogen leak creates a cloud of hydrogen and air mixture. The concentration of hydrogen in the cloud often remains between the minimal and maximal values that determine the occurrence of a fire or an explosion. They may occur if a source of

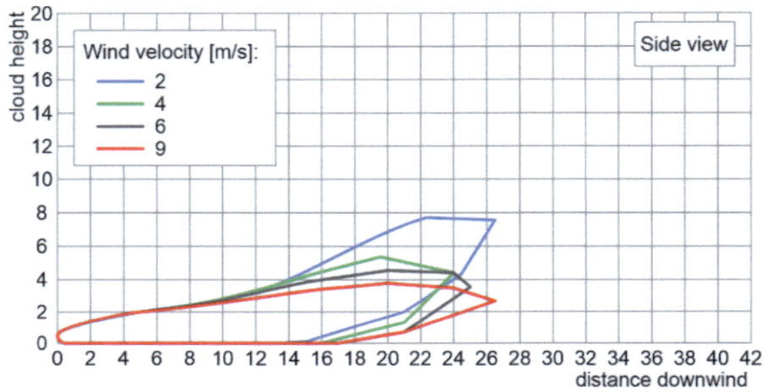

Fig. 6.6 Hydrogen cloud released from a trailer for different wind velocities

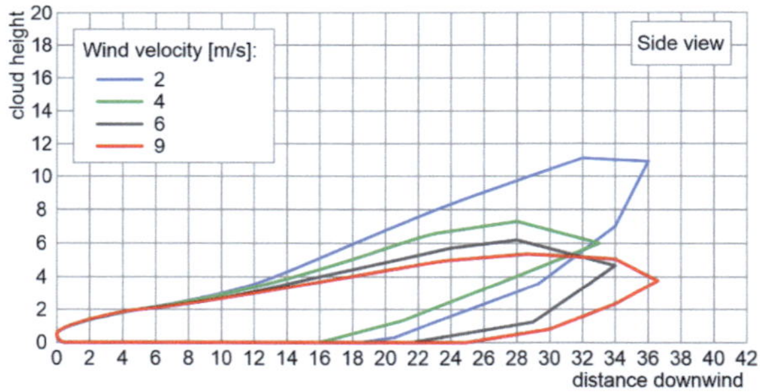

Fig. 6.7 Hydrogen cloud released from a dispenser for different wind velocities

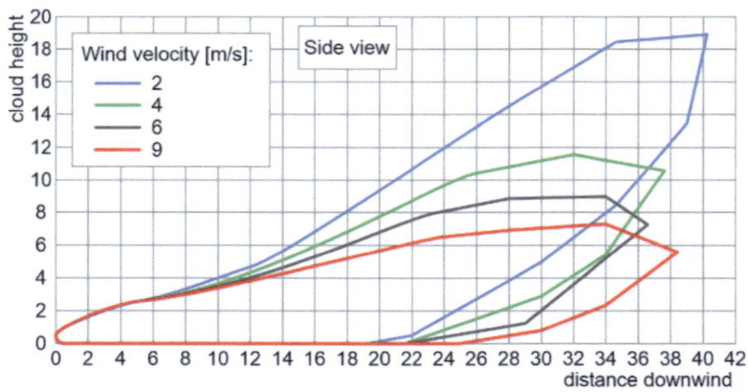

Fig. 6.8 Hydrogen cloud released from a HP tank for different wind velocities

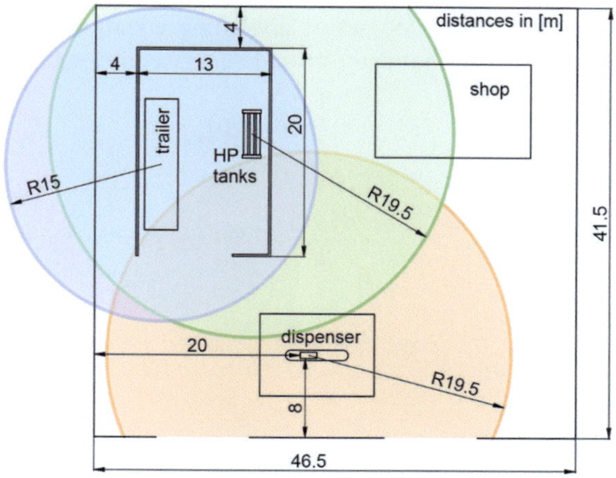

Fig. 6.9 Areas of hydrogen concentration above the lower flammability limit, wind velocity 2 m/s (Rusin et al. 2024b)

ignition lies in the space covered by the cloud. The shape and the dimensions of the cloud are formed by the blowing wind—its velocity and direction. Figure 6.9 shows the largest distances from the sources of leakage a cloud may reach before dispersing in the air. Wind velocity in this case is 2 m/s. The sources of the leakage are placed in the three main structures of a refueling station.

6.3.2 *CFD Model*

The dispersion of hydrogen in the refueling station was also modeled using Computational Fluid Dynamics software, which allowed the investigation of the complex phenomena accompanying the gas release with station structure details and the impact of weather conditions. The software used for this purpose was ANSYS Fluent, which is based on the finite volume method (ANSYS 2024).

(a) Geometry

The computational domain of the station from Fig. 6.2 is presented in Fig. 6.10. It was divided into three parts: storage, shop, and refueling zones. The whole domain was 50 m wide, 50 m long, and 20 m high. The shop zone included a building in which customers can pay for refueling. The storage zone included the open building in which the high-pressure tanks with hydrogen are stored. The refueling zone consisted of one filling terminal with a dispenser. The release area is located in the dispenser and is a circular hole with a diameter equal to 1 cm. In order to calculate the thermodynamic conditions in the puncture, an additional semi-spherical volume is added inside the dispenser to place the inlet boundary inside the terminal.

(b) Numerical Model

The air-hydrogen interaction was simulated with species transport model. For simplicity, it was assumed that air consisted of 79% of nitrogen and 21% of oxygen with no steam. One set of governing equations was solved for the air-hydrogen mixture. Mass, momentum, and energy conservation equations were solved:

$$\frac{\partial(\rho)}{\partial t} + \nabla \cdot (\rho\overline{v}) = 0 \tag{6.1}$$

$$\frac{\partial(\rho\overline{v})}{\partial t} + \nabla \cdot (\rho\overline{v}\overline{v}) = -\nabla p + \nabla \cdot \left(\overline{\overline{\tau}}\right) \tag{6.2}$$

$$\frac{\partial(\rho E)}{\partial t} + \nabla \cdot (\overline{v}(\rho E) + p) = \nabla \cdot \left(k_{\text{eff}}\nabla T - \sum_j h_j J_j + \left(\overline{\overline{\tau}} \cdot \overline{v}\right)\right) \tag{6.3}$$

where

- ρ density,
- v velocity,
- p pressure,
- k_{eff} effective conductivity,
- E energy,
- h specific enthalpy,
- J_j diffusion flux of jth species,
- τ stress tensor,
- T temperature.

The k-omega SST turbulence model (Menter 1993) constituted a turbulence closure.

Peng–Robinson equation of state (Peng and Robinson 1976) was used for each mixture component, and the density of a mixture was calculated based on the molar concentration of components. Mass diffusion coefficients were calculated based on the assumption that Lewis number for all species was unity. Critical pressure, critical temperature, and critical specific volume of mixture were based on the one-fluid

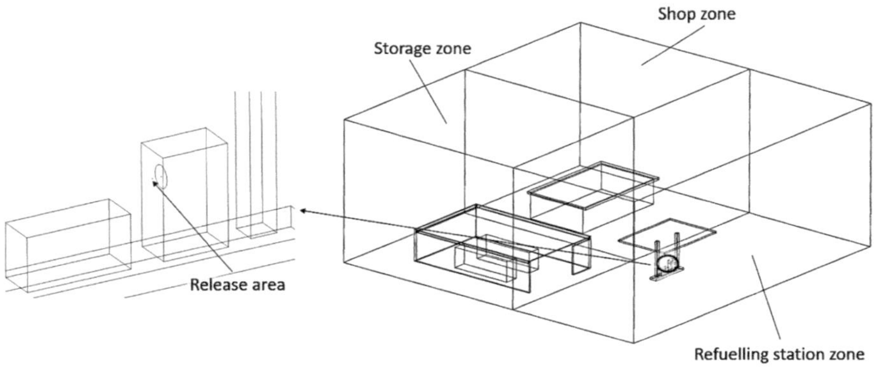

Fig. 6.10 The computational domain

van der Waals mixing rules and acentric factor based on the mole-weighted average. Specific heat capacity at constant pressure was calculated as mass-fraction average of pure species heat capacities.

Hydrogen inlet boundary conditions included total pressure equal to 35 MPa and total temperature of 233 K. Wind was blowing from the eastern side with velocities changed in subsequent cases, and the total temperature was equal to 300 K. The upper surface of the domain was the outlet with static pressure equal to 100 kPa.

The numerical grid was prepared using ICEM software. The mesh selected for the calculations underwent a mesh independence study and consisted of hexahedrons with around 10 M nodes. The mesh was refined in the vicinity of the release zone to capture the flow complexity as it formed the release jet. A fragment of the mesh is presented in Fig. 6.11.

In the numerical simulations, the assumed amount of hydrogen released from the dispenser was 5 kg.

(c) Selected Results

The figures present the molar concentration of hydrogen cloud 2 s after the initialization of the leakage for different weather conditions. In the first case (Fig. 6.12a), the wind was blowing at an angle of 45° to the hydrogen release direction with the velocity equal to 2.5 m/s. In the second case (Fig. 6.12b), the angle was 90° and velocity totaled 9 m/s. It is shown that in both cases the wind impacted the shape of the cloud resulting in non-symmetrical behavior. In both cases the jet impinged against the station column and hydrogen accumulated under the station's roof. Wind blowing westward pushed a portion of hydrogen inside the storage building, but only

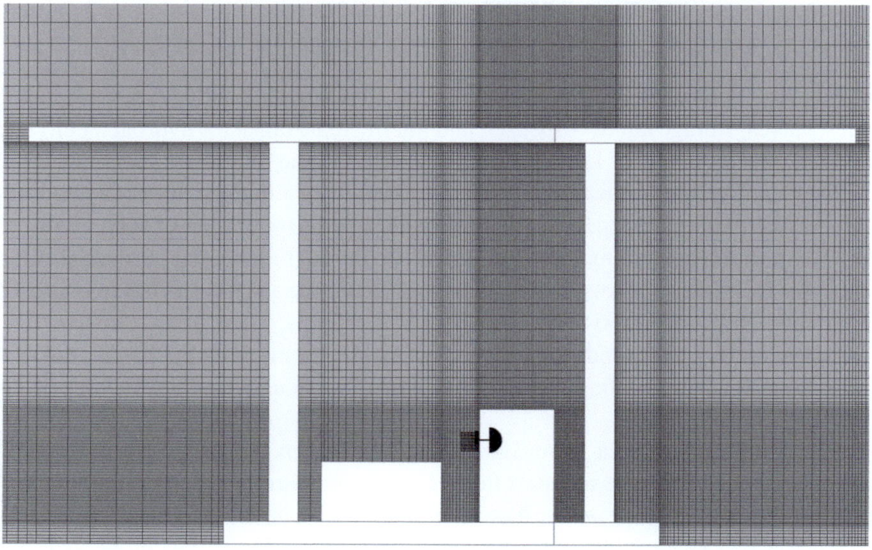

Fig. 6.11 Numerical grid

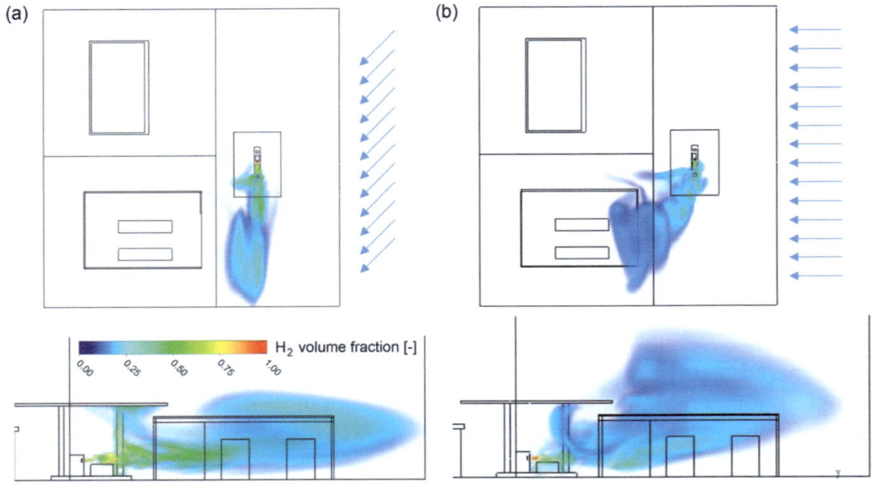

Fig. 6.12 Molar concentration of hydrogen at 2 s: wind speed 2.5 m/s at 45° (**a**) and 9 m/s at 90° (**b**)

in the second case. In the first case, hydrogen managed to move upwards due to buoyancy before it entered the building zone. For this reason, storage building is a risk zone only in the second case.

6.4 Hazards in Hydrogen Systems

An analysis of safety for HRSs must take into consideration the potential locations of leakages. They may occur in the whole infrastructure of a station, which includes pipes, valves, storage tanks, dispenser, and intermediate tanks for hydrogen delivery. The potential locations of leakage are also the cars in the refueling process.

The identification of hazardous scenarios associated with the failure of a refueling station is an important step in the risk assessment of the operation of such a station. An example of a development of hazardous scenarios triggered by an uncontrolled release of hydrogen is shown in the form of an event tree analysis (ETA) in Fig. 6.13 (Rusin et al. 2024a; Suzuki et al. 2021; Pan et al. 2016).

If a leakage is present and the conditions of the fire triangle are satisfied, hazardous events such as a jet fire, a flash fire, and a fireball may occur. An explosion may occur as well.

A jet fire occurs when a gas or liquid is released from a tank or a pipeline and is subsequently ignited. The released flow is under pressure higher than the ambient pressure and creates a long, stable flame with an intensive heat radiation. For relatively small release velocities, the flame is created at the cavity in the punctured area. For

Fig. 6.13 Event tree analysis

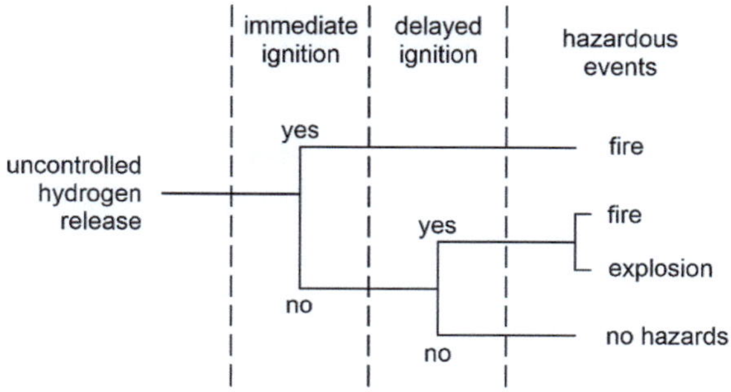

larger velocities the flame is separated from the cavity and is stabilized at a distance from the release location (Bosch and Waterlings 2005; Lin et al. 2024).

A flash fire is triggered when a moving cloud of a mixture consisting of a flammable gas and air is ignited. If this happens, the period when people and the environment are exposed to the fire depends on the concentration of the flammable substance and in a lesser degree on a wind velocity. This period is short and lasts a couple of seconds. The shape of a flame depends on the initial shape of a cloud (CCPS 1984; Shelke and Wen 2021).

A fireball is created when a large number of flammable gases or vapors is rapidly released. A cloud is created that firstly has a shape of a semi-sphere and then a sphere. This is due to the thermal buoyancy forces. A fireball may also be created in a BLEVE phenomenon—Boiling Expanding Liquid Vapor Explosion. This phenomenon occurs when a pressure tank is exposed to an outside flame and heated. The pressure inside a tank increases until the pressure vessel is fractured. The contents of the tank are rapidly thrown out through the created cavities. It forms a cloud, which is immediately ignited (CCPS 1984; Bosch and Waterlings 2005).

All of the above-mentioned events lead to negative consequences for people and structures in a station in the form of death or injury and the destruction of the infrastructure.

The hazards related to fires derive from a direct interaction of a flame and the generation of heat flux. On the other hand, the hazards related to explosions create a pressure wave that affects people and structures. Chosen values of heat fluxes and pressure waves are shown in Figs. 6.14 and 6.15 (LaChance and Tchouvelev 2011; Tretsiakova-McNally 2022; Baraldi et al. 2011; Stolecka and Rusin 2020).

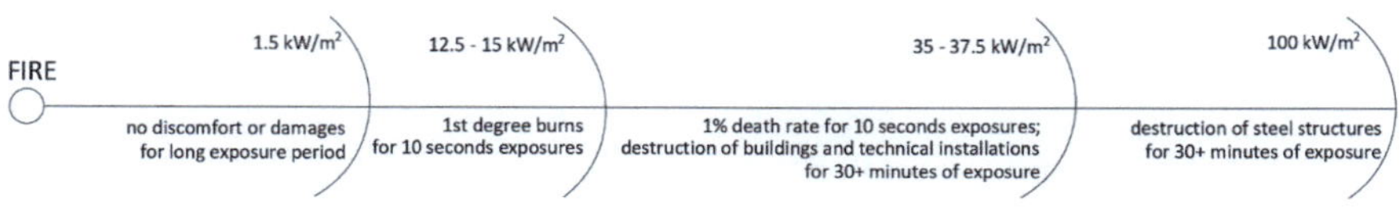

Fig. 6.14 Consequences of heat radiation for humans and the environment

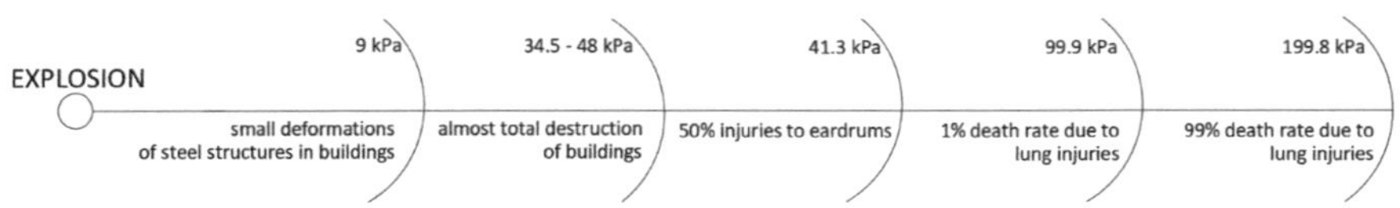

Fig. 6.15 Consequences of overpressure for humans and the environment

6.4.1 Hydrogen Fire

Any failure in a structure containing hydrogen under pressure that breaks the integrity of the structure may lead to an uncontrolled release of the gas. If the hydrogen cloud is ignited, then the leak results in a fire. There are a number of fire types ranging from a jet fire to a fire ball. The specific type of fire that occurs after an ignition depends on the location of the puncture in a structure and the moment of the ignition. In case of pressurized tanks their punctures often evolve into a jet fire. This type of a fire has a long and stable flame sustained by the pressure of the hydrogen inside the punctured structure that keeps on delivering new portions of hydrogen. Nevertheless the progression of this phenomenon is difficult to determine, when wind must also be taken into account (Bosch and Waterlings 2005).

The other type of fire is a fireball followed by a so-called BLEVE process. The magnitude of the fireball and the exposure time to the heat depend on the mass of the leaked hydrogen (Shen et al. 2023).

Figures 6.16, 6.17 and 6.18 show the areas covered by heat fluxes at the level of 12.5 kW/m^2 (blue color) and 37.5 kW/m^2 (red color) for a jet fire initialized with a hydrogen release from the main facilities in a refueling station. The maximal range of the hazard zones in the presented cases are similar. However, it should be noted that a release of different amounts of hydrogen stored in the facilities significantly affects the period it takes for hydrogen to flow out of the structures; therefore it also affects the exposure time to the consequences of a fire.

The hazard zones with the heat flux above 37.5 kW/m^2 shown in Figs. 6.16, 6.17 and 6.18 reach 14.5 m for a trailer puncture, 20 m for a puncture of a pipe in a dispenser, and 26.6 m for an HP tank failure. These are the ranges of the fire that may result in a human death. The size of the hazard zone that may result in skin burns is proportionally larger.

A significant factor that affects the level of hazard related to jet fires is wind blowing with a given velocity. Figures 6.19, 6.20 and 6.21 show the hazardous areas due to heat flux generated from a fire that causes people's death. The results of calculations are shown for wind velocities of 2, 4, 6, and 9 m/s.

Figures 6.22, 6.23 and 6.24 show the results of the analysis for a rapid release of hydrogen and a fire in the shape of a fireball (BLEVE). The resulting heat fluxes are shown in Fig. 6.14. The hazardous areas in Figs. 6.22, 6.23 and 6.24 were derived using the PHAST software. The analysis was conducted for a number of cases with different amounts of hydrogen in the chosen structure. Wind velocity for all these cases is 2 m/s.

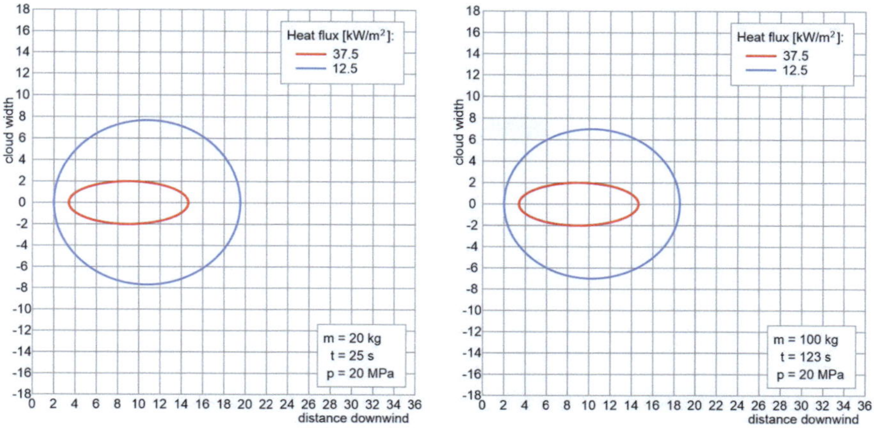

Fig. 6.16 Hazard zone for a jet fire of hydrogen released from a trailer (20 MPa)

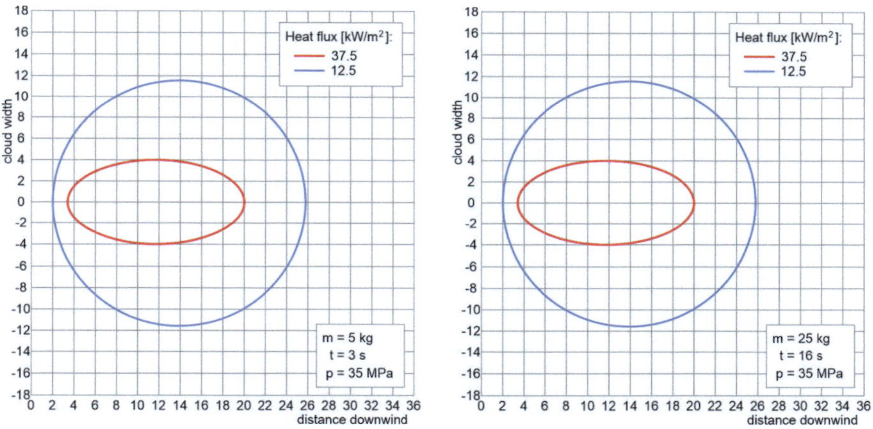

Fig. 6.17 Hazard zone for a jet fire of hydrogen released from a dispenser (35 MPa)

The graphs in Figs. 6.22, 6.23 and 6.24 indicate that in case of a rapid hydrogen release and the occurrence of a fireball/BLEVE, the largest hazardous area related to people's death has a range of around 30 m and appears for a release of 100 kg of hydrogen from a trailer. For a dispenser and an HP tank, the amounts of hydrogen are similar. Therefore, the range of the hazardous areas is practically the same and reaches 19 m.

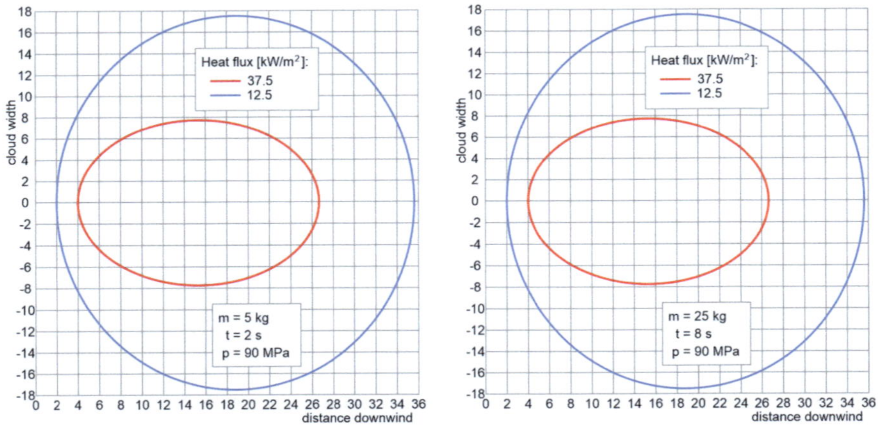

Fig. 6.18 Hazard zone for a jet fire of hydrogen released from a HP tank (90 MPa)

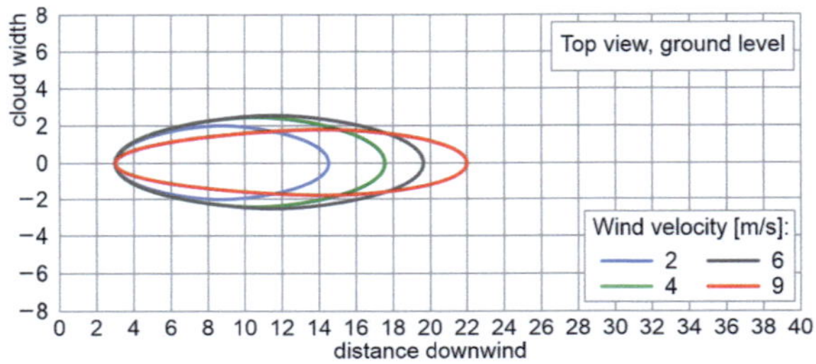

Fig. 6.19 Hazard zone for a jet fire of hydrogen released from a trailer for various velocities of winds

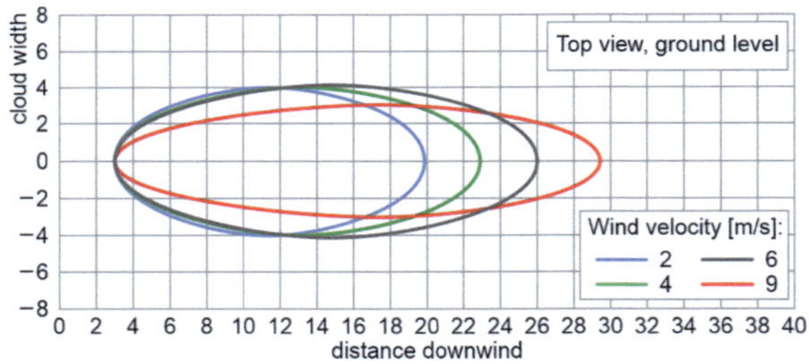

Fig. 6.20 Hazard zone for a jet fire of hydrogen released from a dispenser for various velocities of winds

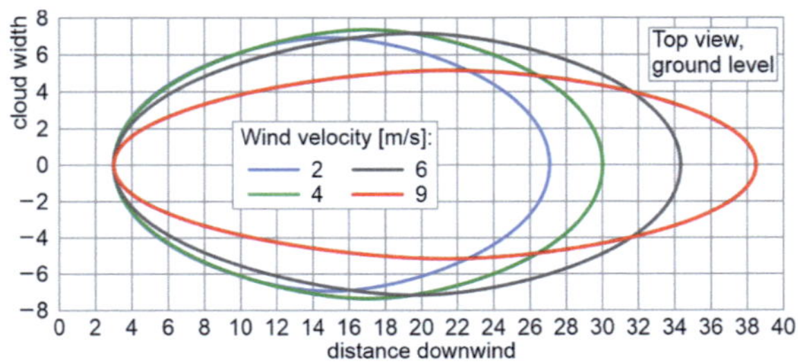

Fig. 6.21 Hazard zone for a jet fire of hydrogen released from an HP tank for various velocities of winds

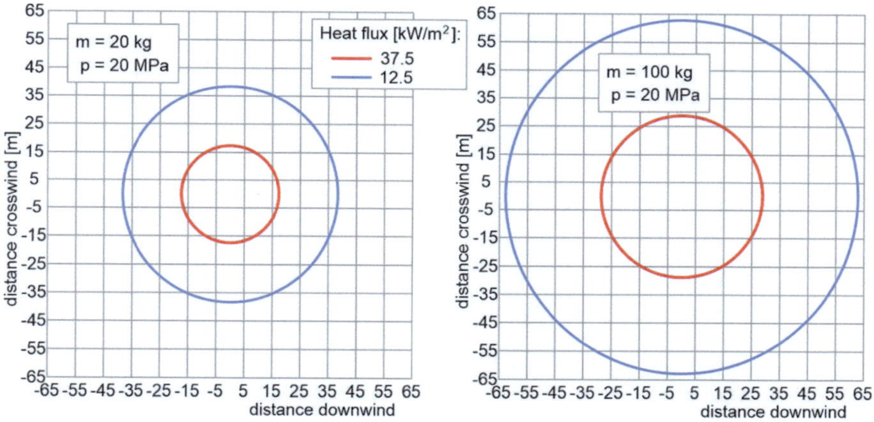

Fig. 6.22 Hazard zone for a fireball of hydrogen released from a trailer (20 MPa)

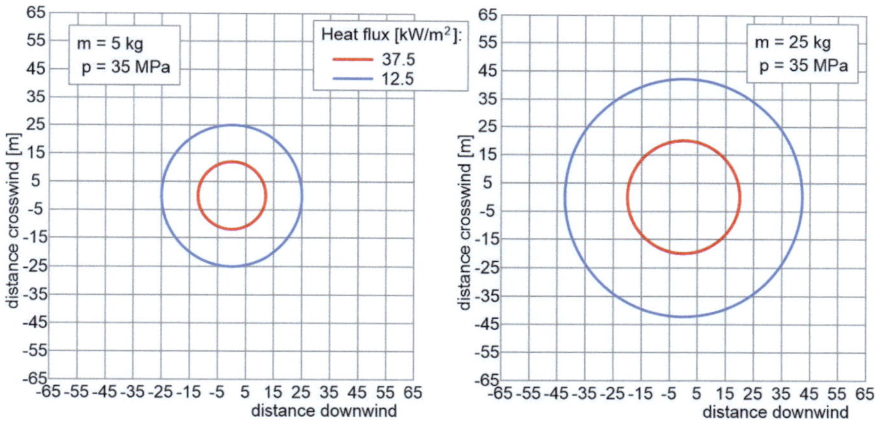

Fig. 6.23 Hazard zone for a fireball of hydrogen released from a dispenser (35 MPa)

6.4.2 Hydrogen Explosion

Uncontrolled hydrogen release in a refueling station may also end in an explosion. Such a hazard scenario may happen when an ignition source is present in a moving cloud of hydrogen at a distance from the release location or at the release point. The level of hazard related to an explosion is affected among others by the amount of hydrogen that is ignited. If an explosion occurs at the location of the release, the amount of hydrogen is larger when compared to a situation, where a cloud moves away from the release location and hydrogen disperses in the air. The dispersion process is additionally intensified by a blowing wind. Negative consequences of an explosion are flying debris from punctured structures and a pressure wave. They cause hazard for people and infrastructure (Sun et al. 2014; Rusin et al. 2024b).

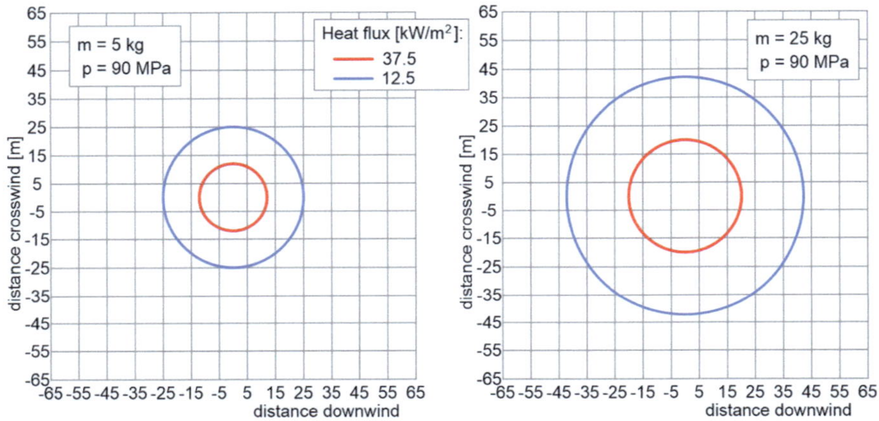

Fig. 6.24 Hazard zone for a fireball of hydrogen released from an HP tank (90 MPa)

The figures present examples of hazard zones related to an explosion of hydrogen released from a trailer, a dispenser, and an HP tank. Figures 6.25, 6.26 and 6.27 depict the situation when the ignition is present at the location of the release, which is an immediate puncture of an installation. Figures 6.28, 6.29 and 6.30 show the ignition at a distance from the release point with a flow of hydrogen through a hole of 10 mm diameter. The areas in the figures contain an overpressure wave of 50 and 100 kPa. These levels of pressure knock people down or cause 1% death rate, respectively. Also, these values of pressure waves cause significant damage to infrastructure. None of the analyzed cases presented here produced a wave of 206.9 kPa, which causes 100% death rate. The results shown in the figures were obtained using the PHAST software for wind velocity of 2 m/s.

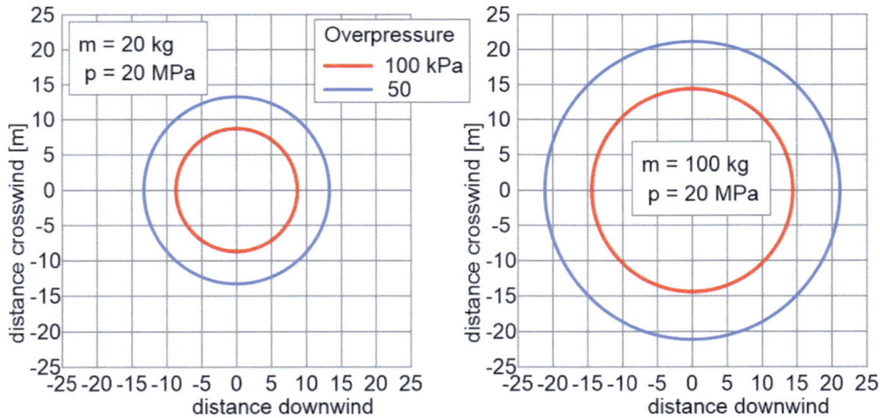

Fig. 6.25 Hazard zone for an explosion of hydrogen released from a trailer (20 MPa, ignition at the point of hydrogen release)

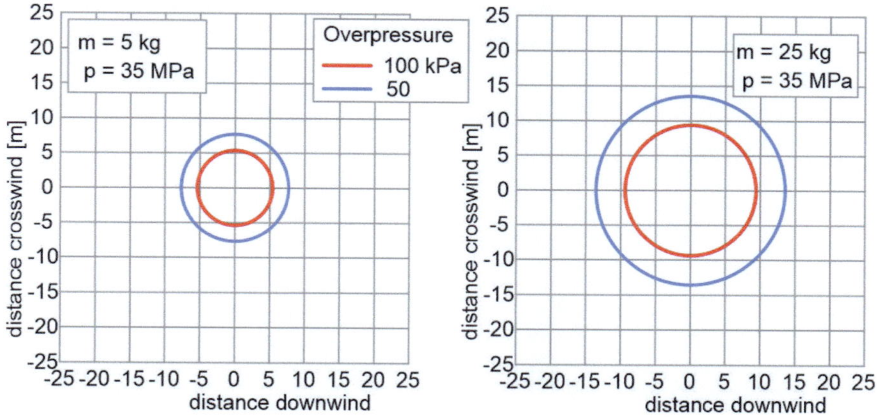

Fig. 6.26 Hazard zone for an explosion of hydrogen released from a dispenser (35 MPa, ignition at the point of hydrogen release)

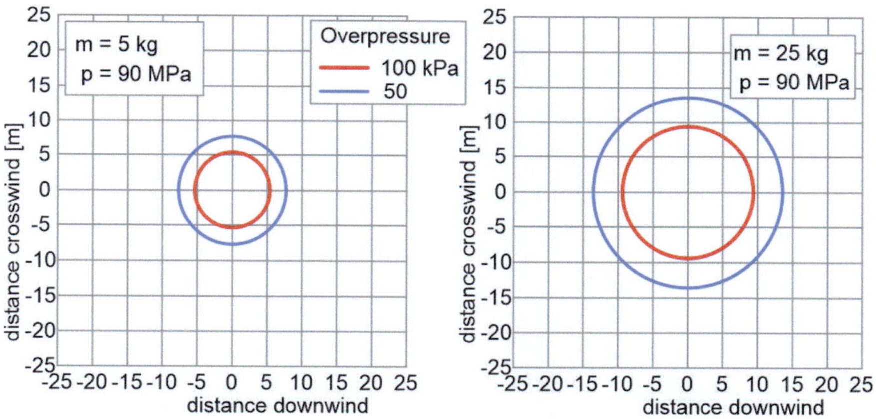

Fig. 6.27 Hazard zone for an explosion of hydrogen released from a HP tank (90 MPa, ignition at the point of hydrogen release)

The locations of the ignition were chosen on the basis of the worst-case scenario. They are located at maximal distances from the release point that the hydrogen concentration is still above the lower flammability limit. For a dispenser this distance is 40 m from the hydrogen release point, and for a trailer it is 30 m. The distance for an HP tank is 40 or 50 m depending on the amount of the released hydrogen.

The results of the calculations presented in the figures indicate that in the case of a released hydrogen explosion, its amount is the dominating factor that affects the level of hazard.

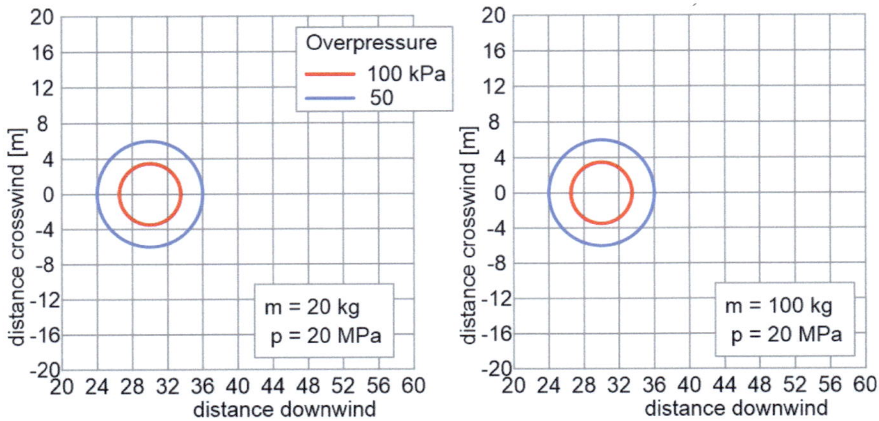

Fig. 6.28 Hazard zone for an explosion of hydrogen released from a trailer (20 MPa, ignition at a distance from the point of hydrogen release)

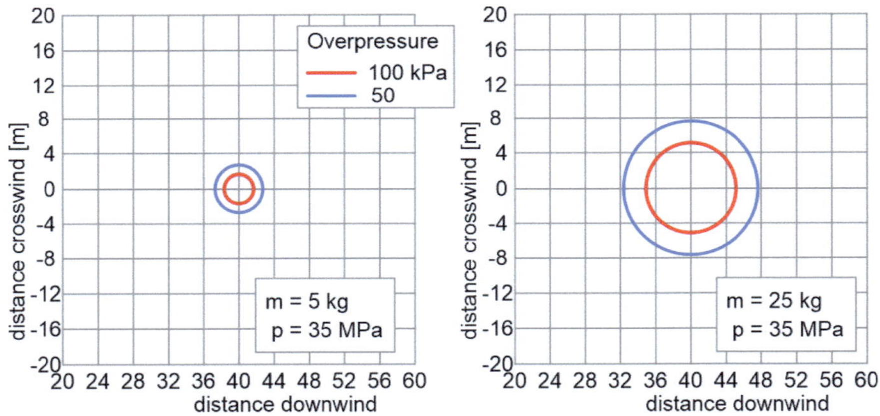

Fig. 6.29 Hazard zone for an explosion of hydrogen released from a dispenser (35 MPa, ignition at a distance from the point of hydrogen release)

6.5 Summary

This chapter describes an analysis of potential scenarios of hazardous events related to uncontrolled hydrogen release from structures located in a refueling station. The analyzed cases included a partial damage and a total rupture of the main structures in a station: a dispenser, a trailer, and an HP tank. Due to hydrogen physicochemical properties, an uncontrolled release may lead to a fire or an explosion, which may affect the people's health or even life. The negative consequences of these events may also destroy infrastructure and lead to significant economic losses.

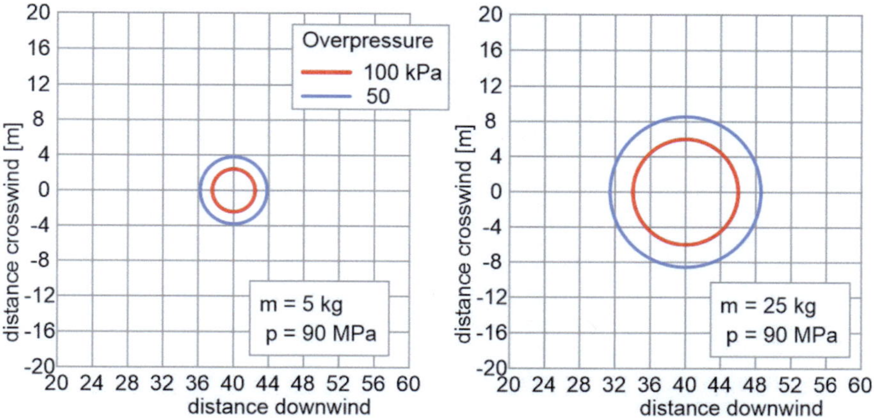

Fig. 6.30 Hazard zone for an explosion of hydrogen released from a HP tank (90 MPa, ignition at a distance from the point of hydrogen release)

The scale of hazards caused by an uncontrolled hydrogen release from station structures depends on the hydrogen parameters, including first of all the gas pressure. It depends also on the amount of the released hydrogen. Other factors also have their impact, including meteorological conditions like the velocity and direction of wind.

References

ANSYS documentation 2024 R1

Baraldi D, Papanikolaou E, Heitsch M, Moretto P, Cant RS, Roekaerts D, Dorofeev S, Kotchourko A, Middha P, Tchouvelev AV, Ledin S, Wen J, Venetsanos A, Molkov VV (2011) Prioritisation of research and development for modelling the safe production, storage, delivery and use of hydrogen. Joint Research Centre Institute for Energy and Transport

Blazquez E, Thorn C (2010) Fire and explosions. Anaesth Intensive Care Med 11(11):455–457

Bosch CJH, Waterlings RAPM (eds) (2005) Yellow book, CPR14E. TNO, Committee for the Prevention of Disasters, The Hague, The Netherlands

CCPS (1984) Guidelines for evaluating the characteristics of vapor cloud explosion, flash fires and BLEVEs. Center of Chemical Process Safety of the American Institute of Chemical Engineers

DNV (2022) Hydrogen forecast to 2050. In: Energy transition outlook. DNV

LaChance J, Tchouvelev A (2011) Development of uniform harm criteria for use in quantitative risk analysis of hydrogen infrastructure. Int J Hydrogen Energy 36:2381–2388

Lin Y, Ling X, Yu A, Liu Y, Liu D, Wang Y, Wu Q, Lu Y (2024) Modeling of hydrogen dispersion, jet fires and explosions caused by hydrogen pipeline leakage. Fire 7(1):8

Menter F (1993) Zonal two equation k-ω turbulence models for aerodynamic flows. In: AIAA 23rd fluid dynamics, plasmadynamics, and lasers conference, AIAA 93-2906

Momirlan M, Veziroglu TN (2005) The properties of hydrogen as fuel tomorrow in sustainable energy system for a cleaner planet. Int J Hydrogen Energy 30:795–802

Moriarty P, Honnery D (2022) When will the hydrogen economy arrive? AIMS Energy 10(6):1100–1121

Pan X, Li Z, Zhang C, Lv H, Liu S, Ma J (2016) Safety study of a wind-solar hybrid renewable hydrogen refueling station in China. Int J Hydrogen Energy 41:13315–13321

Peng DY, Robinson DB (1976) A new two-constant equation of state. Ind Eng Chem Fundam 15(1):59–64

Rusin A, Stolecka-Antczak K, Kosman W, Rusin K (2024a) Influence of driver error on the level of hydrogen refuelling station risk. Int J Hydrogen Energy 73:73–85

Rusin A, Stolecka-Antczak K, Kosman W, Rusin K (2024b) The impact of the configuration of a hydrogen refueling station on risk level. Energies 17:5504

Shelke AV, Wen JX (2021) The burning characteristics and flame evolution of hydrocarbon and hydrogen flash fires. Proc Combust Inst 38(3):4699–4708

Shen Y, Lv H, Hu Y, Li J, Lan H, Zhang C (2023) Preliminary hazard identification for qualitative risk assessment on onboard hydrogen storage and supply systems of hydrogen fuel cell vehicles. Renew Energy 212:834–854

Spiegel C (2018) Characteristics and safety of hydrogen. Fuel Cell Store. https://www.fuelcellstore.com/blog-section/hydrogen-characteristics-safety

Stolecka K, Rusin A (2020) Analysis of hazards related to syngas production and transport. Renew Energy 146:2535–2555

Suleman F, Dincer I, Agelin-Chaab M (2015) Environmental impact assessment and comparison of some hydrogen production options. Int J Hydrogen Energy 40(21):6976–6987

Sun K, Pan X, Li Z, Ma J (2014) Risk analysis on mobile hydrogen refueling stations in Shanghai. Int J Hydrogen Energy 39:20411–20419

Suzuki T, Shiota K, Izato Y, Komori M, Sato K, Takai Y, Ninomiya T, Miyake A (2021) Quantitative risk assessment using a Japanese hydrogen refueling station model. Int J Hydrogen Energy 46:8329–8343

Tretsiakova-McNally S (2022) Harm criteria for people and environment, damage criteria for structures and equipment. Grant agreement No: 325348

Wu Y (2008) Assessment of the impact of jet flame hazard from hydrogen cars in road tunnels. Transp Res Part C Emerg Technol 16:246–254

Chapter 7
Artificial Intelligence in Risk-Based Inspection of Hydrogen Infrastructure

Nicola Paltrinieri

Abstract The global transition to sustainable energy systems has positioned hydrogen as a critical energy carrier. While its benefits, such as high energy density and compatibility with renewables, are clear, integrating hydrogen into existing infrastructure poses significant safety and reliability challenges, particularly related to material degradation mechanisms like hydrogen embrittlement and hydrogen-enhanced fatigue. Traditional risk-based inspection (RBI) methodologies, which optimize maintenance efforts based on component risk levels, fall short in addressing the complexities introduced by hydrogen environments. This chapter introduces HyRIS (Risk-Based Inspection Strategy for Hydrogen Technologies), an innovative framework that integrates artificial intelligence (AI) with conventional RBI practices to address hydrogen-induced degradation mechanisms. By leveraging machine learning models, HyRIS predicts the probability and severity of fatigue-induced failures, enabling precise risk assessments and tailored inspection strategies. The chapter details the methodology's principles, its applications to hydrogen pipeline integrity, and its broader implications for ensuring the safe and reliable operation of hydrogen infrastructure. The integration of AI with RBI frameworks represents a transformative approach to managing the risks associated with hydrogen, supporting its role in the energy transition.

7.1 Introduction

Hydrogen is increasingly recognized as a key component in the global transition toward sustainable energy systems. As a versatile energy carrier, hydrogen offers unique benefits such as high energy density, compatibility with renewable energy sources, and the potential for near-zero emissions when produced through green technologies like water electrolysis (Norwegian Ministry of Petroleum and Energy 2020). It can serve diverse roles, including long-term energy storage, integration of

---

N. Paltrinieri (✉)
Department of Mechanical and Industrial Engineering, Norwegian University of Science and Technology (NTNU), Trondheim, Norway
e-mail: nicola.paltrinieri@ntnu.no

Y. Liu et al. (eds.), *Safety Measures for Hydrogen Fueling Stations*, Risk, Reliability and Safety Engineering, https://doi.org/10.1007/978-981-96-9412-9_7

variable renewable energy sources, and powering fuel cell systems for transportation (Chatzimarkakis et al. 2021). These attributes make hydrogen indispensable for decarbonizing hard-to-abate sectors such as heavy industry, shipping, and aviation, positioning it as a cornerstone of future energy strategies.

However, widespread hydrogen adoption requires the development of safe, efficient, and economical transport systems. Existing natural gas pipeline networks present a promising solution for scaling up hydrogen transport infrastructure with minimal initial investment (Lipiäinen et al. 2023). Despite its promise, integrating hydrogen into existing systems introduces significant technical challenges that must be addressed to ensure reliable operations.

Hydrogen poses unique challenges to the structural integrity of pipelines, stemming from its interaction with pipeline materials. Exposure to hydrogen-rich environments accelerates material degradation processes, including hydrogen embrittlement and hydrogen-enhanced fatigue. These phenomena significantly reduce the load-bearing capacity of pipeline steels, increase crack propagation rates, and shorten the service life of components (Alvaro et al. 2019; Campari et al. 2023). The impact is particularly severe in the presence of pre-existing flaws, geometrical imperfections, and cyclic loading conditions such as those caused by line packing and pressure fluctuations during operation (European Gas Pipeline Incident Data Group 2018; Pipeline and Hazardous Materials Safety Administration 2022).

Hydrogen-induced degradation depends on multiple factors, including operating pressure, temperature, material composition, and the frequency and amplitude of load cycles. For example, ferritic steels commonly used in pipelines exhibit significantly higher fatigue crack growth rates in hydrogen environments compared to inert conditions, with reductions in service life that can reach several orders of magnitude (Yamabe et al. 2016; Takakuwa et al. 2019). Furthermore, welds and heat-affected zones are particularly vulnerable due to residual stresses and uncontrolled microstructures (Drexler et al. 2019). These challenges underscore the need for specialized inspection and maintenance strategies tailored to the unique demands of hydrogen infrastructure.

Inspection and maintenance are critical for ensuring the integrity and safety of hydrogen transport systems. The risk-based inspection (RBI) approach, developed by the American Petroleum Institute (API), prioritizes inspection activities based on component risk levels, optimizing resources while maintaining system safety and reliability (American Petroleum Institute 2019). RBI methodologies have been successfully applied to various industries to mitigate risks associated with mechanical fatigue, corrosion, and other degradation mechanisms. However, the current RBI standards are inadequate for addressing hydrogen-induced degradation due to their reliance on damage factors calibrated for conventional operating environments (Wang et al. 2012; Defteraios et al. 2020).

To address these limitations, an enhanced methodology, named HyRIS (Risk-Based Inspection Strategy for Hydrogen Technologies), has been proposed by Campari et al. (2024a). HyRIS integrates artificial intelligence (AI) techniques with conventional RBI processes to account for hydrogen-specific degradation mechanisms, such as hydrogen-enhanced fatigue crack growth. By leveraging machine learning models, HyRIS predicts the probability and severity of fatigue-induced failures in hydrogen environments. This approach improves the accuracy of risk assessments and facilitates the development of optimized inspection plans for hydrogen infrastructure (Campari et al. 2024a).

This chapter presents a detailed examination of the HyRIS methodology, its underlying principles, and its application to managing hydrogen pipeline integrity. By addressing the unique challenges posed by hydrogen environments, HyRIS offers a significant step forward in ensuring the safety and reliability of hydrogen infrastructure, supporting the broader energy transition.

7.2 Hydrogen-Enhanced Fatigue and Material Challenges

Fatigue is a pervasive failure mechanism in structural materials subjected to cyclic loading. Characterized by the initiation and propagation of cracks, it often stems from surface imperfections, welds, or material inhomogeneities. As cyclic stresses result in crack growth at critical locations, the component's structural integrity can be compromised over time, eventually resulting in catastrophic failure. Fatigue failure accounts for 80–90% of mechanical failures in industrial systems (Fetters 1964). Environmental factors, such as exposure to reactive or corrosive gases, further accelerate degradation by interacting with the material at a microscopic level.

The presence of hydrogen introduces unique and severe challenges to fatigue performance. Hydrogen atoms penetrate the metal lattice, weaken interatomic bonds, and exacerbate fatigue degradation through processes such as hydrogen embrittlement and hydrogen-enhanced fatigue crack growth (Barnoush and Vehoff 2010). These mechanisms significantly increase the crack propagation rate, particularly in pipeline steels. Hydrogen's impact is most pronounced in welded zones, where residual stresses and microstructural inhomogeneities create favorable conditions for embrittlement and crack growth (Drexler et al. 2019).

The extent of hydrogen-induced degradation depends on a complex interplay of parameters. Among these, operating pressure is a critical factor. Higher hydrogen partial pressures increase the concentration of hydrogen atoms within the metal lattice, creating conditions that accelerate crack propagation. This relationship is particularly evident in ferritic steels, where fatigue crack growth rates rise significantly under elevated pressures (Slifka et al. 2018).

Temperature also plays a pivotal role, as hydrogen-enhanced fatigue crack growth is most severe near-ambient temperatures. Ferritic steels, commonly used in pipelines, exhibit peak degradation rates between 20 and 40 °C, while martensitic steels show similar vulnerabilities in slightly lower temperature ranges (Takakuwa et al. 2019; Yamabe et al. 2016). These observations highlight the importance of temperature control in mitigating fatigue degradation.

The purity of hydrogen and the presence of impurities further influence fatigue performance. Certain impurities, such as oxygen and sulfur dioxide, inhibit hydrogen uptake by forming protective surface films. In contrast, methane and carbon dioxide, frequently present in natural gas-hydrogen blends, exhibit negligible effects on hydrogen embrittlement under specific operating conditions (Komoda et al. 2019; Somerday et al. 2013). These results underscore the need for detailed compositional analysis when assessing pipeline risk, as the effects of impurities may vary depending on pressure, temperature, and material characteristics.

Material microstructure is another determinant of hydrogen-induced fatigue. Ferritic and martensitic steels, while widely used for pipelines, vary significantly in their compatibility with hydrogen environments. Microstructural features such as grain boundaries, inclusions, and phase distributions affect hydrogen diffusion and crack propagation. Not all steels within these classifications exhibit the same degree of susceptibility; for instance, finer-grained microstructures with controlled residual stresses tend to perform better in hydrogen environments compared to coarser structures (Campari et al. 2024b).

Finally, loading conditions profoundly affect fatigue performance. Cyclic stresses, characterized by their amplitude and frequency, directly influence crack growth rates. Low-frequency cycles allow hydrogen more time to accumulate at the crack tip, exacerbating its detrimental effects. Similarly, high-stress amplitudes increase the stress intensity range, further accelerating crack propagation. These effects combine to reduce the fatigue life of pipeline materials, making precise monitoring and control of operational loads essential (Yamabe et al. 2016; Nanninga et al. 2010).

The influence of these environmental parameters, material characteristics, and loading conditions on hydrogen-enhanced fatigue is summarized in Table 7.1.

The complex interactions among these parameters create significant challenges for designing, operating, and maintaining hydrogen pipelines. Understanding the

Table 7.1 Key parameters influencing hydrogen-enhanced fatigue

Parameter	Characteristics	References
Operating pressure	Higher pressures increase hydrogen concentration in metal lattices, accelerating fatigue crack growth	Slifka et al. (2018)
Temperature	Maximum degradation occurs at near-ambient temperatures (20–40 °C for ferritic steels)	Takakuwa et al. (2019) and Yamabe et al. (2016)
Hydrogen purity	Impurities like oxygen can mitigate hydrogen uptake; methane and CO_2 have negligible effects under specific conditions	Komoda et al. (2019) and Somerday et al. (2013)
Microstructure	Ferritic and martensitic steels exhibit varying susceptibilities depending on grain boundaries, phase distributions, and residual stresses	Campari et al. (2024b)
Loading conditions	Low frequencies and high-stress amplitudes exacerbate crack propagation in hydrogen environments	Yamabe et al. (2016) and Nanninga et al. (2010)

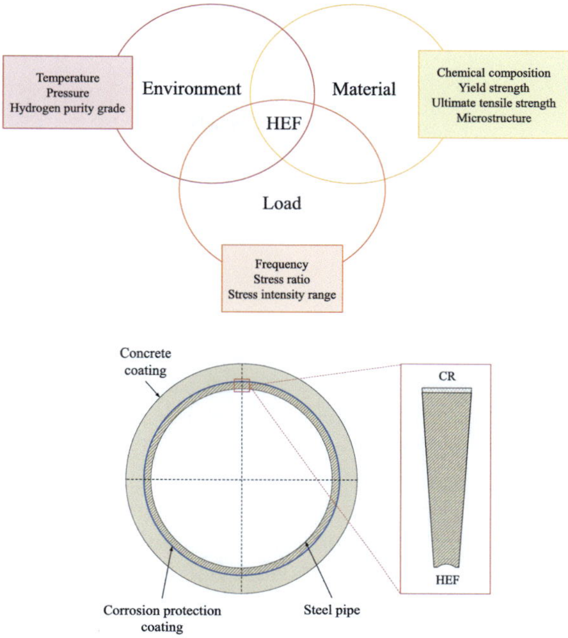

Fig. 7.1 Synergy of material, environment, and mechanical loading for the occurrence of hydrogen-enhanced fatigue (HEF)

interplay of material properties and operating conditions is critical for developing inspection strategies that address hydrogen's unique degradation mechanisms, thereby ensuring the safety and reliability of hydrogen infrastructure. The synergistic interaction between there factors is illustrated in Fig. 7.1.

7.3 Risk-Based Inspection Framework

RBI is a structured methodology used to optimize the inspection and maintenance of industrial systems by focusing efforts on the most critical components. Its primary objective is to balance safety, reliability, and cost-effectiveness by prioritizing inspection activities based on the assessed risk of failure for each component. Risk, in this context, is evaluated as a function of two factors: the probability of failure (P_f) and the consequence of failure (C_f). By systematically identifying high-risk components, RBI allows resources to be allocated efficiently, reducing the likelihood of catastrophic failures while minimizing downtime and inspection costs (American Petroleum Institute 2019).

The RBI process typically involves several key steps, as shown in Fig. 7.2:

1. **Data Collection**: Gathering historical and operational data for the plant, including design specifications, operating conditions, and inspection history.

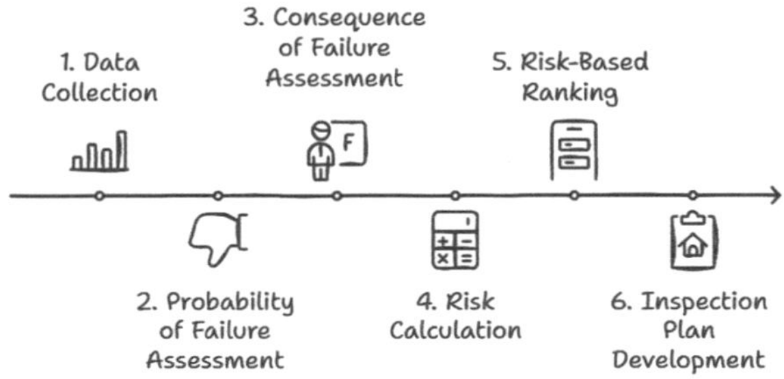

Fig. 7.2 Risk-based inspection key steps

2. **Probability of Failure Assessment**: Evaluating the likelihood of failure for each component. As indicated in Eq. 7.1, the probability of failure can be calculated as the product of three factors: generic failure frequency (gff), damage factor (D_f), and management system factor (F_{MS}). The damage factor, a function of time (t) and inspection effectiveness (I_E), is the most crucial parameter and accounts for the characteristics of each component. It depends on the susceptibility to various damage mechanisms, which is influenced by various environmental parameters, loading conditions, and material characteristics:

$$P_f(t, I_E) = \text{gff} \cdot D_f(t, I_E) \cdot F_{MS} \quad (7.1)$$

3. **Consequence of Failure Assessment**: Estimating the potential impacts of failure, including safety hazards, environmental consequences, and financial losses. In this case, the impact areas of flammable releases are evaluated through integral models and consolidated consequence analysis techniques. The release rates are calculated for each hole size, weighting the flammable fatality areas ($CA_{flam,n}$) on the event probability (gff_n), as per Eq. 7.2:

$$CA_{flam} = \frac{\sum_{i=1}^{4} gff_n \cdot CA_{flam,n}}{\sum_{i=1}^{4} gff_n} \quad (7.2)$$

4. **Risk Calculation**: Combining P_f and C_f to quantify the risk for each component, according to the definition of Kaplan and Garrick (1981) in Eq. 7.3.

$$R_f(t, I_E) = P_f(t, I_E) \cdot C_f \quad (7.3)$$

5. **Risk-Based Ranking**: Prioritizing components based on their calculated risk levels.
6. **Inspection Plan Development**: Designing an inspection strategy that addresses high-risk components while optimizing resource allocation.

RBI methodologies are widely used in industries such as oil and gas, chemical processing, and power generation. For example, the API recommended practices, including API 580 and API 581, provide detailed guidelines for implementing

RBI in various contexts. However, these frameworks were developed for traditional industrial environments and do not account for the unique degradation mechanisms introduced by hydrogen-rich operating conditions.

Hydrogen technologies present significant challenges for traditional RBI frameworks due to the complex and often unpredictable nature of hydrogen-induced material degradation. Conventional RBI methodologies rely on predefined damage factors and historical data to assess P_f. However, these assumptions are not readily applicable to hydrogen environments, where degradation mechanisms such as hydrogen embrittlement and hydrogen-enhanced fatigue are highly context-dependent and not well-represented in existing standards (Wang et al. 2012; Defteraios et al. 2020).

For example, hydrogen-induced fatigue degradation depends on factors such as pressure, temperature, material composition, and cyclic loading conditions. These variables interact in nonlinear ways, making it difficult to predict failure probabilities using conventional methods. Furthermore, the lack of comprehensive datasets on hydrogen-metal interactions exacerbates uncertainties in risk assessments, leading to either overly conservative or insufficiently cautious inspection strategies.

Another challenge lies in quantifying the C_f for hydrogen-related failures. Hydrogen leaks can result in fire hazards, explosions, or environmental impacts, but the severity of these consequences depends on the specific operating conditions and mitigation measures in place. Traditional consequence models, designed for hydrocarbons or other industrial substances, often underestimate the rapid dispersal and flammability of hydrogen, leading to inaccuracies in risk calculations (Giannini et al. 2023).

These limitations necessitate the development of enhanced RBI methodologies that explicitly account for the unique risks associated with hydrogen technologies.

7.4 Integrating Data-Driven Approaches

To address the challenges posed by hydrogen environments, integrating data-driven approaches into RBI frameworks offers a promising pathway for improving risk assessment accuracy and reliability. Advanced tools such as machine learning (ML) and AI can augment traditional RBI methodologies by providing predictive insights into failure probabilities and degradation mechanisms.

Data-driven models can analyze complex interactions among variables such as pressure, temperature, material properties, and loading conditions to predict the likelihood of hydrogen-induced failures. For instance, Campari et al. (2024b) proposed HyRIS, which integrates AI techniques with conventional RBI processes. HyRIS proposes the use of machine learning models to predict hydrogen-enhanced fatigue crack growth rates based on operating and material conditions, providing a more nuanced understanding of P_f in hydrogen environments.

While these predictive capabilities offer significant potential, their effectiveness depends on the availability and quality of hydrogen-specific datasets. Additionally, data-driven models may face challenges in generalizing predictions across different operating contexts or material systems. Nonetheless, when used in conjunction with traditional RBI frameworks, AI algorithms can dynamically adjust inspection strategies, ensuring resources are allocated effectively to high-risk components. For example, ML models can identify trends in material degradation over time, enabling preemptive inspections and reducing the likelihood of unexpected failures.

By combining the systematic framework of RBI with the analytical power of AI, data-driven approaches address the key limitations of traditional methodologies. They enhance the ability to assess risks in hydrogen technologies, accounting for the unique challenges posed by hydrogen-induced degradation.

7.5 Machine Learning-Enhanced RBI

The integration of ML into RBI frameworks represents a significant advancement in addressing the challenges associated with hydrogen-induced material degradation. ML models excel in identifying complex, nonlinear relationships between multiple variables, making them particularly suited to predict hydrogen-enhanced fatigue crack growth (HEFCG) and other degradation mechanisms.

The HyRIS methodology (Campari et al. 2024b) demonstrates the potential for ML models to enhance fatigue risk assessment. Figure 7.3 illustrates a schematic of the HyRIS approach. The different steps of the methodology are presented in the following. In this case, a Gradient Boosting Machine (GBM) model was adopted. This ensemble learning algorithm combines the predictive capabilities of multiple decision trees, iteratively improving accuracy by correcting previous errors. In the context of hydrogen technologies, ML models predict parameters such as the fatigue acceleration factor (FAF), quantifying the increase in crack growth rate in hydrogen environments relative to inert conditions. These predictions are used to prioritize inspection activities within the RBI framework.

7.6 ML-Based Prediction: Data Preparation, Training, and Evaluation

Developing an ML model for RBI involves several key steps, each critical to ensuring accurate and reliable predictions.

1. **Data Preparation**: The first step involves compiling and cleaning data from various sources, including experimental results and material testing repositories. For example, the Technical Reference for Hydrogen Compatibility of Materials (San Marchi and Somerday 2012), maintained by Sandia National Laboratories, provides extensive datasets on the mechanical behavior of materials in hydrogen environments. Attributes such as material composition, microstructure, operating pressures, temperatures, and cyclic loading conditions are collected,

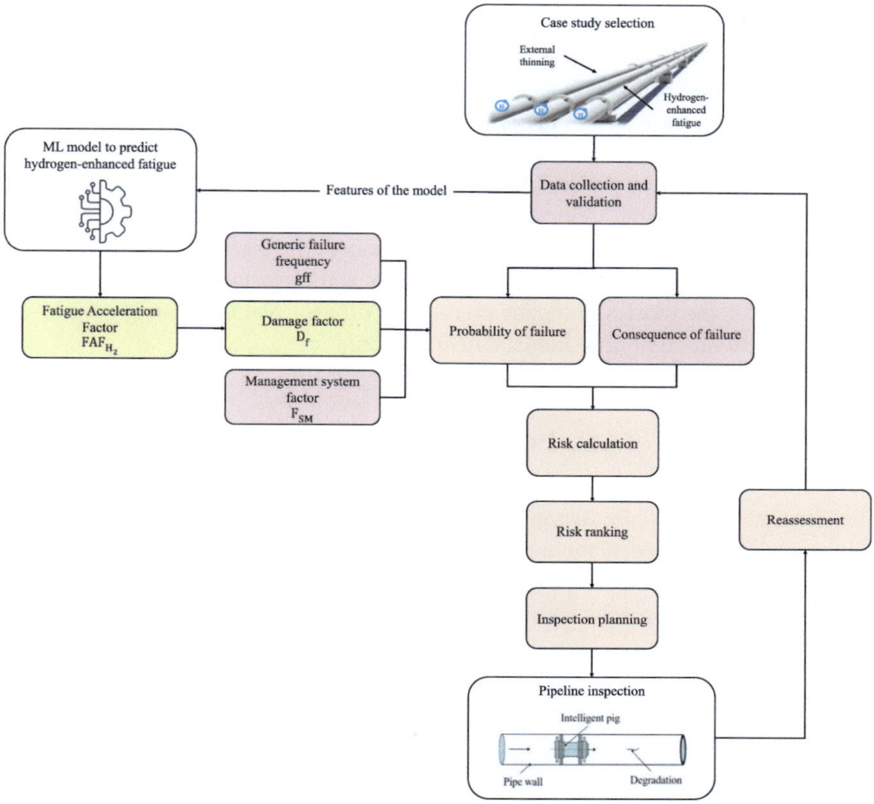

Fig. 7.3 Flow diagram of the HyRIS methodology, derived from the integration of the risk-based inspection framework and a data-driven model. Adapted from Campari et al. (2024b)

preprocessed, and merged into a unified repository. Features with low variability or redundancy are removed, while categorical attributes, such as steel grades and microstructures, are encoded for compatibility with ML algorithms. Continuous variables, such as stress intensity ranges and hydrogen partial pressures, are normalized to ensure uniform scaling.

2. **Model Training**: The dataset is divided into training and evaluation subsets, typically in a 70:30 ratio. The training set is used to fit the GBM model, optimizing hyperparameters such as the number of decision trees, learning rate, and tree depth. The model iteratively adjusts its predictions by minimizing errors between predicted and actual FAF values. During training, the model learns to identify patterns and relationships among input features, such as the correlation between microstructure and hydrogen-induced crack growth rates or the influence of cyclic loading conditions on material degradation.

3. **Model Evaluation**: Once trained, the model is tested on the evaluation subset, a previously unseen portion of the data, to assess its generalization capability. Performance metrics such as accuracy, precision, recall, and F_1-score are used to evaluate classification outcomes. Additionally, the area under the precision-recall curve (AUC-PR) is analyzed to measure the model's ability to handle imbalanced datasets, particularly when predicting severe FAF categories like high and extreme. Feature importance analysis is conducted to determine the relative contribution of each input variable to the model's predictions. Parameters such as stress intensity range, operating pressure, and material microstructure often emerge as the most significant predictors, offering valuable insights into the factors driving hydrogen-induced degradation.

7.7 Integration of ML Predictions into RBI Processes

The integration of ML predictions into RBI frameworks enhances the accuracy and efficiency of inspection planning for hydrogen technologies. The HyRIS methodology incorporates ML-derived FAF predictions into the calculation of the P_f. This integration encompasses the following steps:

1. **Adjusting Damage Factors**: The ML model's FAF predictions are used to adjust the damage factor for mechanical fatigue within the RBI framework. For example, depending on the predicted severity class (e.g., negligible, low, medium, high, extreme), the damage factor may be scaled by predefined multipliers to reflect the accelerated degradation caused by hydrogen environments.
2. **Dynamic Risk Assessments**: The revised damage factors are combined with traditional P_f metrics (i.e., generic failure frequencies and management system factors) to calculate updated P_f values. These dynamic risk assessments enable the prioritization of inspection efforts for high-risk components, ensuring a targeted and cost-effective approach to maintenance.
3. **Inspection Planning and Optimization**: Based on the updated risk profiles, inspection plans are developed to address components with the highest risk levels. For example, pipeline segments exposed to severe cyclic loading and high hydrogen pressures may be scheduled for immediate inspection using advanced non-destructive testing (NDT) techniques, such as ultrasonic testing or magnetic flux leakage.
4. **Continuous Improvement**: The ML model's predictions are continuously refined using new data collected from inspections and operational monitoring. This iterative process reduces uncertainties over time and enhances the model's ability to generalize to diverse scenarios, further improving the robustness of the RBI framework.

By incorporating ML-based predictions, the RBI process becomes more adaptive, addressing the unique challenges posed by hydrogen technologies. This integration not only improves the reliability of hydrogen infrastructure but also supports the safe and efficient deployment of hydrogen as a critical energy carrier in the global energy transition.

7.8 Case Studies on Pipeline Integrity

To evaluate the application of RBI methodologies enhanced with ML, this section explores case studies involving pipelines subjected to various operational scenarios. The selected pipelines represent typical configurations used in energy transport, including those designed for natural gas and those repurposed or newly constructed for hydrogen or hydrogen-natural gas blends. Each case study examines pipelines with differences in materials, operating conditions, and degradation mechanisms, as summarized in Table 7.2.

- **Pipeline A**: A conventional natural gas pipeline constructed from API 5L X70 steel. It operates at a pressure of 70 bar and a temperature range of 10–30 °C. The pipeline is selected as a baseline for assessing the risks associated with traditional transport systems.
- **Pipeline B**: A repurposed natural gas pipeline used for transporting pure hydrogen. This pipeline, also constructed from API 5L X70 steel, experiences operating pressures of 60 bar and a temperature range of 15–40 °C.
- **Pipeline C**: A newly constructed pipeline designed specifically for hydrogen transport, made from a API 5L X60 steel with enhanced resistance to hydrogen embrittlement. It operates at 80 bar and 20–35 °C.
- **Pipeline D**: A pipeline used for transporting a 30:70 hydrogen-natural gas blend. This system, constructed from API 5L X65 steel, operates at 75 bar and 10–40 °C.

Each scenario includes simulated operational profiles incorporating cyclic loading conditions representative of real-world operation, such as compressor start-stop cycles and pressure variations during gas delivery.

The case studies reveal distinct degradation behaviors for pipelines under different transport scenarios (Table 7.2).

Pipeline A, carrying natural gas, experiences degradation mechanisms primarily driven by mechanical fatigue and external corrosion. Fatigue crack growth rates remain moderate, as hydrogen-induced effects are absent, allowing conventional RBI frameworks to provide reliable inspection planning.

In Pipeline B, repurposed for hydrogen transport, an increased rate of fatigue crack growth is observed due to hydrogen embrittlement and hydrogen-enhanced fatigue. Weld zones and pre-existing defects are particularly vulnerable, necessitating adjustments to inspection intervals. Advanced NDT techniques, such as phased-array ultrasonic testing (PAUT) or acoustic emission monitoring, are recommended to detect early signs of material degradation.

Pipeline C, designed for hydrogen transport, illustrates the benefits of material optimization. The use of low-carbon steel reduces susceptibility to hydrogen

Table 7.2 Summary of pipeline characteristics, operating conditions, predicted degradations, and inspection recommendations

Pipeline	Material	Transported gas	Operating conditions	Observed degradation mechanisms	Inspection recommendations
Pipeline A	API 5L X70 steel	Natural gas	Pressure: 70 bar; temperature: 10–30 °C; cyclic loading: moderate	Mechanical fatigue and external corrosion	Conventional RBI sufficient; focus on external corrosion and geometric defects
Pipeline B	API 5L X70 steel	Pure hydrogen	Pressure: 60 bar; temperature: 15–40 °C; cyclic loading: high	Hydrogen embrittlement, hydrogen-enhanced fatigue, external corrosion	Shortened inspection intervals; use PAUT and acoustic emission monitoring
Pipeline C	API 5L X60 steel	Pure hydrogen	Pressure: 80 bar; temperature: 20–35 °C; cyclic loading: moderate	Reduced hydrogen-induced effects due to optimized material properties, external corrosion	Longer inspection intervals feasible; use predictive models from ML-enhanced RBI frameworks
Pipeline D	API 5L X65 steel	Hydrogen-natural gas blend	Pressure: 75 bar; temperature: 10–40 °C; cyclic loading: moderate	Intermediate degradation from partial hydrogen effects, external corrosion	Frequent recalibration of risk models; monitor hydrogen concentration and inspect high-stress zones

embrittlement, resulting in lower fatigue crack growth rates compared to Pipeline B. RBI frameworks integrating predictive models, such as those proposed in the HyRIS methodology, enable targeted inspection planning while balancing safety and cost-efficiency.

For Pipeline D, carrying a hydrogen-natural gas blend, fatigue crack growth rates are intermediate, reflecting the partial mitigation of hydrogen's effects by the presence of natural gas. Inspection strategies must account for the variable behavior of hydrogen content, using monitoring techniques and predictive insights to refine risk assessments over time.

The risk matrix in Fig. 7.4 qualitatively illustrates the risk level associated with each pipeline. As shown, the probabilities and consequences of failure equally impact the risk of component breakdown. In general, pipelines for natural gas transport exhibit higher consequences in the case of unintended release but a lower probability of such release occurring.

Additionally, Fig. 7.5 shows the iso-risk plot, which has the consequences and probabilities of failure on the x-axis and y-axis, respectively. This graph is in double logarithmic scale and allows to identify the most critical components in a plant and develop risk-based inspection plans.

The comparative analysis reveals distinct differences in degradation mechanisms between natural gas and hydrogen transport. For Pipeline A, traditional RBI frameworks are sufficient to manage risks associated with mechanical fatigue. The absence of hydrogen allows reliance on established damage factors, and inspection strategies can focus on external corrosion and geometric defects.

In contrast, Pipeline B highlights the challenges of repurposing existing infrastructure for hydrogen transport. Hydrogen embrittlement and hydrogen-enhanced fatigue

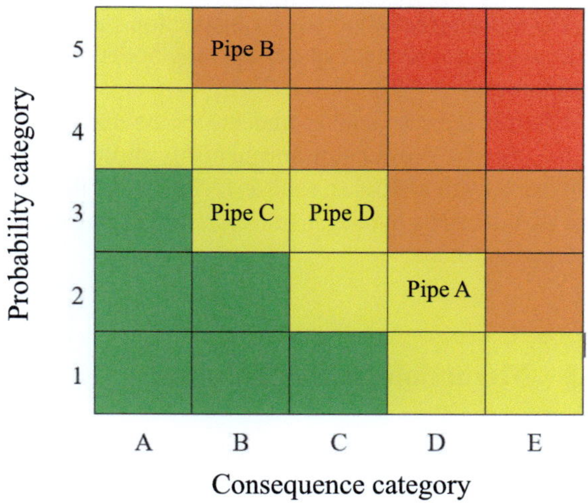

Fig. 7.4 Risk matrix for the four pipelines

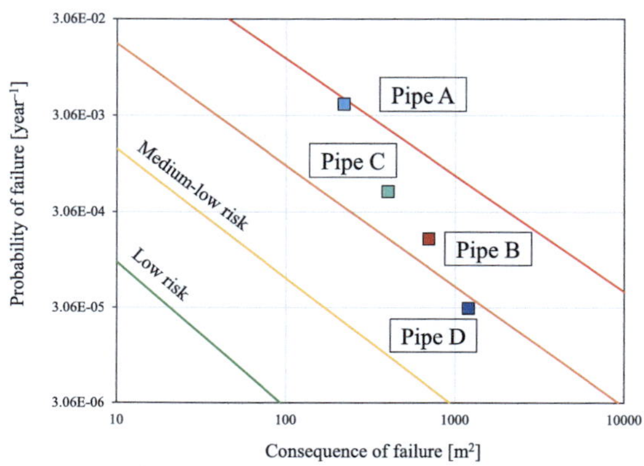

Fig. 7.5 Iso-risk plot for the pipelines carrying natural gas (Pipe A), hydrogen (Pipe B and C), or natural gas-hydrogen blends (Pipe D)

significantly reduce material strength, particularly in regions of stress concentration. Inspection planning for such pipelines requires advanced RBI frameworks that integrate ML predictions for hydrogen-induced degradation. Techniques such as PAUT and magnetic particle inspection are essential for detecting microstructural damage early.

Pipeline C illustrates the benefits of material optimization in mitigating hydrogen-induced effects. Although this pipeline is less vulnerable to hydrogen embrittlement, inspection intervals remain critical for addressing potential issues such as micro-cracking and weld defects. Predictive models, such as those proposed in the HyRIS framework, enable precise planning by quantifying risks under specific operating conditions.

For Pipeline D, the presence of a hydrogen-natural gas blend introduces additional variability in degradation mechanisms. Fatigue crack growth rates are influenced by the hydrogen content, necessitating frequent recalibration of damage factors. Monitoring hydrogen concentration and its effects on material behavior is critical to ensuring safe operations. Inspection strategies must balance the intermediate risks associated with blends, employing both traditional and hydrogen-specific NDT techniques.

Overall, the case studies underscore the importance of tailoring RBI frameworks to the specific degradation mechanisms and operational profiles of pipelines. ML-enhanced methodologies, such as HyRIS, offer the potential to address the complexities introduced by hydrogen transport, improving safety and reliability in the evolving energy landscape.

7.9 Methodological Highlights

The integration of ML with RBI frameworks represents a pivotal advancement in addressing hydrogen-specific challenges, such as hydrogen embrittlement and hydrogen-enhanced fatigue crack growth. These degradation mechanisms depend on complex interactions among material properties, operating conditions, and environmental factors, which conventional RBI methodologies struggle to capture.

ML models, such as those proposed in the HyRIS framework (Campari et al. 2024b), provide predictive insights by analyzing large and complex datasets. These models can quantify the fatigue acceleration factor in hydrogen environments, enabling dynamic adjustments to damage factors and improving the precision of risk assessments. By prioritizing high-risk components, ML-based RBI frameworks optimize inspection planning and resource allocation, enhancing safety and operational efficiency. However, these benefits are contingent on the quality and availability of data, as well as the continuous refinement of predictive accuracy through regular updates.

7.10 Limitations and Opportunities for Improving RBI Practices

The integration of ML with RBI for hydrogen infrastructure offers both significant opportunities and notable challenges (Table 7.3). On the one hand, ML-enhanced RBI frameworks can revolutionize inspection planning and risk management by leveraging advanced predictive capabilities. On the other hand, the practical implementation of such systems faces limitations that must be carefully addressed to ensure their effectiveness.

One of the primary challenges in integrating ML with RBI is the limited availability of comprehensive and high-quality datasets on hydrogen-induced degradation mechanisms. ML models often depend on experimental data or historical records to make predictions. However, datasets that capture critical variables such as hydrogen pressure, temperature, and cyclic loading effects on various pipeline materials remain limited. This lack of data hinders the ability of ML models to generalize their predictions across diverse scenarios, reducing their reliability in practical applications.

Another challenge lies in validating and generalizing ML models to ensure their predictions are applicable across a wide range of operating conditions. While frameworks like HyRIS have demonstrated promising results in specific contexts, their effectiveness in other hydrogen applications, such as storage tanks or distribution

Table 7.3 Limitations and opportunities in ML-enhanced RBI

Aspect	Limitations	Opportunities
Data	Availability of datasets on hydrogen-induced degradation mechanisms is currently limited	Development of comprehensive datasets through standardized testing protocols and open-access repositories
Model validation	Challenges in generalizing ML models to diverse materials, environments, and operational profiles	Integration of hybrid models combining ML with physics-based simulations for improved accuracy and reliability
Workflow integration	Need for significant modifications to existing RBI workflows, requiring specialized expertise	Integration into decision support systems for streamlined inspection planning and actionable insights
Regulatory acceptance	Adopting ML-based risk assessment methodologies may require thorough safety evaluations, delaying approval	Collaborative development of regulatory standards that incorporate data-driven methodologies
Uncertainty	Probabilistic nature of ML introduces uncertainties in predictions, especially with sparse or noisy data	Real-time data integration from sensors to refine predictions and enable proactive maintenance strategies
Scalability	Initial applications may focus on pipelines, with scalability to other hydrogen infrastructure requiring additional validation	Adaptation of methodologies to other hydrogen infrastructure (e.g., storage tanks, compressors, valves, and heat exchangers)

networks, is yet to be fully established. Variability in material properties, environmental conditions, and operational profiles introduces complexity, making it difficult for a single ML model to perform well in all situations without extensive validation.

Adapting traditional RBI frameworks to incorporate ML predictions also requires significant modifications to workflows and decision-making processes. Integrating ML into existing systems demands specialized expertise, investment in computational infrastructure, and regulatory approval. These factors can create barriers to adoption, particularly for industries with established practices that are not yet designed to accommodate data-driven methodologies. Furthermore, the probabilistic nature of ML models introduces uncertainties in their predictions, especially when dealing with limited or noisy data. These uncertainties must be carefully managed to avoid overconfidence in the outputs or errors in risk assessments.

Despite these challenges, the opportunities for enhancing RBI practices through ML integration are substantial. For instance, combining ML models with physics-based simulations, such as finite element analysis (FEA), can improve predictive accuracy and reliability. Hybrid modeling approaches enable the prediction of crack propagation in hydrogen environments with greater precision, offering valuable insights for targeted inspection planning.

The use of real-time data from sensors embedded in pipelines and storage systems further supports dynamic risk assessments. Advanced ML models trained on this data can detect early signs of degradation, allowing for proactive maintenance and reducing the likelihood of unexpected failures. This capability transforms traditional inspection strategies into adaptive systems that respond to evolving risks in real time.

Efforts to expand datasets also present significant opportunities. Collaborative initiatives between research institutions and industry stakeholders can focus on developing comprehensive datasets for hydrogen technologies. Standardized testing protocols and open-access data repositories will improve the training and validation of ML models, addressing current data limitations and supporting the development of more robust predictive tools.

Integrating ML-enhanced RBI frameworks into decision support systems offers another avenue for improvement. These systems can streamline the inspection planning process by providing actionable insights, such as optimized inspection intervals and recommended techniques based on real-time risk profiles. This integration facilitates more efficient resource allocation and supports the safe operation of hydrogen infrastructure.

Finally, the methodologies developed for pipelines have the potential to scale to other hydrogen applications, such as production plants, storage infrastructures, and refueling stations. Expanding these techniques across the hydrogen value chain enhances the relevance and utility of ML-enhanced RBI frameworks, contributing to the broader adoption of hydrogen as a critical energy carrier.

By addressing these challenges and leveraging the outlined opportunities, ML-enhanced RBI frameworks can become a cornerstone of hydrogen infrastructure management. Their ongoing evolution will play a crucial role in supporting the global transition to a hydrogen-based economy, ensuring safe, efficient, and reliable operations across the sector.

7.11 Conclusions

The integration of ML with RBI frameworks represents a transformative advancement in managing the safety and reliability of hydrogen infrastructure. Hydrogen's unique degradation mechanisms, including hydrogen embrittlement and hydrogen-enhanced fatigue, pose significant challenges to conventional inspection methodologies. ML-enhanced RBI frameworks address these challenges by leveraging predictive capabilities to assess complex synergies among material properties, operating conditions, and environmental factors.

This chapter has highlighted several key insights. First, the adoption of ML models, such as those proposed in the HyRIS methodology, enhances the precision of risk assessments by dynamically adjusting damage factors to reflect hydrogen-specific risks. This allows for more targeted and cost-effective inspection planning, particularly for pipelines transporting hydrogen or hydrogen-natural gas blends. Second, the use of hybrid modeling approaches, combining ML with physics-based simulations, offers a pathway to improved predictive accuracy, bridging gaps in knowledge and addressing the variability of hydrogen-induced degradation across diverse materials and operational scenarios. Third, the potential for real-time risk assessment, driven by sensor data and continuous model updates, underscores the role of ML in enabling proactive maintenance strategies that mitigate the likelihood of unexpected failures.

Despite these advances, challenges remain. The limited availability of high-quality datasets, the need for extensive model validation, and barriers to integrating ML into traditional RBI workflows highlight areas requiring focused effort. Addressing these limitations is essential to fully realizing the benefits of ML-enhanced RBI frameworks. Collaborative initiatives involving research institutions, industry stakeholders, and regulatory bodies will play a pivotal role in overcoming these barriers and establishing robust, scalable solutions.

The implications of these advancements for the hydrogen economy are profound. As hydrogen transitions from a niche application to a cornerstone of global energy systems, ensuring the safety and reliability of its infrastructure becomes paramount. ML-enhanced RBI frameworks provide the tools to meet this challenge, enabling the efficient management of pipelines, storage tanks, compressors, and other critical components. By reducing risks and optimizing inspection strategies, these methodologies support the deployment of hydrogen at scale, contributing to the decarbonization of energy systems and the realization of climate goals.

Looking ahead, the prospects for risk-based inspection in the hydrogen economy are both promising and dynamic. Continued advancements in ML algorithms, the development of comprehensive datasets, and the integration of real-time monitoring technologies will further enhance the effectiveness of RBI frameworks. Additionally, expanding these methodologies to encompass the entire hydrogen value chain will ensure that safety and reliability remain central to hydrogen's growth as a sustainable energy carrier.

In conclusion, ML-enhanced RBI represents a pivotal step forward in hydrogen infrastructure management. By addressing existing challenges and embracing emerging opportunities, these frameworks are poised to play a critical role in the safe and efficient transition to a hydrogen-based economy.

Acknowledgements The author acknowledges the contribution of Dr. Alessandro Campari in editing the final draft and the figures.

References

Alvaro A, Wan D, Olden V, Barnoush A (2019) Hydrogen enhanced fatigue crack growth rates in a ferritic Fe-3 wt% Si alloy and an X70 pipeline steel. Eng Fract Mech 219:106641. https://doi.org/10.1016/j.engfracmech.2019.106641

American Petroleum Institute (2019) API RP 581—risk-based inspection methodology

Barnoush A, Vehoff H (2010) Recent developments in the study of hydrogen embrittlement: hydrogen effect on dislocation nucleation. Acta Mater 58:5274–5285. https://doi.org/10.1016/j.actamat.2010.05.057

Campari A, Nakhal Akel AJ, Ustolin F, Alvaro A, Ledda A, Agnello P, Moretto P, Patriarca R, Paltrinieri N (2023) Lessons learned from HIAD 2.0: inspection and maintenance to avoid hydrogen-induced material failures. Comput Chem Eng 173:108199. https://doi.org/10.1016/j.compchemeng.2023.108199

Campari A, Vianello C, Ustolin F, Alvaro A, Paltrinieri N (2024a) Machine learning-aided risk-based inspection strategy for hydrogen technologies. Process Saf Environ Prot 191(Part A):1239–1253. https://doi.org/10.1016/j.psep.2024.09.031

Campari A, Ustolin F, Alvaro A, Paltrinieri N (2024b) Inspection of hydrogen transport equipment: a data-driven approach to predict fatigue degradation. Reliab Eng Syst Saf 251:110342. https://doi.org/10.1016/j.ress.2024.110342

Chatzimarkakis J, Levoyannis C, van Wijk A, Wouters F (2021) Hydrogen act—towards the creation of the European hydrogen economy

Defteraios N, Kyranoudis C, Nivolianitou Z, Aneziris O (2020) Hydrogen explosion incident mitigation in steam reforming units through enhanced inspection and forecasting corrosion tools implementation. J Loss Prev Process Ind 63:104016. https://doi.org/10.1016/j.jlp.2019.104016

Drexler ES, Slifka AJ, Amaro RL, Sowards JW, Connolly MJ, Martin ML, Lauria DS (2019) Fatigue testing of pipeline welds and heat-affected zones in pressurized hydrogen gas. J Res Natl Inst Stand Technol 124:1–19. https://doi.org/10.6028/jres.124.008

European Gas Pipeline Incident Data Group (2018) 10th report of the European gas pipeline incident data group (1970–2016)

Fetters KL (1964) The nature of fatigue. J Pet Technol 16:869–872

Giannini L, Jafarzadeh S, Campari A, Ustolin F, Paltrinieri N (2023) Inspection planning in the marine sector, a case study of a hydrogen-fueled fishing vessel. In: Proceedings of the international conference on offshore mechanics and Arctic engineering—OMAE 2023, pp 1–9. https://doi.org/10.1115/OMAE2023-100914

Kaplan S, Garrick BJ (1981) On the quantitative definition of risk. Risk Anal 1:11–27. https://doi.org/10.1111/j.1539-6924.1981.tb01350.x

Komoda R, Yamada K, Kubota M, Ginet P, Barbier F, Furtado J, Prost L (2019) The inhibitory effect of carbon monoxide contained in hydrogen gas environment on hydrogen-accelerated fatigue crack growth and its loading frequency dependency. Int J Hydrogen Energy 44:29007–29016. https://doi.org/10.1016/j.ijhydene.2019.09.146

Lipiäinen S, Lipiäinen K, Ahola A, Vakkilainen E (2023) Use of existing gas infrastructure in European hydrogen economy. Int J Hydrogen Energy 48:31317–31329. https://doi.org/10.1016/j.ijhydene.2023.04.283

Nanninga N, Slifka A, Levy Y, White C (2010) A review of fatigue crack growth for pipeline steels exposed to hydrogen. J Res Natl Inst Stand Technol 115:437–452. https://doi.org/10.6028/jres.115.030

Norwegian Ministry of Petroleum and Energy (2020) The Norwegian Government's hydrogen strategy

Pipeline and Hazardous Materials Safety Administration (2022) Pipeline incident 20 year trends [WWW document]

San Marchi C, Somerday BP (2012) Technical reference for hydrogen compatibility of materials. Report SAND2012-7312

Slifka AJ, Drexler ES, Amaro RL, Hayden LE, Stalheim DG, Lauria DS, Hrabe NW (2018) Fatigue measurement of pipeline steels for the application of transporting gaseous hydrogen. J Press Vessel Technol Trans ASME 140:1–12. https://doi.org/10.1115/1.4038594

Somerday BP, Sofronis P, Nibur KA, Marchi CS, Kirchheim R (2013) Elucidating the variables affecting accelerated fatigue crack growth of steels in hydrogen gas with low oxygen concentrations. Acta Mater 61:6153–6170. https://doi.org/10.1016/j.actamat.2013.07.001

Takakuwa O, Ogawa Y, Okazaki S, Matsunaga H, Matsuoka S (2019) Temperature dependence of fatigue crack growth in low-carbon steel under gaseous hydrogen. Am Soc Mech Eng Press Vessel Pip Div PVP 6B-2019:1–6. https://doi.org/10.1115/PVP2019-93451

Wang Y, Cheng G, Hu H, Wu W (2012) Development of a risk-based maintenance strategy using FMEA for a continuous catalytic reforming plant. J Loss Prev Process Ind 25:958–965. https://doi.org/10.1016/j.jlp.2012.05.009

Yamabe J, Yoshikawa M, Matsunaga H, Matsuoka S (2016) Effects of hydrogen pressure, test frequency, and test temperature on fatigue crack growth properties of low-carbon steel in gaseous hydrogen. Procedia Struct Integr 2:525–532. https://doi.org/10.1016/j.prostr.2016.06.068

Chapter 8
Beyond Accidents: Human and Organizational Factors

Eulàlia Badia, Roser Sala, and Joaquín Navajas

Abstract Current research emphasizes the need for examining safety through a sociotechnical lens, particularly in high-risk organizations, where safety emerges from the complex interplay of social and technical factors. This chapter pretends to stress out the significance of human behaviour and organizational practices specifically in the context of hydrogen fuelling stations (HRSs). By recognizing the vital contributions of these factors, we can better understand how to enhance safety measures and foster a culture of safety in this emerging industry. The understanding of the interaction between human actions, organizational structures, and technological systems is essential for preventing incidents and promoting a safer operational environment.

8.1 Safety Approach

8.1.1 Safety Considerations of Hydrogen

Hydrogen is a versatile element used in many industries that presents unique safety challenges due to its properties. While not inherently more dangerous than other fuels, its characteristics demand careful handling.

One of the primary safety concerns of hydrogen is its high flammability and low ignition energy, which allow it to ignite easily across a wide range of concentrations, posing a significant risk of accidental fires. Additionally, hydrogen's small molecular size and low viscosity contribute to its high leakage potential, making it difficult to contain. Its low viscosity alone makes the escape rate of liquid hydrogen 50 times greater than water and 10 times greater than liquid nitrogen. In its gaseous state, hydrogen also has the lowest viscosity at room temperature among all other gases.

E. Badia (✉) · R. Sala · J. Navajas
Socio-Technical Research Centre, Energy, Environmental, and Technological Research Centre (CIEMAT), Barcelona, Spain
e-mail: eulalia.badia@ciemat.es

Y. Liu et al. (eds.), *Safety Measures for Hydrogen Fueling Stations*, Risk, Reliability and Safety Engineering, https://doi.org/10.1007/978-981-96-9412-9_8

Therefore, it easily crosses porous walls and escapes through the smallest gaps (Barpi 2009; Lanz et al. 2013).

Another critical safety aspect is hydrogen's ability to embrittle metals and alloys. Prolonged exposure can degrade the mechanical properties of materials, increasing the risk of equipment failure and leaks (Wei et al. 2022). Furthermore, hydrogen's reducing properties can lead to violent reactions with certain compounds, posing additional hazards in industrial settings involving reactive substances. Finally, it's important to note that hydrogen's invisibility and lack of odour make it difficult to detect, increasing the risk of undetected leaks and potential explosions. This concern is especially significant in enclosed spaces where hydrogen can accumulate.

After going through hydrogen's properties and safety aspects, it's clear that accidents involving this gas can lead to severe human and environmental consequences. Understanding these risks is crucial for industries that work with hydrogen. By proactively addressing safety concerns—such as proper handling, effective leak detection, and the implementation of robust safety protocols—companies can foster a safer working environment. This not only protects employees and the surrounding community but also significantly reduces the likelihood of hydrogen-related incidents. Ultimately, a comprehensive approach to safety will help industries harness hydrogen's potential while mitigating its associated risks.

8.1.2 Systemic Approach to Safety

Safety management in high-risk organizations has evolved significantly over the years linked to the occurrence of industrial disasters (Hollnagel 2009; Reason 1990). Early safety efforts primarily centred on technological advancements, operating under the belief that reliable equipment could inherently prevent accidents. Thus, measures were aimed at eliminating equipment failures, seen as the main cause of incidents. However, in the mid-twentieth century, as technological reliability improved, human errors became more evident, shifting the focus to human factors such as ergonomic design, human–machine interfaces, and mistakes. Later, major accidents such as Seveso (1976), Chernobyl (1986), Challenger (1986), and even Fukushima (2011) challenged the view that accidents stemmed solely from engineering failures and human error, highlighting the critical role of organizational deficiencies. Issues such as tolerated degradation of safety systems, non-compliance with regulations, complacency, and poor leadership were identified as key contributors to these catastrophic events. It became clear that accidents were rarely caused by isolated equipment failures or human errors alone, but instead stemmed from deeper organizational weaknesses related to less tangible areas of safety. In this context, human errors were seen as symptoms of deeper organizational and managerial gaps, emphasizing the impact of organizational factors on human performance. This period also marked the emergence of the concept of "safety culture," which seeks to understand how cultural elements within an organization influence safety, a topic that will be discussed further later on.

Beyond the previous phases included in Reason's framework, some authors argued that accidents should be evaluated by considering not only internal organizational relationships but also interactions with key external stakeholders (Wilpert et al. 1999). This broader inter-organizational perspective recognizes the importance of considering the relationships between organizations and external stakeholders in accident analysis.

Lately, several scholars (Glendon et al. 2016) have advocated for a holistic approach to safety management that integrates previous theories and recognizes the complex interactions among technology, humans, and organizational structures. Also, the importance of understanding and influencing multiple organizational subcultures to effectively promote safety is highlighted (Borys et al. 2009; Martin 2002). This viewpoint challenges the idea of a singular organizational culture and emphasizes the importance of acknowledging the existence of socially constructed subcultures (Gherardi and Nicolini 2000) in managing performance variability and uncertainty. Addressing these issues requires a multifaceted approach that considers various perspectives to identify and tackle the root causes of incidents.

In summary, operating experience has shown that analysing technological, human, and organizational variables separately is insufficient for understanding the multiple causes of accidents. A comprehensive and systemic approach to safety is essential from a preventive standpoint. This perspective aligns with Rasmussen's sociotechnical model (Rasmussen 1997) which recognizes that the unit of analysis—whether an individual, group, or organization—comprises interconnected systems that influence one another. The model shows up the importance of human and organizational factors in the safety of complex systems, identifying key actors—involving different research disciplines—within and outside the organization who influence safety through control mechanisms and decision-making. These actors are arranged hierarchically, with the decisions of some limiting the options of others through laws, standards, and rules. Figure 8.1 illustrates Rasmussen's sociotechnical model, showing the interactions between all actors in a system and Table 8.1 describes the roles of each actor.

Fig. 8.1 Rasmussen's sociotechnical model (Rasmussen 1997)

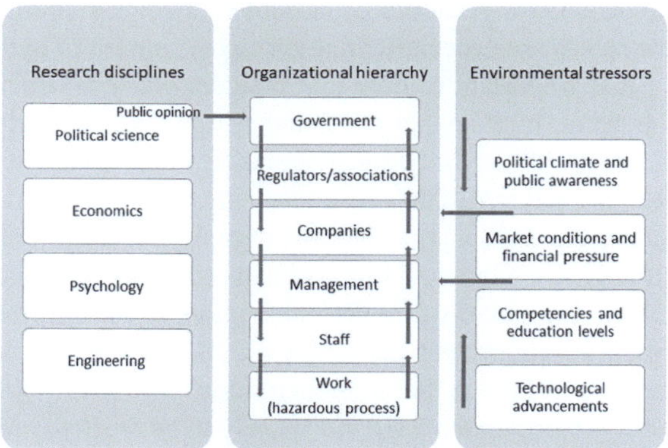

Table 8.1 Actors involved in a system (original table based on Rasmussen's sociotechnical model)

Actors from the system	Roles in the overall safety
Government	Regulates hazardous processes through the development of laws and regulations
Regulators/ associations	Interpret the laws and translate them into norms and standards to be followed and they establish the necessary requirements for the operation of a facility
Company/ organization	Responsible for managing the hazardous process and allocating human and economic resources Creates policies and procedures to ensure the safe management of hazardous processes while adhering to regulations
Management	They lead, manage, and control work. They translate policies into practices (rules, procedures) that will in part guide human performance
Staff	They focus on the individuals who work directly with the hazardous process implementing the safety rules and procedures in their daily activities and making informed decisions based on their observations and data
Work	It pertains to the hazardous process itself, including the equipment and the work environment

As said, these various actors interact and influence one another, emphasizing their collective responsibility in ensuring the safe and prudent operation of the processes. For instance, research findings can shape government regulations, which subsequently affect company policies and employee practices. Furthermore, environmental stressors, such as shifting political landscapes, public awareness, fluctuating market conditions, financial pressures, and rapid technological advancements, also influence the different actors.

The model also takes into account the role of public opinion within the system, recognizing its power to shape policies, drive regulatory changes, and influence organizational decision-making, by pressuring companies to prioritize sustainability, safety, and ethical practices in response to growing societal expectations.

This holistic approach allows for a deeper understanding of how various factors influence safety outcomes, particularly in high-risk environments such as HRS, where the stakes are especially high due to the nature of hydrogen as a fuel.

It is important to note that incident analysis often focuses on factors closest to the event, both spatially and temporally, while overlooking the significance of the actors previously mentioned and their interdependence (Mogles et al. 2018; Santos-Reyes and Beard 2008). By addressing the interdependence of human, organizational, technical, and external factors, organizations can implement more effective safety strategies, leading to fewer accidents and improved operational reliability. Therefore, safety assessments for HRS must move beyond surface-level analysis of individual errors or equipment malfunctions, to focusing on the broader organizational and contextual factors that influence safety outcomes. Without this holistic perspective, safety improvements are more likely to be reactive rather than proactive, and critical risks may remain unaddressed.

8.1.3 Organizational and Safety Culture

Culture develops in environments where people live and work together, requiring a level of shared understanding. This shared understanding is the essence of culture, which enables individuals to interpret their surroundings, perform, and comprehend the behaviours of others. Thus, culture is essential not only in society but also within organizations (Guldenmund 2018).

The literature offers a variety of definitions of organizational culture, reflecting its broad scope. These definitions suggest that culture includes many elements, such as behaviours, beliefs, values, norms, attitudes, expectations, and shared symbols and meanings. One of the most widely used is that of Schein who defines culture as "a pattern of shared basic assumptions that the group learned as it solved its problems of external adaptation and internal integration, that has worked well enough to be considered valid and, therefore, to be taught to new members as the correct way to perceive, think, and feel in relation to those problems" (Schein 2004, p. 17). Thus, organizational culture helps the organization adapt to its environment and integrates employees by shaping interpersonal dynamics, power structures, and the accepted language. Its knowledge provides insights into how members behave in their daily activities and offers a deeper understanding of their thought processes and actions within the organization. The philosophy of management, the mission of the organization, and the strategic choices management determine the culture of the organization (Cooke and Burack 1987).

When focusing on safety, organizational culture is seen as a crucial factor in influencing human performance and determining safety management within organizations. In this regard, safety culture, as a component of organizational culture (Antonsen 2009), encompasses the norms, values, and practices that promote safety. In high-risk organizations, where safety is paramount, the way people think and behave regarding safety become a critical aspect of the overall organizational culture. Safety culture can be best understood as the organizational processes that shape and reinforce a culture prioritizing safety within the organization (Guldenmund 2010).

The term "safety culture" was first used in the Chernobyl nuclear power plant accident investigation reports when it was shown that the lack of safety culture had been one of the key causes of the accident. It was defined as "that assembly of characteristics and attitudes in organizations and individuals which establishes that, as an overriding priority, nuclear plant safety issues receive the attention warranted by their significance" (IAEA 1991, p. 1). It is noteworthy that the IAEA's perspective on safety culture has evolved considerably over time as its complexities have become more fully understood.

Some scholars have conceptualized safety culture as an interplay of a psychological and sociological component that encompasses perceptions and values related to safety; a behavioural component that reflects the attitudes and behaviours of both workers and managers; and a structural component that includes both formal and informal organizational elements (Guillaume et al. 2018).

It is important to note that despite the well-documented importance of safety culture in preventing accidents, events like the Fukushima Daiichi disaster reveal that it is still not prioritized. Post-incident analyses highlighted significant lapses in safety awareness from both the plant owner and regulator, who relied too heavily on the nuclear industry's technical expertise when crafting regulations. A widespread belief that nuclear power plants were infallible went unchallenged by operators, regulators, or the government, leaving Japan unprepared for a severe accident that could have been anticipated and prevented. The Japanese nuclear industry failed to learn from the lessons of Three Mile Island and Chernobyl, fostering a culture that normalized resistance to regulatory pressure and the concealment of minor incidents. This underscores the ongoing relevance of safety culture in high-risk industries, reminding us that major accidents can still occur in its absence (US National Research Council 2014).

Since every organization has a culture, or potentially several subcultures, that can influence safety, understanding the specific safety culture within an organization is crucial for identifying potential risks and areas for improvement (Hopkins 2006). By examining safety culture, organizations can identify the necessary adjustments to strengthen their focus on safety, ensuring that safety becomes a shared priority at all levels. There are different validated instruments and frameworks available to assess safety culture, providing organizations with concrete tools to measure and track the effectiveness of their safety initiatives. This proactive approach not only reduces the likelihood of accidents but also fosters a work environment where safety is seamlessly integrated into everyday practices and decision-making. With the use of these assessment tools, organizations can continuously monitor their safety culture, identify areas for improvement, and make data-driven decisions to further enhance safety outcomes.

8.2 Human and Organizational Factors

As previously mentioned, investigations of past accidents worldwide have shown that human error and organizational factors contribute to incidents across all sectors and countries. In sectors such as health, aviation, nuclear, oil, gas, and HRS, accidents can result in devastating losses in terms of lives, property, and reputation. These settings, where the consequences of an error can be severe and where high-performance standards are essential, are referred to as complex systems (Perrow 1999). Therefore, recognizing human and organizational factors is crucial to preventing accidents and ensuring safe operations.

In the case of hydrogen installations, addressing these factors can significantly reduce accident risks, protect lives, and support the long-term success of hydrogen as a clean energy source. However, despite the acknowledged importance of human and organizational factors in event occurrence there is still some way to go. For instance, an analysis of 211 hydrogen fuelling station incidents revealed a significant under-emphasis on these elements in post-event corrective actions (Badia et al.

2024a). Most corrective actions prioritize technical solutions, such as equipment modifications, material enhancements, and process changes. In contrast, human-organizational and inter-organizational factors are often overlooked in addressing hydrogen accidents. Notably, only 13% and 11% of countermeasures implemented after an accident focus on human-organizational aspects and inter-organizational aspects respectively (Fig. 8.2).

When examining human and organizational corrective actions, the study reveals that the most prevalent countermeasures focus on additional personnel training in emergency response, inspections, and specific equipment use (such as piping connections and jigs), as well as the utilization of operating experience related to O-rings, piping, and operating manuals (Fig. 8.3 and Table 8.2).

When examining corrective actions that extend beyond the boundaries of the HRS (inter-organizational), results show that the most prevalent countermeasures include further investigations of the incident, leakage tests, torque tightening, and periodic inspections by the manufacturer (see Fig. 8.3 and Table 8.3).

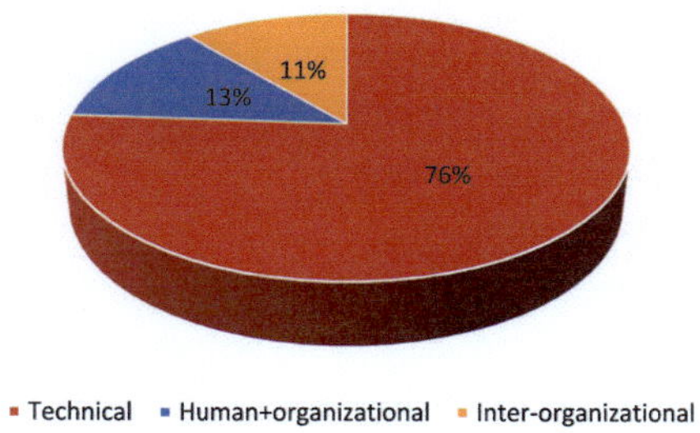

Fig. 8.2 Distribution of the countermeasures considering three different sociotechnical levels (Badia et al. 2024a)

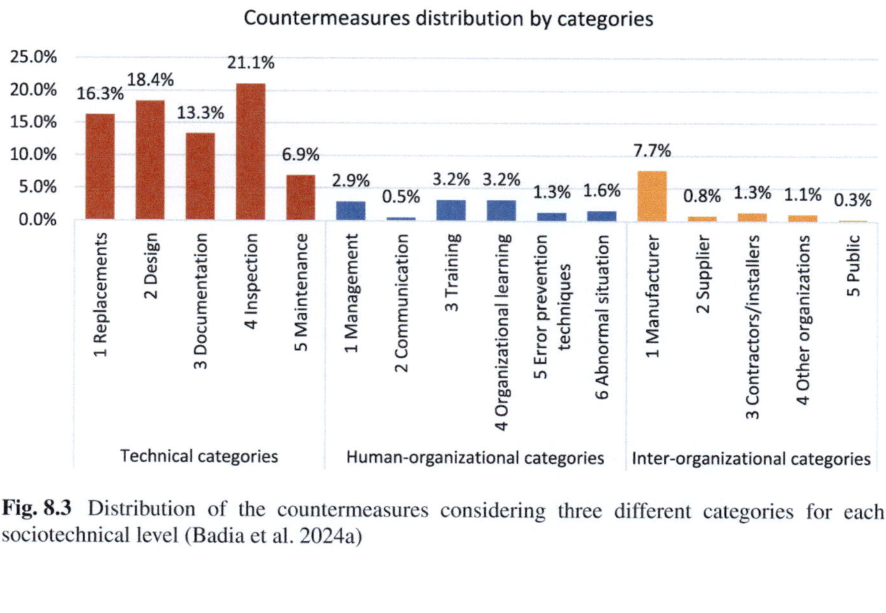

Fig. 8.3 Distribution of the countermeasures considering three different categories for each sociotechnical level (Badia et al. 2024a)

Table 8.2 Human and organizational countermeasure category distribution (Badia et al. 2024a)

Human-organizational categories	No. events	% Overall countermeasures
1. Management	11	2.9
Change management systems	1	0.3
Establishment of management guidelines	1	0.3
Resources	3	0.8
Roles	1	0.3
Supervision	3	0.8
Work planning	2	0.5
2. Communication	2	0.5
Communication of information	1	0.3
Communication plan modification	1	0.3
3. Training	12	3.2
Conduct training	7	1.9
Refresh/recurrent training	2	0.5
Training documentation modification	1	0.3
Training instructions/periodicity	2	0.5
4. Organizational learning	12	3.2
Accident investigation	3	0.8
Implementation of operating experience	8	2.1
Safety evaluation instructions/periodicity	1	0.3
5. Error prevention techniques	5	1.3
Check procedures before work/approve operations before work	2	0.5
Check works	3	0.8
6. Abnormal situations	6	1.6
Conduct emergency drills	3	0.8
Emergency drills instructions/periodicity	2	0.5
Emergency manual improvement/implementation	1	0.3
Total human-organizational countermeasures	48	12.8
Total overall countermeasures	375	100.0

Moreover, the study highlights that the recurring nature of similar incidents across different HRS indicates a failure to effectively leverage external operational experiences and best practices.

It is worth noting that while the literature and incident analysis acknowledge human factors and organizational issues as contributors to hydrogen accidents, these factors are not prioritized in corrective actions compared to technical issues. This disconnection between the recognized importance of these factors and their actual

Table 8.3 Inter-organizational countermeasure categories distribution (Badia et al. 2024a)

Inter-organizational categories	No. events	% Overall countermeasures
1. Manufacturer	29	7.7
Design modification	1	0.3
Design review	1	0.3
Equipment recall	1	0.3
Fixings	3	0.8
Instructions/documentation	5	1.3
Investigation/inspection/tests	16	4.3
Replacements	2	0.5
2. Supplier	3	0.8
Design modification	2	0.5
Improve qualification process	1	0.3
3. Contractors/installers	5	1.3
Communication system review/communication of information	2	0.5
Instructions	2	0.5
Review training	1	0.3
4. Other organizations	4	1.1
Communication of information	1	0.3
Conduct emergency drills	1	0.3
Conduct training	1	0.3
Improve communication between companies	1	0.3
5. Public	1	0.3
Communication with public	1	0.3
Total inter-organizational countermeasures	42	11.2
Total overall countermeasures	375	100.0

implementation is particularly concerning, especially given the potential risks within the hydrogen fuelling sector.

8.2.1 Human Factors

Human factor research examines the relationship between individuals, the tools they use, and the physical and social environments in which they live and work (Kohn et al. 2000). Human factors comprise a variety of elements that influence human performance, primarily stemming from physical, physiological, psychological, and socio-psychological factors. This broad range of elements includes aspects such as human abilities (skills, knowledge/training, and competency), restraints and constraints, work environment, motivation, emotions and thoughts, interpersonal relations, communication, attitudes, behaviour, creativity, initiative, self-drive, and stress management, among others (Ramprasad and Kumar 2018).

The connection between human factors and performance is evident in several ways. For instance, inadequate human abilities for the work can result in errors and diminish overall performance. Stress and high-pressure situations can impair decision-making abilities, while fatigue can reduce alertness and increase reaction times, negatively impacting both performance and safety. Additionally, deficiencies in interpersonal relationships and unsupportive team environments can hinder collaboration and communication, resulting in decreased productivity and increased stress. Disengagement stemming from unclear goals and lack of recognition can also adversely affect performance. Furthermore, the design of workspaces can influence physical strain and discomfort, ultimately impacting both performance and productivity.

In other words, human actions are influenced by factors beyond an individual's direct control. This can make it challenging to avoid unintended actions. Errors can result from a combination of personal, task-related, situational, and organizational factors. For example, personal factors can include a lack of knowledge or skill, while task-related factors can lead to complex or poorly designed procedures. Situational factors, such as environmental conditions or distractions, can also contribute to errors. Moreover, organizational factors, like inadequate training or poor communication, can create an environment that increases the likelihood of mistakes (Reason 1997).

Human performance can result in human failures which can be either unintentional (errors) or deliberate (violations) (Reason 1990). There are two types of errors: First, execution errors (slips and lapses) occur when an individual's actions do not align with the plan; although the intentions may be correct, the execution may be flawed or omitted entirely. Second, planning errors (mistakes) happen when an individual follows a plan that is inadequate for achieving the desired goals. While errors relate to an individual's cognitive processes, violations (routine or exceptional) are deviations from established practices designed to ensure the safe operation of a hazardous system. These violations are generally committed intentionally, although they can sometimes be unintentional.

Rasmussen further explains that human behaviour is divided into three levels based on how people interact with a system: skill-based (slips and lapses), rule-based (mistakes), and knowledge-based (mistakes). These levels are ordered by decreasing familiarity with the task, reflecting different ways in which people apply their cognitive abilities, leading to varying performance outcomes. Consequently, human errors occur at these different performance levels (Rasmussen 1986). Human failure types and levels are shown in Table 8.4.

In a study aimed at describing improved approaches for identifying human failures in process hazard analysis, it was stated that human failures are typically responsible for 50–90% of process accidents (Baybutt 2002). These human failures can be related to actions that are not performed, performed incorrectly, performed in the wrong place, performed in the wrong sequence, performed at the wrong time, or performed when they are not required. The author identifies several aspects related to incorrect

Table 8.4 Human failure types and causes for each performance level (author's elaborated table, based on Reason's and Rasmussen's model)

Human failure (Reason)	Description	Level (Rasmussen)	Type	Causes	Examples at HRS
Execution errors (action not as planned)	Occur when the individual's actions do not align with the plan (correct intention, incorrect execution, or no action performed)	Skill-based: Actions rely on well-established routines and memory developed through repetition or practice	Slips: Associated with attention failure during routine tasks that demand minimal conscious effort	Unintentional • Double-capture slips • Omissions following interruptions • Reduced intentionality • Perceptual confusions • Interference errors • Reversals (action in the wrong order) • Omissions • Repetitions	An operator at a hydrogen storage facility, distracted by a conversation, mistakenly reads the temperature gauge instead of the pressure gauge, resulting in incorrect system assessments
			Lapses: Associated with memory failure		An operator forgets to close the valve after completing the refuelling process, leaving the system open, which could result in a potential gas leak or safety risk later on
Planning errors (action as planned)	Occur when actions conform to the plan, but the plan itself is inadequate to achieve the desired goals	Rule-based: It involves using stored rules or procedures, often learnt from experiences or communicated instructions	Mistakes: Associated with familiar situations	Unintentional • Misapplication of good rules • Application of bad rules	An operator begins refuelling without inspecting the hose for proper connection, as specified in the procedure based on the assumption that the hose is always secure

(continued)

Table 8.4 (continued)

Human failure (Reason)	Description	Level (Rasmussen)	Type	Causes	Examples at HRS
		Knowledge-based: It occurs when there are no existing rules or prior experience to rely on, requiring the individual to use available knowledge or newly acquired information	Mistakes: Associated with unfamiliar situations	Unintentional • Selectivity • Workspace limitations • Out of sight/mind • Confirmation bias • Overconfidence • Biased reviewing • Illusory correlation • Halo effect • Problems with causality • Problems with complexity (delayed feedback, vagabonding, lack of time awareness …)	A technician at a hydrogen refuelling station mistakenly configures the refuelling hose to the wrong pressure setting by failing to account for the temperature variations in the hydrogen storage tank
Violations	Deviations from established safe practices		Routine, exceptional non-compliance violations	Generally deliberate actions	A driver intentionally exceeds speed limits while transporting hazardous materials to reach the destination faster, despite knowing the procedure is meant to ensure safe handling and prevent accidents

actions by operators that organizations should consider to reduce the likelihood of human errors and enhance overall safety in process operations. This list, which can be customized for specific processes and companies in the context of process hazard analysis, includes various human failures that may serve as initiating events. A detailed list of these human failures is provided in Table 8.5.

Numerous research studies have focused on human factor assessment methodologies tailored to the specific challenges associated with hydrogen fuelling station

Table 8.5 Types of human failures and descriptions (author's elaborated table based on Baybutt's study)

Human factor contributors to failure	Description
Work overload/underload	Excessive workload or time pressure can lead to errors due to stress, fatigue, and reduced attention Insufficient work or a monotonous routine can result in boredom, lowered motivation, and higher likelihood of errors
Insufficient training	Lack of knowledge to perform a task safely and effectively
Inadequate skills	Insufficient ability or competency to carry out a task, even with adequate training
Inadequate resources	Lack of necessary tools, equipment, or support to perform a task safely and efficiently
Inadequate procedures	Poorly defined or unclear procedures can create confusion and mistakes
Inadequate labelling	Insufficient or unclear labelling of equipment, controls, or materials, which can contribute to errors
Equipment not easily operable	Complex or difficult-to-use equipment that can increase the likelihood of errors
Displays/controls not visible/heard	Limited visibility or audibility of displays and controls can make it difficult to monitor and operate equipment safely
Displays/controls confusing	Complex or confusing displays and controls that can lead to misunderstandings and mistakes
Displays/controls not accessible/usable	Displays and controls that are difficult to reach or use, increasing the risk of errors
Inadequate communications	Poor communication between individuals or teams can lead to misunderstandings and errors
Environmental issues	Factors like temperature, humidity, light, noise, or distractions can impair performance and increase the risk of errors
Error	A deviation from the intended action, often caused by slips, lapses
Mistake	A wrong action resulting from a misunderstanding or misjudgement

incidents. It's the case of a study which examines the role of human factors in operational errors during the maintenance of safety venting devices in HRS, utilizing first- and second-generation human reliability assessment (HRA) techniques (Castiglia and Giardina 2013). The results show that human errors substantially raise the risk of accidental overpressure in hydrogen storage systems. The findings underscore the need to enhance safety standards by identifying and addressing human errors, recommending measures such as worker certification and training, along with the use of redundant supervisory controls and checklists during maintenance activities.

Another study, aimed at improving the safety and reliability of hydrogen refuelling station operations and maintenance, focuses on assessing and reducing human error through an integrated methodology. This methodology combines the Human Error Assessment and Reduction Technique (HEART) with Bayesian Networks (BN) and the Best–Worst Method (BWM) to evaluate key human error factors (Chauhan et al. 2023). The study identifies several types of human errors in the context of HRS operations and maintenance, including (1) Errors caused by adverse environmental conditions, such as extreme weather, which significantly increase the likelihood of human error; (2) Errors linked to experience and training, where inadequate experience and insufficient training contribute to a higher probability of mistakes; and (3) Task-specific errors, such as mistakes in filling the storage unit, which—despite normal weather conditions and adequate training—can increase the risk of errors in critical systems like emergency ventilation.

Additionally, a research evaluating the influence of driver awareness on refuelling risk identified three main triggers for dangerous events: valve failure, human error, and external factors such as disasters or equipment wear (Rusin et al. 2024). The probability of human error, particularly during refuelling, was assessed using the HEART method, highlighting its critical importance. Drivers were categorized as either cautious or careless, with results indicating that the likelihood of human error is significantly higher for careless drivers. Overall, the findings suggest that human error is the most significant contributor to refuelling risk.

8.2.2 Organizational Factors

Error theories suggest that in complex systems, human errors arise from antecedents or pathways that existed within the system at the time the errors were made (Strauch 2017). As conceptualized by Turner and Pidgeon, human errors are outcomes of systemic issues rather than isolated faults. These errors are symptoms indicating the existence of underlying latent conditions within the system that contribute to their occurrence (Turner and Pidgeon 1997). Therefore, human factors extend beyond the actions of an individual worker performing a specific task; they also encompass higher level elements such as safe systems of work, procedures, training and competence, work organization, management systems, and business culture. It's essential not to overlook potential human factor issues and the risk of human errors also in the design process itself.

According to Reason, there are two types of accidents: those affecting individuals and those impacting organizations (Reason 1997). While individual accidents are more common, organizational (or system) accidents are less frequent but can have devastating consequences, particularly in critical sectors like nuclear power, commercial aviation, and the chemical and petrochemical industries. Unlike individual accidents, which result from operators' mistakes and have immediate consequences (known as active errors), organizational accidents arise from deeper systemic issues (referred to as latent conditions) that exist at higher levels within the organization. These latent conditions often remain dormant and largely undetected until they interact with other factors, ultimately compromising the system's defences. Additionally, this author considers that active errors are symptoms or indicators of a flawed system.

Organizational factors cover everything from management practices to workplace culture. They include elements such as vision and objectives, targets and goals, strategies and priorities, integrated management systems, commitment and involvement, responsibility and authority, delegation and decision-making, performance assessment and review, experience and feedback, continuous improvement, work environment and culture, knowledge management, communication, and contract management, among others (Ramprasad and Kumar 2018).

A research project identified and defined a set of organizational dimensions (Table 8.6) that contribute to nuclear power plants' safety (Haber and Barriere 1998). While this study was initially focused on the nuclear industry, its findings may also be applicable to other high-risk industries, such as HRS.

As noted in the previous section, human errors significantly contribute to many accidents at HRS. However, it is crucial to go beyond merely examining individual performance. Focusing solely on human factors can obscure the larger organizational influences that contribute to these errors. By addressing these higher level elements, a more robust safety environment can be established that not only mitigates the risk of human error but also enhances the overall effectiveness of safety systems in hydrogen refuelling operations.

8.2.3 Human and Organizational Factors Involved in Incidents

Alongside human factor research, findings from incident analysis emphasize the importance of prioritizing human factors during maintenance, operations, and other activities to prevent accidents related to human error.

Perrow's research from Three Mile Island nuclear power plant accident suggests that human error is a significant contributor to industrial accidents, accounting for an estimated 60–80% of incidents. However, it is important to note that identifying human error as a contributing factor to an accident does not equate to assigning blame. A no-blame culture acknowledges that errors are inevitable and shifts the focus from

Table 8.6 Organizational dimensions related to safety and their description (author's elaborated table based on Haber and Barriere study)

Organizational dimensions	Description
Culture	
Attention to safety	The work environment's characteristics, including norms, rules, and shared understandings, that shape how personnel perceive the organization's commitment to safety
Organizational culture	Personnel's shared perceptions of the organization including its traditions, values, goals, and practices, which define the "personality" of the organization
Management and leadership	
Goal setting/prioritization	The degree to which facility personnel comprehend, accept, and align with the goals' purpose and importance
Resource allocation	The way the facility allocates its financial resources, including both the actual distribution and personnel's perceptions of it
Decision-making	The way decisions that affect the operation are made and who makes the decisions
Organization and processes control	
Formalization	The adequacy and clarity of rules, procedures, and methods for both routine and unusual activities
Roles and responsibilities	The extent to which personnel's positions and work activities are defined and performed
Coordination of work	The planning, integration, and implementation of the work activities of individuals or groups
Organizational knowledge	Personnel's understanding about the interactions of the organizational subsystems and the way in which work is accomplished
Communication	
Intradepartmental communication	The exchange of information, both formal and informal, within a given department or unit
Interdepartmental communication	The exchange of information, both formal and informal, between the different departments or units
External communication	The exchange of information, both formal and informal, between the organization and external organizations
Learning and improvement	
Organizational learning	The extent to which personnel and the organization use past experience to improve future performance
Problem identification	The extent to which the organization uses knowledge, experience, and information to identify potential problems
Human resources management	
Personnel selection	The degree to which the organization selects personnel capable of handling both routine and unique job demands

(continued)

Table 8.6 (continued)

Organizational dimensions	Description
Training	The extent to which personnel are equipped with the necessary skills and knowledge for safe, effective tasks, and their view of the training's value
Performance evaluation	The degree to which personnel are provided with fair assessments of their work-related behaviours
Performance quality	The extent to which personnel follow job requirements, perform correctly, and take responsibility for their actions
Time urgency	The degree to which personnel perceive schedule pressures while completing various tasks

personal accountability to organizational responsibility. In fact, even though errors should be minimized, they can also serve as valuable learning opportunities. Reason argues, a "no-blame culture" is ineffective and undesirable. Instead, a culture of trust with clear accountability is essential and where boundaries between acceptable and unacceptable behaviour are established (Reason 1997).

Post-event analysis reports and data from incident databases have been instrumental in identifying the human and organizational contributing factors to these events, as well as in deriving key lessons learnt. Below are several examples of hydrogen-related accidents from both sources.

Incident Reports: In 2019, a hydrogen tank explosion at a testing site in South Korea, where renewable hydrogen was being generated through a water electrolyser powered by solar panels, was triggered by a static spark in the hydrogen buffer tank, compounded by oxygen contamination in the hydrogen stream (Kim 2019). The incident resulted in two fatalities, six injuries, and substantial damage to buildings within a 100-m radius of the tanks. The accident investigation revealed that two contributing factors to the incident were the safety management team's failure to adhere to protocols requiring daily testing of hydrogen stream quality, and production's disregard of warnings that the oxygen level had exceeded 3%, yet continuing with the test validation. This incident highlights the critical role of rigorous adherence to procedures, effective communication and awareness within the organization, and proper training and understanding among staff regarding safety measures. Overall, it demonstrates the interconnectedness of human behaviour and organizational practices in ensuring safety in high-risk environments.

In the context of hydrogen fuelling station operations, the Kjørbo incident in Norway in 2019, caused by an assembly error in a hydrogen tank's high-pressure storage unit where inner bolts were improperly torqued due to human error, also highlights the critical role of human factors (Nel ASA, n.d.). This error led to a hydrogen leak, resulting in the formation of a flammable hydrogen-air mixture that ignited. In response, a key post-incident measure was the implementation of a dual-witness procedure to ensure proper torquing and minimize human error. This incident, along with others, has drawn attention to the importance of human factors in hydrogen station operations, emphasizing the need for enhanced training, supervision, and procedural rigor to prevent such failures in the future.

In the same year, a major uncontrolled release of high-pressure hydrogen occurred at the Air Products hydrogen trailer transfill facility in Santa Clara, California (Hydrogen Safety Panel 2021). The incident took place during the gaseous hydrogen filling of a modular multi-cylinder trailer, when hydrogen was unintentionally released from an open pipe. This was triggered by an unauthorized attempt (without adhering to standard) to repair a leaking valve and compounded by a miscommunication between the two drivers overseeing the fill operation. This incident further emphasizes the need for strict adherence to safety protocols and proper communication during hydrogen handling operations. As follow-up actions, improvements were made to the trailer filling procedures, an enhanced driver training programme was implemented, and control functions were more clearly labelled.

These outcomes align with findings from some authors who identify human factors as key contributors to hydrogen leak accidents at refuelling stations. They recommend that companies prioritize safety promotion and operational training to reduce unsafe behaviours (Wang and Gao 2023).

Incident Databases: An analysis of 215 hydrogen-related incidents between 1992 and 2008, recorded in France's ARIA public database, revealed that over 70% of these accidents were primarily driven by organizational and human factors mainly related to maintenance or upkeep operations and errors in process management (Barpi 2009). The document emphasizes the importance of continuous vigilance at all levels within the facility—management, supervisory staff, technicians, and subcontractors—besides recognizing the ongoing risk of ignition in the presence of hydrogen.

The examination of the H2Tools database with more than 200 documented incidents and accidents, primarily occurring in Europe and the USA over the past two decades, revealed that the primary causes of these incidents include human error, equipment failures, and design flaws. In addition, factors such as inadequate protocols, training deficiencies, changes in procedures and equipment, and poor situational awareness have also contributed (Salehi et al. 2022).

Other studies, including those based on the HIAD database, show that the main contributors to incidents are a lack of situational awareness, unclear safety procedures and policies, and inappropriate human behaviour. Specifically, a study using network theory to analyse hydrogen logistics incidents from the HIAD database highlighted the interdependencies between various factors and their effects. It found that inappropriate operations are often linked to human or recognition errors, which can be mitigated by implementing operation checklists, establishing confirmation steps before completing tasks, and fostering collaboration to reduce individual mistakes (Lam et al. 2019).

A study of hydrogen-related events also from the HIAD 2.0 database shows that over half were caused by job, individual/human, and organizational/management factors. The main issues include inadequate maintenance and inspection, unclear instructions, failure to confirm safety after repairs, improper handling of safety devices, incorrect chemical identification, incomplete degasification, insufficient staff training, improper equipment installation, and non-compliance with procedures. These findings highlight the need for better maintenance, clearer instructions, and improved safety practices to prevent incidents (Wen et al. 2022).

An analysis of the HIAD database reveals that incidents and accidents attributed to human errors accounting for a 9.4% of the total frequently stem from factors such as insufficient training and competence among operators, fatigue, stress, boredom, underlying medical conditions, and unfavourable working conditions. In contrast, events with organizational root causes representing 9.0% of the total often relate to deficiencies in safety barriers, a failure to learn from previous incidents, inadequate coordination and planning, and a weak health and safety culture. Additionally, a substantial number of these undesirable events were directly connected to outdated maintenance guidelines (Campari et al. 2023).

A study examining lessons learnt from hydrogen events, with a particular focus on maintenance-related incidents, underscores the critical importance of conducting proper maintenance activities (Collina et al. 2023). Among the contributing factors to these undesired events, human error emerges as a significant issue. Even minor mistakes can have severe consequences, highlighting the need for meticulous attention to detail and robust maintenance procedures in the hydrogen sector The authors present nine performance shaping factors that significantly influence human performance and the likelihood of error. These include the need for adequate time for detection, diagnosis, decision-making, and action; the impact of perceived danger on stress levels; and the complexity of tasks, which affects error rates. Additionally, an operator's experience and training play crucial roles, along with the clarity of procedures and adherence to them. The quality of the human–machine interface, positive attitudes towards safety and managerial support, effective teamwork, and the physical working environment also contribute to performance outcomes. Together, these factors highlight the multifaceted nature of human error in complex tasks.

Another research focusing on lessons learnt from hydrogen incident analysis found that maintenance operations were a primary cause of many incidents. Human errors, such as failure to follow procedures, inadequate training, and a lack of situational awareness, were frequently identified as contributing factors. Based on these findings, it's reasonable to conclude that organizational and human factors were responsible for approximately 87% of the analysed hydrogen-related events (Alfasfos et al. 2024).

An investigation focused on the analysis of events in the hydrogen value chain highlights the significant role of human factors in contributing to incidents and accidents within this sector. It suggests that incidents linked to human performance can have a very damaging impact. In particular, human activities—such as operations conducted by personnel in the storage area—are a major risk factor in hydrogen storage. Furthermore, in industrial applications of hydrogen, leaks from pipes within facilities, primarily due to human actions and welding failures, represent critical safety concerns. In processes directly related to HRS, operator-induced failures can lead to serious consequences, including equipment damage, fires, explosions, and even personal injuries (Badia et al. 2024b).

8.2.4 Factors That Enhance Safety from a Human and Organizational Perspective

Building on the insights from scientific literature and lessons to be learnt from hydrogen refuelling station incidents, this section highlights some key human and organizational factors that the authors believe are crucial for enhancing safety in HRS operations.

Safety Management Systems: Safety management systems are the procedures, components, and activities designed to enhance the safety performance within an organization (Li and Guldenmund 2018). Their implementation must begin in the design phase in order to identify, assess, manage, and mitigate risks in HRS. To be effective, they must include the following: (1) Safety Policy, outlining goals and responsibilities with a clear commitment to safety; (2) Risk Management, for identifying and mitigating risks through hazard analysis and regular safety audits; (3) Safety Assurance, to monitor and improve safety processes; (4) Safety Promotion, fostering a safety culture through training and communication; and (5) Documentation, for record-keeping and continuous improvement.

To support this, investing in safety is key to the long-term success and reliability of HRS. Adequate resources—time, personnel, and funding—must be allocated to safety programmes and equipment, ensuring safety is an integral part of operations, not an afterthought.

Design Phase: Human Factors Engineering: Many hydrogen-related incidents stem from deficiencies in the design phase (Navajas et al. 2023), where safety features and risks are insufficiently addressed. Human-centred design principles are crucial in HRS to ensure systems are safe, efficient, and minimize human error. By considering human limitations and capabilities, designs can be made more intuitive and user-friendly. Clear instructions and effective feedback reduce cognitive load, helping operators make better decisions. Usability testing and iterative design are essential for continuous improvement, allowing systems to evolve based on real-world feedback. Prioritizing human-centred design enhances safety, efficiency, and performance while reducing the risk of operational errors.

Leadership and Management Practices: The role of leadership in fostering an error-resilient organization is paramount. Leadership practices and attitudes towards safety can influence the overall safety culture of an organization. Leaders who visibly prioritize safety through actions, policies, and transparent communication set the tone for the organization. They can empower teams by providing clear strategies, promoting open communication about safety expectations, and encouraging involvement in safety decisions and change processes. Employee empowerment is further reinforced through feedback mechanisms that allow concerns to be raised without fear of retaliation. Recognizing and rewarding safe behaviours, along with celebrating safety milestones, also helps reinforce a culture of continuous improvement and motivates ongoing safety efforts.

Haber and Barriere recognized specific leadership practices (within organizational factors) that can influence safety outcomes (Haber and Barriere 1998). These practices are shown in Table 8.7 and could help improve safety outcomes in HRS.

Hazard Identification and Risk Assessment: Hazard identification and risk assessment are essential for ensuring safety in hydrogen refuelling stations as they help to develop strategies to mitigate risks. Different methods, such as task analysis and hazard identification techniques, can be employed to assess potential risks associated with specific tasks. These approaches help determine hazards and their potential consequences, particularly those arising from human error. Risk assessments must be continuous, as safety needs to be regularly reviewed and adapted to new challenges, emerging hazards or operational changes. Case studies related to human and organizational factors in HRS offer valuable lessons by demonstrating how they impact safety, operational efficiency, and the organization's overall risk profile. These real-world examples underline the importance of a systematic approach to hazard identification and the ongoing need for adaptation in safety practices.

Organizational Learning: Organizations that excel at learning are more likely to detect and correct errors, ultimately improving safety and performance (Argyris 1999). Learning within an organization involves the collective adoption of new skills, attitudes, values, and behaviours that contribute to continuous improvement. One critical mechanism for fostering organizational learning is the establishment of a robust system for reporting safety incidents and near misses. Such a system provides valuable data that, when properly analysed, enables organizations to learn from both accidents and close calls, rather than merely penalizing mistakes. By

Table 8.7 Clusters and leadership clusters

Clusters	Leadership practices
Motivating and influencing	• Setting goals • Recognizing • Rewarding
Encouraging development	• Mentoring • Supporting • Giving feedback
Making decisions	• Planning and organizing • Problem solving • Consulting • Innovating
Giving-seeking information	• Monitoring • Informing • Clarifying goals and objectives • Delegating
Building relationships	• Networking • Teambuilding managing conflict

implementing systems to track errors and performance, coupled with a non-punitive incident reporting system, organizations can cultivate a culture of trust and openness.

Another important practice that supports organizational learning is continuous assessment. In dynamic environments, such as hydrogen refuelling stations, where technology, human interaction, and organizational practices intersect, systematic evaluation and improvement are essential. Conducting regular audits, establishing feedback loops, and investigating incidents with a focus on both human and organizational factors are critical to identifying weaknesses and driving ongoing improvements.

Not forgetting that feedback is central to preventing future errors, organizations can learn from past mistakes to refine their processes and systems, ultimately reducing risk. Sharing lessons to be learnt across the organization ensures that everyone benefits from collective knowledge, strengthening the overall safety culture. Transparent communication about safety policies and incidents fosters trust, while a non-punitive environment encourages proactive reporting. Recognizing and celebrating those who report safety concerns further reinforces a culture of learning and continuous improvement.

Training: As human behaviour plays a significant role in the safe handling and operation of hydrogen systems, employees must be well-trained in safety protocols, emergency response procedures, and the specific properties of hydrogen, such as its flammability and leakage potential. Recognizing that human error can occur at any stage, especially during inspection and maintenance activities, effective training programmes that address human error must be designed.

These programmes should include scenario-based simulations to prepare employees for real-life situations, with regular updates and repeated sessions to reinforce key practices. Additionally, regular drills and emergency response planning, along with training for external emergency services, ensure readiness in critical situations. To be effective, training must be updated and to reflect procedural changes.

Finally, clear expectations regarding roles and responsibilities, along with a strong emphasis on accountability, will ensure that employees understand their safety obligations and the consequences of neglecting protocols. This understanding helps create a culture of safety where employees are more aware of potential hazards and better prepared to act responsibly.

In summary, the factors discussed all emphasize the prioritization of safety and the promotion of a strong safety culture within the organization. Safety must be deeply embedded in the organization's core values and mission. To achieve this, safety practices must be proactive and comprehensive. This includes conducting frequent and random inspections and updating procedures for start-up, inspection, operation, maintenance, and shutdown to adapt to changing conditions. Operating procedures must be carefully aligned with both maintenance and operational requirements, while ensuring that equipment and materials meet the specifications required for safe and efficient operation. Deviations from established procedures should only be allowed after thorough evaluation and approval, and any process changes should be implemented only after a detailed risk assessment. Furthermore, safety procedures should be regularly updated to reflect changes in operations, with relevant documentation kept readily accessible and communicated to all personnel. It is also critical to define and communicate safe operating parameters—such as temperature, pressure, and flow rates—to all relevant staff, while ensuring that safety management procedures remain current and effective. Adequate supervision must be provided for critical repairs and maintenance, and any procedural changes should be communicated promptly to both staff and subcontractors. Lastly, lessons to be learnt from past incidents should be actively incorporated into the safety plan to continuously improve safety practices and reduce the risk of future events. Through these measures, safety becomes not just a regulatory obligation, but a central and ongoing commitment within the organization (Wen et al. 2022).

8.3 Conclusions

This chapter highlights the key role of human and organizational aspects in ensuring the safety of HRS. It advocates for a systemic approach to safety, which transcends the narrow focus on individual errors or equipment failures to focus on broader organizational and contextual influences. Without this holistic perspective, safety improvements are likely to be reactive rather than proactive, and critical risks may remain unaddressed. Specifically, organizational culture and safety culture are essential drivers of human performance and are central to effective safety management.

By drawing on evidence from other high-risk organizations, this chapter offers a broad perspective on how human and organizational issues can impact the safe operation of hydrogen systems. It also summarizes recent research on how human error and organizational deficiencies have contributed to hydrogen-related incidents and accidents.

The need to consider these aspects in incident reports and databases, but most crucially the procedures and daily practice of organizations responsible for HRS is argued. Some key human and organizational factors that could significantly improve HRS safety, such as adequate resources allocated to safety programmes and equipment, human factor engineering in the design phase, adequate leadership, and management practices or organizational learning, are pointed out. Ultimately, prioritizing safety at all levels is paramount. Fostering a strong safety culture within HRS organizations is essential to achieving and maintaining a high level of safety performance.

References

Alfasfos R, Sillman J, Soukka R (2024) Lessons learned and recommendations from analysis of hydrogen incidents and accidents to support risk assessment for the hydrogen economy. Int J Hydrogen Energy 60:1203–1214. https://doi.org/10.1016/j.ijhydene.2024.02.226

Antonsen S (2009) Safety culture and the issue of power. Saf Sci 47(2):183–191. https://doi.org/10.1016/j.ssci.2008.02.004

Argyris C (1999) On organizational learning. Wiley. https://books.google.es/books?id=JmTIWYs3yhwC

Badia E, Navajas J, Sala R, Paltrinieri N, Sato H (2024a) Analysis of hydrogen value chain events: implications for hydrogen refueling stations' safety. Safety 10(2):1–26. https://doi.org/10.3390/safety10020044

Badia E, Navajas J, Sala R, Sato H, Liu Y, Paltrinieri N (2024b) Safety lessons to learn from HRSs events analysis, 2022–2023

Barpi (2009) Accidentology involving hydrogen. In: Bureau for analysis of industrial risks and pollutions. https://www.aria.developpement-durable.gouv.fr/wp-content/files_mf/SY_hydrogen_GB_2009.pdf

Baybutt P (2002) Layers of protection analysis for human factors (LOPA-HF). Process Saf Prog 21(2):119–129. https://doi.org/10.1002/prs.680210208

Borys D, Else D, Leggett S (2009) The fifth age of safety: the adaptive age. J Health Serv Res Policy 1(1):19–27

Campari A, Nakhal Akel AJ, Ustolin F, Alvaro A, Ledda A, Agnello P, Moretto P, Patriarca R, Paltrinieri N (2023) Lessons learned from HIAD 2.0: inspection and maintenance to avoid hydrogen-induced material failures. Comput Chem Eng 173. https://doi.org/10.1016/j.compchemeng.2023.108199

Castiglia F, Giardina M (2013) Analysis of operator human errors in hydrogen refuelling stations: comparison between human rate assessment techniques. Int J Hydrogen Energy 38(2):1166–1176. https://doi.org/10.1016/j.ijhydene.2012.10.092

Chauhan A, Golestani N, Yazdi M, Njue JCW, Abbassi R, Salehi F (2023) A novel integrated methodology for human reliability assessment in hydrogen fuelling stations. Int J Hydrogen Energy 48(34):12917–12933

Collina G, Subedi A, Campari A, Thapa BS, Paltrinieri N (2023) Lesson learned from H2-related incidents: criticality of maintenance operations. In: Institution of chemical engineers symposium series No. 170

Cooke RA, Burack EH (1987) Measuring norms and expectations with the OCI. In: Organizational culture inventory, leader's guide

Gherardi S, Nicolini D (2000) To transfer is to transform: the circulation of safety knowledge. Organization 7(2):329–348. https://doi.org/10.1177/135050840072008

Glendon AI, Clarke S, McKenna E (2016) Human safety and risk management. CRC Press. https://books.google.es/books?id=u9O1bblQHFEC

Guillaume O, Herchin N, Neveu C, Noël P (2018) An industrial view on safety culture and safety models. In: Gilbert C, Journé B, Laroche H, Bieder C (eds) Safety cultures, safety models taking stock and moving forward

Guldenmund FW (2010) (Mis)understanding safety culture and its relationship to safety management. Risk Anal 30(10):1466–1480. https://doi.org/10.1111/j.1539-6924.2010.01452.x

Guldenmund FW (2018) Understanding safety culture through models and metaphors. In: Gilbert C, Journé B, Laroche H, Bieder C (eds) Safety cultures, safety models taking stock and moving forward

Haber SB, Barriere MT (1998) Development of a regulatory organizational and management review method. Research report RSP-0060

Hollnagel E (2009) The ETTO principle: efficiency-thoroughness trade-off: why things that go right sometimes go wrong. Ashgate. https://books.google.es/books?id=srYNX2OgagYC

Hopkins A (2006) Studying organisational cultures and their effects on safety. Saf Sci 44:875–889. https://doi.org/10.1016/j.ssci.2006.05.005

Hydrogen Safety Panel (2021) Report on the hydrogen explosion and fire incident in Santa Clara, California, June

IAEA (1991) Safety culture. Safety series 75-INSAG-4. IAEA. https://www.iaea.org/publications/3753/safety-culture

Kim Y (2019) Review: hydrogen tank explosion in Gangneung, South Korea. In: 2019 center for hydrogen safety conference. https://www.aiche.org/chs/conferences/international-center-hydrogen-safety-conference/2019/proceeding/paper/review-hydrogen-tank-explosion-gangneung-south-korea

Kohn LT, Corrigan JM, Donaldson MS (eds) (2000) To err is human. Building a safer health system. https://doi.org/10.17226/9728

Lam CY, Fuse M, Shimizu T (2019) Assessment of risk factors and effects in hydrogen logistics incidents from a network modeling perspective. Int J Hydrogen Energy 44(36):20572–20586. https://doi.org/10.1016/j.ijhydene.2019.05.187

Lanz A, Heffel J, Messer C (2013) Hydrogen fuel cell engines—module 6: fuel cell engine safety. J Chem Inf Model 53(9):1689–1699

Li Y, Guldenmund FW (2018) Safety management systems: a broad overview of the literature. Saf Sci 103:94–123. https://doi.org/10.1016/j.ssci.2017.11.016

Martin J (2002) Organizational culture: mapping the terrain. Sage. https://doi.org/10.4135/9781483328478

Mogles N, Padget J, Bosse T (2018) Systemic approaches to incident analysis in aviation: comparison of STAMP, agent-based modelling and institutions. Saf Sci 108:59–71. https://doi.org/10.1016/j.ssci.2018.04.009

Navajas J, Badia E, Sala R, Sato H, Paltrinieri N (2023) Development of a protocol for the systematic analysis of events at hydrogen refuelling stations. Chem Eng Trans 105:127–132. https://doi.org/10.3303/CET23105022

Nel ASA (n.d.) Status and Q&A regarding the Kjørbo incident. https://nelhydrogen.com/status-and-qa-regarding-the-kjorbo-incident/

Perrow C (1999) Normal accidents (REV-Revi). Princeton University Press. https://doi.org/10.2307/j.ctt7srgf

Ramprasad SK, Kumar P (2018) 3 factor—HOT (human, organizational and technical) model for construction safety culture. Int J Civ Eng Technol (IJCIET) 9(8):530–541. http://www.iaeme.com/IJCIET/index.asp530http://www.iaeme.com/ijciet/issues.asp?JType=IJCIET&VType=9&IType=8http://www.iaeme.com/IJCIET/index.asp531http://www.iaeme.com/IJCIET/issues.asp?JType=IJCIET&VType=9&IType=8

Rasmussen J (1986) Information processing and human-machine interaction: an approach to cognitive engineering. https://api.semanticscholar.org/CorpusID:60765343

Rasmussen J (1997) Risk management in a dynamic society: a modelling problem. Saf Sci 27(2):183–213. https://doi.org/10.1016/S0925-7535(97)00052-0

Reason J (1990) Human error. Cambridge University Press. https://books.google.es/books?id=WJL8NZc8lZ8C

Reason J (1997) Managing the risks of organizational accidents. Ashgate. https://books.google.es/books?id=ZnhRAAAAMAAJ

Rusin A, Stolecka-Antczak K, Kosman W, Rusin K (2024) Influence of driver error on the level of hydrogen refuelling station risk. Int J Hydrogen Energy 73:73–85. https://doi.org/10.1016/j.ijhydene.2024.05.436

Salehi F, Abbassi R, Asadnia M, Chan B, Chen L (2022) Overview of safety practices in sustainable hydrogen economy—an Australian perspective. Int J Hydrogen Energy 47(81):34689–34703. https://doi.org/10.1016/j.ijhydene.2022.08.041

Santos-Reyes J, Beard AN (2008) A systemic approach to managing safety. J Loss Prev Process Ind 21(1):15–28. https://doi.org/10.1016/j.jlp.2007.06.009

Schein EH (2004) Organizational culture and leadership, 3rd edn. Jossey-Bass. https://books.google.es/books?id=THQa4txcMl4C
Strauch B (2017) Investigating human error: incidents, accidents, and complex systems. Routledge. https://doi.org/10.4324/9781315251851
Turner BA, Pidgeon NF (1997) Man-made disasters, 2nd edn. Butterworth-Heinemann. https://books.google.es/books?id=7Hq6AAAAIAAJ
US National Research Council (2014) Lessons learned from the Fukushima nuclear accident for improving safety of U.S. nuclear plants. The National Academies Press. https://doi.org/10.17226/18294
Wang X, Gao W (2023) Hydrogen leakage risk assessment for hydrogen refueling stations. Int J Hydrogen Energy 48(91):35795–35808. https://doi.org/10.1016/j.ijhydene.2023.05.340
Wei R, Lan J, Lian L, Huang S, Zhao C, Dong Z, Weng J (2022) A bibliometric study on research trends in hydrogen safety. Process Saf Environ Prot 159:1064–1081. https://doi.org/10.1016/j.psep.2022.01.078
Wen JX, Marono M, Moretto P, Reinecke EA, Sathiah P, Studer E, Vyazmina E, Melideo D (2022) Statistics, lessons learned and recommendations from analysis of HIAD 2.0 database. Int J Hydrogen Energy 47(38):17082–17096. https://doi.org/10.1016/j.ijhydene.2022.03.170
Wilpert B, Fahlbruch B, Miller R, Baggen R, Gans A (1999) Inter-organisational development in the German nuclear safety system. In: Misumi J, Wilpert B, Miller R (eds) Nuclear safety. A human's factors perspective. Taylor & Francis, pp 127–140

Chapter 9
Risk Analysis: Qualitative and Quantitative Methods

Andrzej Rusin, Katarzyna Stolecka-Antczak, Wojciech Kosman, and Krzysztof Rusin

Abstract This chapter deals with the risk assessment for hydrogen installations. Two methods are presented. The first one aims to describe the risk in qualitative terms. Probabilities of failures and severities of the failure consequences are described using relative scales. The latter method is more complex and tends to determine the risk for the area with hydrogen structures. Risk evaluated for a specific location is determined by all risk-generating structures according to hazardous scenarios. Once each hazard is accounted for, the total risk is the combination of risk from different structures depending on the distance from that structure. This approach allows to determine the most hazardous area and then rearrange the layout of a hydrogen installation in order to decrease the resulting risk.

9.1 Introduction

Hydrogen physicochemical properties result in the gas applications exposed to the presence of a potential risk of hazardous events for people and the environment. This concerns every stage of hydrogen industry.

Hazardous events may also happen in hydrogen refueling stations. An example of such an incident happened in Norway in 2019. A failure started with a hydrogen leak from a high-pressure storage unit and resulted in a fire and an explosion. The fire was suppressed after 3 h, and there were no injuries to the station maintenance crew. However, the explosion and the following pressure wave triggered an airbag in a nearby car, and the passengers in that car suffered minor injuries. The cause of the failure was determined to be an assembly error of a plug in a high-pressure storage tank. The screws in the plug were not properly fastened. It led to hydrogen leakage and created a mixture of hydrogen and air, which was then ignited. The source of the ignition was not identified.

A. Rusin · K. Stolecka-Antczak · W. Kosman (✉) · K. Rusin
Silesian University of Technology, Gliwice, Poland
e-mail: wojciech.kosman@polsl.pl

Y. Liu et al. (eds.), *Safety Measures for Hydrogen Fueling Stations*, Risk, Reliability and Safety Engineering, https://doi.org/10.1007/978-981-96-9412-9_9

Another example of a hazardous incident at a hydrogen refueling station is the failure of an excess valve located on a high-pressure pipeline leading to a storage area. This incident took place in the USA in 2012. The failure led to a leakage of approximately 300 kg of hydrogen. Firemen evacuated residents in the surrounding areas. There were no human casualties, and the material losses were insignificant: a roof over a dispenser was mildly burned by a hydrogen flame. The safety system in the station responded correctly and the equipment and fuel storages were immediately cut off. The pressure and the temperature levels in storage tanks were within their allowable ranges before and during the incident. After a detailed analysis of the event, there was a decision to exchange the safety valves with new ones, elevate the ventilation ducts, modify the first response procedures, and improve the communication channels with the first responders. A significant amount of time was spent on a thorough overview of the station design, evaluation of the plans and first response procedures, meetings with the local society, additional first aid training, and test drills involving station employees together with first responders.

The events described above prove that an unfortunate course of a failure resulting from an uncontrolled hydrogen leak may cause a significant danger for people and the environment. The evaluation presented in Chap. 5 shows the potential hazard zones in a refueling station and is an exemplary analysis of dangerous events related to hydrogen processing. The safety of hydrogen processes is improved, when all persons involved are aware of the accompanying risk. The safety is further improved, when appropriate technology is applied. This includes hydrogen leak sensors, correct arrangement of cutoff valves, and firewalls (Tian et al. 2024).

9.2 Methodology for Risk Analysis Related to Hydrogen Refueling Station

A risk related to the presence of different technical objects is defined as a product of hazardous scenario probability and the consequences of the scenario. The consequences are for example a threat of human death or injury. If a number of scenarios exist, then the risk is combined (Witkowski et al. 2017; Yoo et al. 2021):

$$R = \sum_{i}^{n} P_i C_i \tag{9.1}$$

where

P_i probability or frequency of occurrence of a hazardous scenario
C_i consequences of the scenario
n number of hazardous scenarios related to an object or a process.

Results of a qualitative evaluation of risk are risk diagrams, while results of quantitative evaluation are values of a risk assigned to each scenario for the whole systems

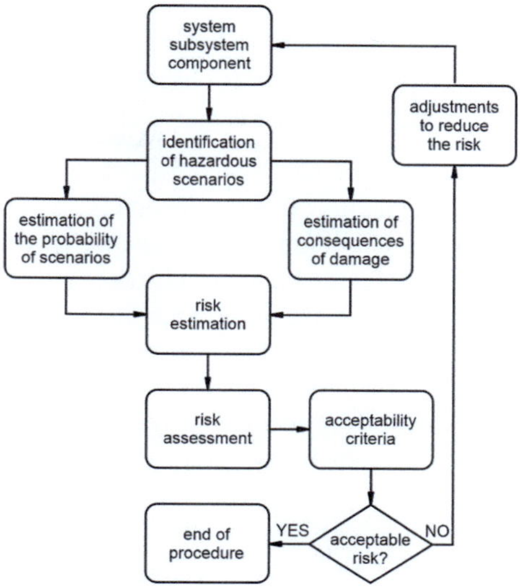

Fig. 9.1 Risk analysis procedure

or subsystems. They are described in more details in the following sections of this chapter.

Risk analysis is a formalized procedure and includes among others the following steps: definition of a system, subsystems and their objects, identification of hazards, qualitative risk evaluation, quantitative risk evaluation, risk evaluation according to assumed criteria, and actions resulting from the conducted risk analysis—see Fig. 9.1 (Suzuki et al. 2021).

The first essential step in a risk analysis is the identification of a system and its structure. Complex systems like a hydrogen refueling station are divided into a number of subsystems, which are then divided into objects. As described in Chap. 5, a typical arrangement of a station includes a refueling area with a dispenser, a storage system, and a shop/cashier. An example of station layout is shown in Chap. 6 of this book in Fig. 6.2.

The second step in a risk analysis is the identification of hazardous scenarios related to the operation of the system or object under the investigation (Rusin et al. 2024a, b).

The next step of a risk analysis is the determination of consequences of the hazardous events defined in an event tree. In order to determine the rate for people exposed to hazardous events, so-called probit functions are used. These functions allow to determine the probability of injuries due to effects like fires, explosions, and reaction to toxic substances. In general, the probit functions have the following form (Rusin et al. 2024a, b):

$$P_r = a + b \cdot \ln V \quad (9.2)$$

where

a, b empirical constants basing on data from recorded real events
V a measure of the negative effect.

The measure of the negative effect may be a single value or a set of values. For example, an effect of an explosion is an overpressure wave described by the value of the pressure as a function of distance from the center of the explosion. On the other hand, the release of a toxic substance is described by a combination of the substance concentration in the air and the time of exposure to the substance.

Depending on the risk analysis results, further actions may be taken to decrease the level of the risk. This is a part of risk management.

The following sections provide a description of a comprehensive risk analysis of a hydrogen refueling station.

9.3 Analysis of the Risk Related to Hydrogen Release at a Refueling Station

9.3.1 Qualitative Risk Analysis

Qualitative risk analysis often relies on professional experience and good practice. The risk is expressed in relative categories with subjective measures and ratings. These are descriptive values used to differentiate the levels of risk, for example: low, average, and high. Qualitative methods may also employ dimensionless numbers that allow to compare the risks related to different objects and sort them from the highest to the lowest (Graves 2000; Nakayama et al. 2022).

An example of a risk matrix is shown in Fig. 9.2. The matrix was created for the hydrogen refueling station introduced in Chap. 6. The analysis focused on partial (a puncture) and total (a rupture) damages to the chosen structures located in the station: a dispenser, a compressor, a trailer, and a high-pressure tank. Failures of these types may lead to an uncontrolled release of hydrogen and result in negative consequences like fires or explosions.

The probabilities of hazardous scenarios and the intensity of the consequences are evaluated in five-point scales. The probability levels include almost certain, likely, possible, unlikely, and rare, while for the consequences, extreme, major, moderate, minor, and insignificant. The following hazard scenarios are considered:

- S1—ruptured dispenser
- S2—punctured dispenser
- S3—ruptured trailer
- S4—punctured trailer
- S5—ruptured HP tank
- S6—punctured HP tank

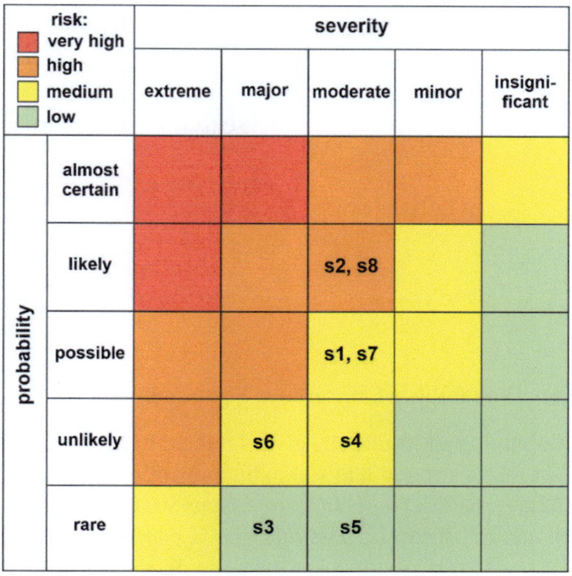

Fig. 9.2 Risk matrix

- S7—ruptured compressor
- S8—punctured compressor.

The severity of the two failure types—a rupture or a puncture—is related to different amounts and flow rates of the released hydrogen and to the duration of the leak.

The colors of the areas in the matrix refer to different risk levels according to the legend given in the figure. Depending on the levels, actions should be taken to change the level of risk especially for the very high and high levels. The particular risk diagram given for the refueling station indicates that two out of eight considered scenarios generate risk at the high level. This is due to likely probability of the scenarios and moderate consequences. These two scenarios must be analyzed in detail, and actions should be taken to decrease the risk.

9.3.2 Quantitative Risk Analysis

The first step of the quantitative risk analysis is the evaluation of the hazardous scenario probability. In the second step, the consequences of the scenarios are also evaluated. The measure of the consequences is the human death probability. Then the risk may be calculated according to Eq. (9.1) for any area in a hydrogen installation, where risk is affected by more than one object or more than one hazardous scenario. The obtained results allow to rearrange the structures in a hydrogen system to minimize the human death probability and material losses to buildings and facilities.

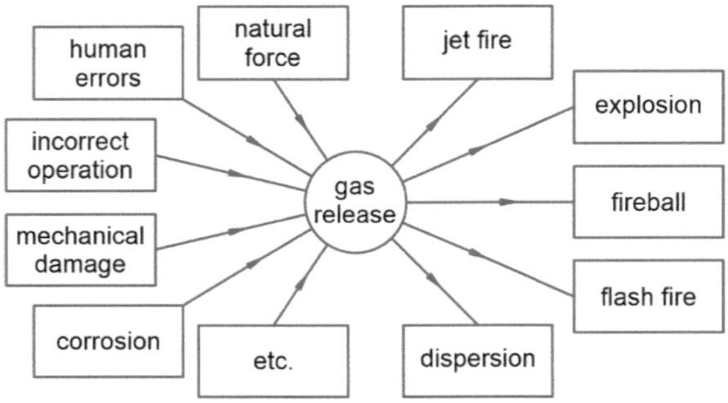

Fig. 9.3 Causes of failures and their consequences

(a) Probability of Hazardous Scenarios

Probability of hazardous events related to hydrogen release depends on a number of factors. Hydrogen leaks may be a result of a human error, valve or pipeline fittings failure, material defects, components wear due to, for example, corrosion or fatigue, or natural disasters. The last one is especially difficult to predict. Also, problems may arise that are linked to mistakes done at a design stage of a hydrogen installation or to mistakes made during the operation of the system. An important factor that determines the effects of failures is also the first response to a failure including the response of the automatic safety systems. The factors are depicted in Fig. 9.3 (Rusin et al. 2021).

The initial event for every object considered in a risk analysis is an uncontrolled hydrogen leak. This is the root of the event tree shown for example in Fig. 9.4 for a dispenser, trailer/HP tank. In the examples provided in this chapter, the probabilities of leakages are taken from the available technical publications. In each tree, the first division into branches located right after the root splits the potential events between ruptures and punctures. This division implicates the amount of the released hydrogen, which in turn affects the level of hazard caused by a potential fire or explosion if the hydrogen is ignited. The tree in Fig. 9.4 presents also the branches related to the time between the initial gas release and the ignition, which correspond to an immediate ignition or a delayed ignition. The final events are fireballs, jet fires, explosions, or dispersion of the released hydrogen with the last one having no negative consequences (Pan et al. 2016; Timmers and Stam 2017). The event trees provide probabilities for each branch (Li et al. 2011).

Further analysis uses the hazard zones described in Chap. 6. In those zones, a heat flux or an overpressure wave may occur as a result of a potential fire or explosion (Matthijsen and Kooi 2006).

(b) Consequences of Hazardous Scenarios

In order to evaluate the negative consequences of hazardous events in a refueling station, one may use probit functions. As written earlier in this chapter, the functions allow to determine the probability of injuries or death for given measures of effects related to negative consequences. In a refueling station, a fire is measured

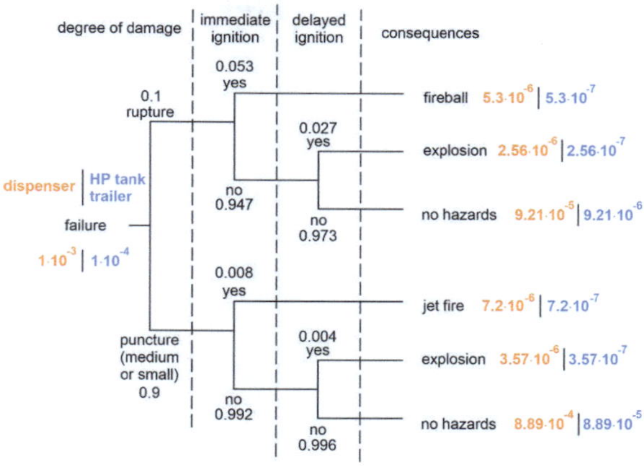

Fig. 9.4 Event tree analysis for hydrogen release from a dispenser, trailer, or an HP tank

through a generated heat flux, and an explosion is measured through the value of the overpressure. The corresponding probit functions are

- for a heat radiation (Groth et al. 2017; Eisenberg et al. 1975)

$$P_r = -38.48 + 2.56 \cdot \ln\left(q^{\frac{4}{3}} \cdot t\right) \tag{9.3}$$

- for an overpressure (Baraldi et al. 2011)

$$P_r = -77.1 + 6.91 \cdot \ln(\Delta p) \tag{9.4}$$

where

q heat flux
t time of exposure to the heat flux
Δp overpressure.

The values from the probit functions are later converted to the death probabilities using conversion tables. Since the heat flux and the overpressure depend on the distance between the center of a fire or an explosion, the probit function and the death probability resulting from the probit functions are also dependent on the distance from the center of a negative effect. This is shown in the graphs in Figs. 9.5, 9.6, 9.7 and 9.8. The figures depict the death probabilities for the tree's main structures in a refueling station: a dispenser, a trailer, and an HP tank. Figure 9.5 corresponds to a jet fire, and Fig. 9.6 to a fireball. Both phenomena are related to a heat flux and their effects are calculated using Eq. (9.3).

Figure 9.7 shows the death probability for the explosion that occurs at a distance from the location of the hydrogen release. The distance is measured here not from the point of the release but from the center of the explosion. The figure may be compared to Fig. 9.8, where the death probability is shown for the explosion triggered at the point of the hydrogen release. It is worth emphasizing here that the vertical axes in

Fig. 9.5 Probability of human death for a jet fire

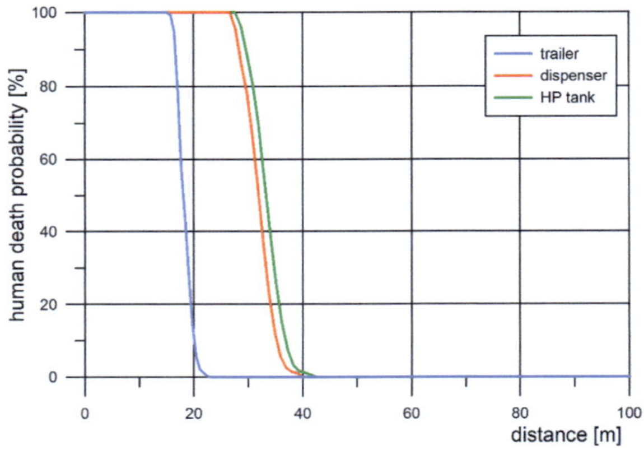

Fig. 9.6 Probability of human death for a fireball

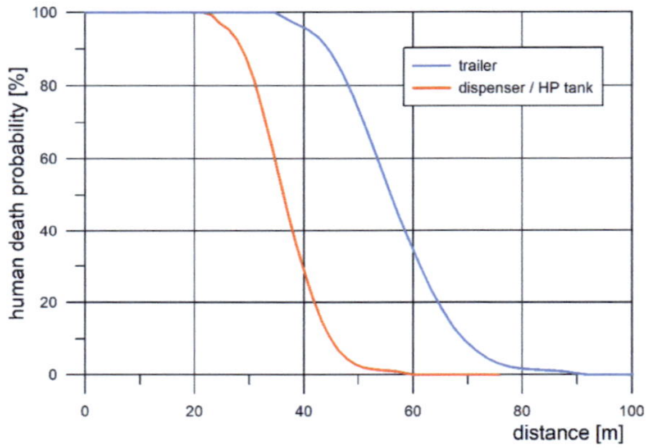

Fig. 9.7 Probability of human death for an explosion away from the point of release

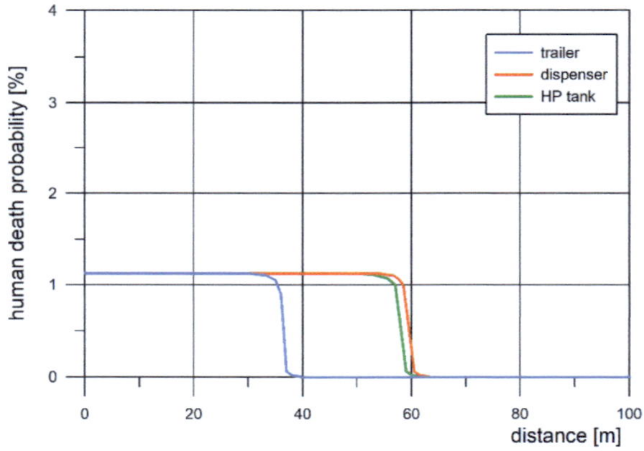

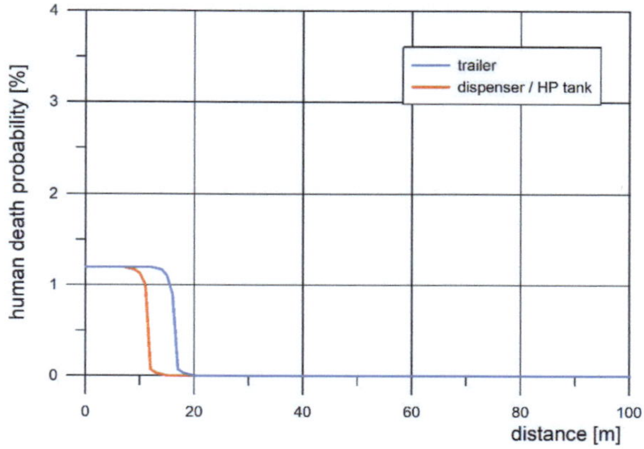

Fig. 9.8 Probability of human death for an explosion at the location of the hydrogen release

Figs. 9.5 and 9.6 have different ranges than the vertical axes in Figs. 9.7 and 9.8. This is due to differences between fires and explosions.

(c) Risk Analysis

The probabilities of hazardous scenarios from the event tree shown in Fig. 9.4 together with the probabilities of human death presented in Figs. 9.5, 9.6, 9.7 and 9.8 allow to calculate the risk according to Eq. (9.1) related to the main structures in a refueling station. Figures 9.5, 9.6, 9.7 and 9.8 prove that the probability of human death depends on the distance from the point, which is the center of explosion or the source of fire. This probability is multiplied by scenario probability according to Eq. (9.1) in order to calculate the risk. For this reason, the calculated risk is also a function of distance from the center or the source of a negative consequence of uncontrolled hydrogen release. The distance may be treated as a range of a risk and may be plotted over an aerial view of a refueling station layout. This concept is shown in Fig. 9.9.

Death probability is shown earlier in this chapter for four consequences that may occur at the end of a hazardous scenario. In order to keep the consequences uniform, the delayed explosion will be omitted in the further analysis. In this way a jet fire, a fireball, and an immediate explosion translate to the risk ranges shown in Fig. 9.10 with centers in a common point.

According to Eq. (9.1), the risks due to different scenarios may be added to determine the total risk. A single structure like a pressure tank that is subject to a

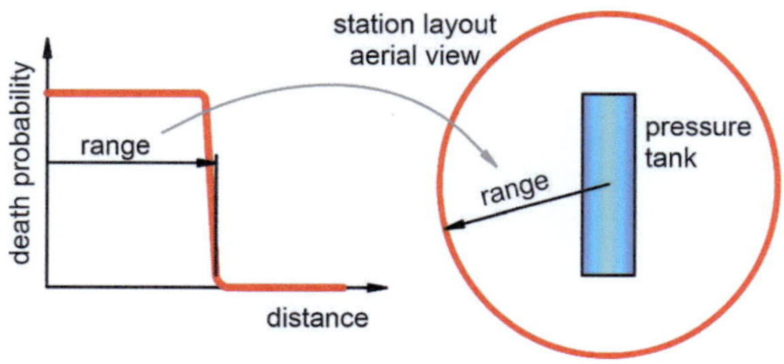

Fig. 9.9 Risk range from death probability dependency on the distance from a hydrogen leak

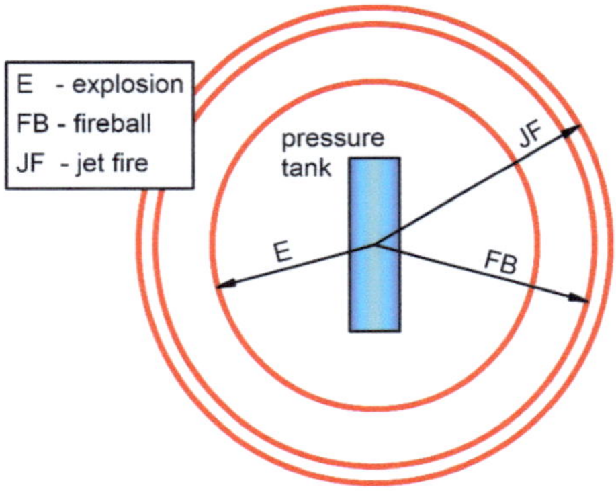

Fig. 9.10 Risk ranges from three potential consequences of hydrogen leak

possible hydrogen leak "generates" the risk area around itself. The highest value of the risk occurs in the zone that is the intersection of the areas corresponding to different consequences. In Fig. 9.10, this area is the circle with the smallest radius, because it is the common area (the intersection area) of all three circles.

In a refueling station or any other hydrogen installation, there are a number of structures that are prone to leaks and therefore "generate" risk areas. These areas may overlap in a limited space of a station. Each overlap increases the total risk in the intersection area. This is shown in Fig. 9.11 for two structures.

In the figure, red curves correspond to basic risk areas from a single structure. Green color denotes locations, where at least two basic risk areas intersect. There is also the area bounded by violet curves, where the highest number of risk areas intersect and the total risk reaches the maximal value.

Due to different mass and pressure of the stored hydrogen, the risk ranges may be and usually are different for different structures.

In a risk analysis, the ranges and the values of the risks from single structures are determined first. The ranges are taken from the death probability versus distance relations and the risk values from Eq. (9.1). After that, the task becomes a geometric

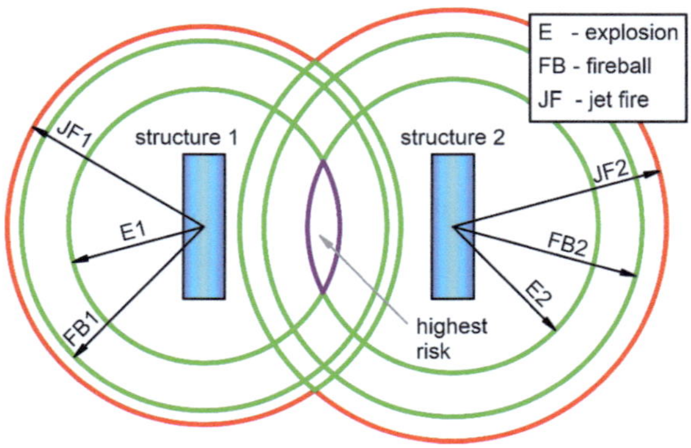

Fig. 9.11 Risk ranges for two structures and risk area intersections

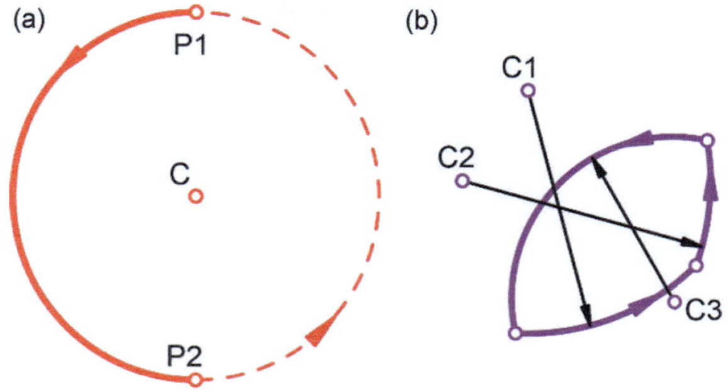

Fig. 9.12 Representation of a basic risk range (**a**) and an intersection area (**b**)

analysis with the goal of establishing the intersection area with the highest value of the risk. The examples shown later in this chapter base on the following algorithm.

Step 1

The basic risk areas for a single structure are represented as a combination of two circular arcs as shown in Fig. 9.12a. Each arc is described by a starting point P1, a center point C, and an ending point P2. All the shapes of the intersection areas that are established later are also represented by a combination of circular arcs as shown in Fig. 9.12b. The arcs are defined in a counterclockwise order.

Step 2

The intersection area with the largest risk is found. The search must avoid adding the value of a basic risk more than once. For that reason, a list is generated that initially includes the basic risk areas for single structures. In subsequent steps, each area is tested against other basic risk areas that are up the list, that is toward the end of the list. If an intersection is found, it is added to the end of the list with an index that identifies the second basic area in a pair that was tested. The intersections are then themselves tested against the basic risk areas. The index determines against which basic area the tests should start. The basic risk areas are always taken starting from the one located right after the indexed area, and the tests are conducted moving toward the end of the list of the basic areas. Again, if an intersection is found then it is placed at the end of the list with the index describing the basic area from the last test. The steps are repeated until the end of the list is reached. The algorithm allows to find multiple intersection areas that may cover separate space in the analyzed hydrogen station or other installation. For each intersection, the basic risk values are added.

Step 3

Among all the intersections found, the one with the highest total risk is the result of the search. The surface area is calculated for that shape. Curved edges are approximated by short line segments. Then the surface area is calculated using the equation for a surface area of a flat polygon, which has the form:

Fig. 9.13 Intersection of circles

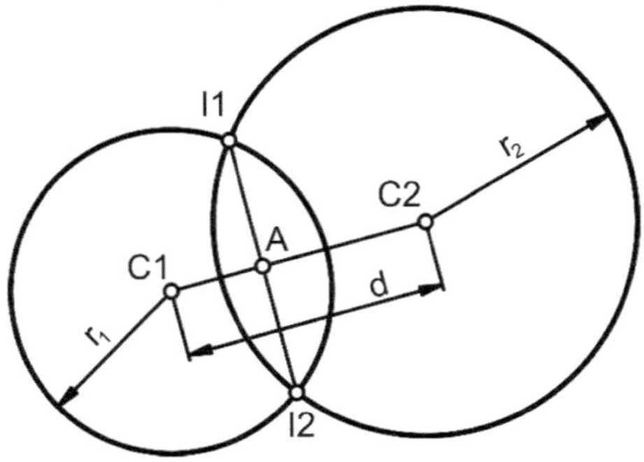

$$A = \frac{1}{2} \sum_{i}^{n} (x_i y_{i+1} - x_{i+1} y_i) \tag{9.5}$$

where x and y are coordinates of consecutive approximation points along the edge.

Step 2 described above involves the search for intersections between shapes that consist of circular arcs. A pair of areas is tested for points of intersection using comparisons between pairs of edges. Two arcs may have one, two, or no point of intersection. In a general case, the coordinates of an auxiliary point A are calculated first; see Fig. 9.13.

The auxiliary point coordinates are

$$\begin{aligned} x_A &= x_{C1} + a \frac{x_{C2} - x_{C1}}{d} \\ y_A &= y_{C1} + a \frac{y_{C2} - y_{C1}}{d} \end{aligned} \tag{9.6}$$

with

$$a = \frac{r_1^2 - r_2^2 + d^2}{2d} \tag{9.7}$$

where d is the distance between the centers of the circles, $(x_{C1}; y_{C1})$ and $(x_{C2}; y_{C2})$ are the coordinates of circle centers, r_1 and r_2 are circle radii. The coordinates of the intersection points if any are defined as

$$\begin{aligned} x_i &= x_A \pm h \frac{y_{C2} - y_{C1}}{d} \\ y_i &= y_A \mp h \frac{x_{C2} - x_{C1}}{d} \end{aligned} \tag{9.8}$$

with

$$h = \sqrt{r_1^2 - a^2} \tag{9.9}$$

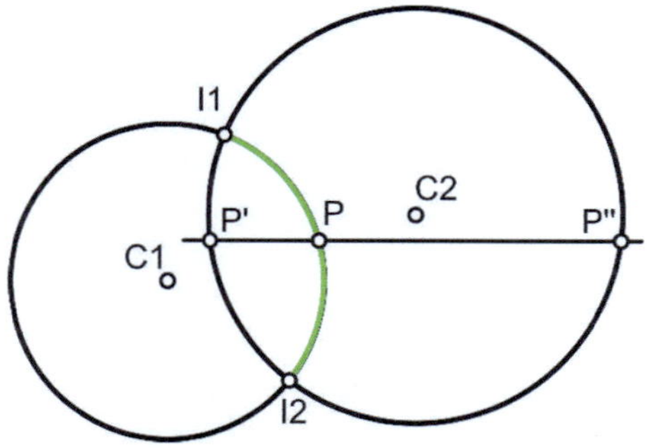

Fig. 9.14 Internal edge testing

No points of intersection exist if the expression under the square root in Eq. (9.9) is negative. Also, the intersection points must lie on the arcs between their starting and ending points. One way to verify the location of the intersection points between the arc ends is the use of a cross product using the vectors formulated by the endpoints and the center point and the tested intersection point.

The shapes that are tested for intersection are then divided at the intersection points. A new shape is built from the edges of the tested shapes. The intersection area includes only the internal edges meaning for example an edge of one of the tested shape that is located completely inside the other shape. The test of the internal/ external location is done using a midpoint on an edge and casting a horizontal ray through that midpoint. If the number of intersections between the ray and the border of an area on one side of the point is odd, then the edge is located inside that area. This is shown in Fig. 9.14. The green arc is tested against the circle with center C2. The horizontal line going through point P intersects the circle only once on the left side and only once on the right side of the point P. Therefore, the green arc is located inside the circle with the center in C2.

The methodology described above allows to identify the zone that corresponds to the highest risk "generated" from multiple structures in a refueling station. The surface area of this zone is a value that may be calculated as explained above. This value may be used as a measure in the evaluation of the level of risk in a hydrogen installation. It is strictly correlated to the location of the main structures in an installation.

9.4 Refueling Station Layout Risk Analysis

A number of station layouts is chosen here to analyze the risk in a hydrogen facility. The layouts base on the real station design as well as projects presented in the literature. However, in order to allow a comparison between the analyzed cases, the layouts are unified in the following way. Firstly, all the layouts are fit into a rectangular area of 46.5 by 41.5 m. Secondly, the analysis is restricted to three main structures in each layout: a trailer that delivers gaseous hydrogen to the station, a high-pressure storage tank, which represents a pack of tanks, and a dispenser for cars or trucks refilling. The range of a risk area for each single structure and the risk magnitude are taken from the scenario probabilities shown in Fig. 9.4, and the death probability relations are presented in Figs. 9.5, 9.6, 9.7 and 9.8. The analyzed cases are shown in Figs. 9.15, 9.16, 9.17 and 9.18. The figures include the main structures, a shop/cashier, and intersection of the risk areas "generated" by the main structures. In each drawing, the area with the highest risk is also presented.

In the first case in Fig. 9.15, the trailer is parked behind the utility shop/cashier. High-pressure tank storage is also located behind the shop. Dispenser for refilling cars and trucks is located in front of the shop. The surface area of the highest risk zone is 7.09 m^2 with the risk value of 1.503×10^{-5} per year. In this configuration, the zone of the highest risk is located at the shop.

In Case 2 (Fig. 9.16), the trailer and storage section are located in the north-west part of the station, while the shop is on the opposite end of the station diagonally. The surface area of the highest risk zone is 52.51 m^2 with the risk value of 1.503×10^{-5} per year.

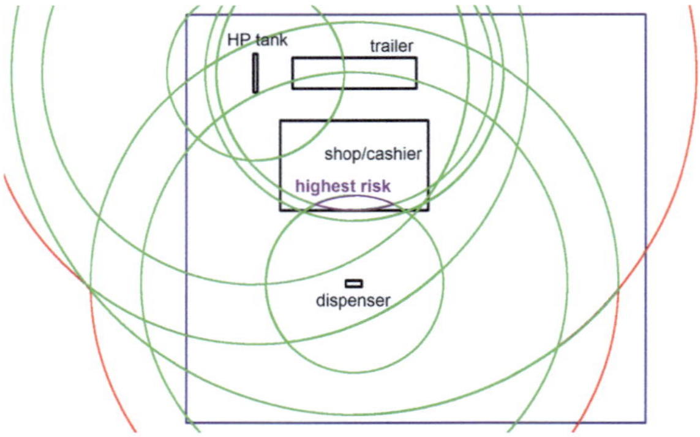

Fig. 9.15 Refueling station risk analysis—Case 1

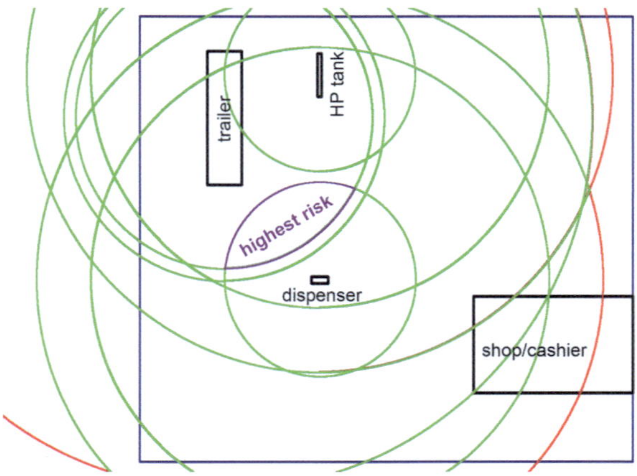

Fig. 9.16 Refueling station risk analysis—Case 2

Fig. 9.17 Refueling station risk analysis—Case 3

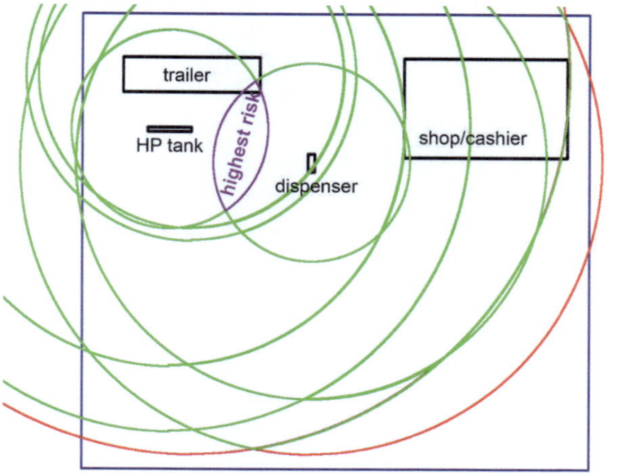

Fig. 9.18 Refueling station risk analysis—Case 4

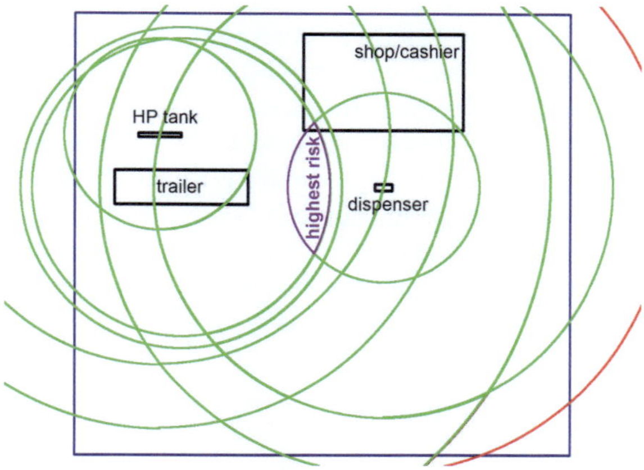

Case 3 in Fig. 9.17 corresponds to a station with the main structures located in the northern part. The dispenser is between the storage area and the shop. The surface area of the highest risk zone is 37.37 m^2 with the risk value of 1.503×10^{-5} per year.

A similar case is shown in Fig. 9.18. The dispenser is located in the center. Storage area is to the west from the dispenser and the shop is to the north. The surface area of the highest risk zone is 32.63 m^2 with the risk value of 1.503×10^{-5} per year.

The analysis of the four cases in Figs. 9.15, 9.16, 9.17 and 9.18 leads to the following conclusions. Firstly, the highest value of the total risk in the area, where most basic risks intersect, is the same in each case. This is because the hydrogen refueling station analyzed here occupies a relatively small space. The area of the highest risk is the intersection of the basic risk areas for jet fire and fireball for each of the three main structures (trailer, HP tank, dispenser). The risk corresponding to explosion is one order lower than the risks for jet fire and fireball. So, even if the intersection area with the highest risk does not include the basic area for an explosion, the risk value is not significantly affected.

Secondly, the arrangement and the distances between the structures affect the size of the area with the highest risk and its location. In Case 1, the concentration of the risk is located in the shop/cashier section. This should be avoided because this section is visited by untrained customers and is occupied by staff during the working hours. The appropriate layout of a refueling station should avoid putting the shop between the storage and the dispenser sections.

It may be seen from the presented layouts with the highest risk areas that the risk concentration is located between the storage and the dispenser section. Since the size of the highest risk area depends on the distances between the structures, then it may be decreased by moving the structures away from each other, as long as they remain inside the station boundary.

An example of modification of the preliminary layout is shown in Figs. 9.19 and 9.20. The dispenser seen in Fig. 9.18 is moved south, that is, away from the shop/cashier. The area of the highest risk follows this change. Figure 9.19 presents a dependency between the position of the dispenser and the surface area of the highest risk zone. The location of the dispenser is measured from the position in Fig. 9.18 and toward the vertical direction to the south.

Figure 9.20 presents the layout of the hydrogen refueling station from Case 4 after modification. The dispenser is relocated 14 m to the south. The highest risk area highlighted in the figure is smaller than in Fig. 9.18. It is also located further away from the shop/cashier. The shop/cashier facility is relocated itself—it is moved 7 m to the east boundary. This modification puts the shop further away from the storage section and the dispenser.

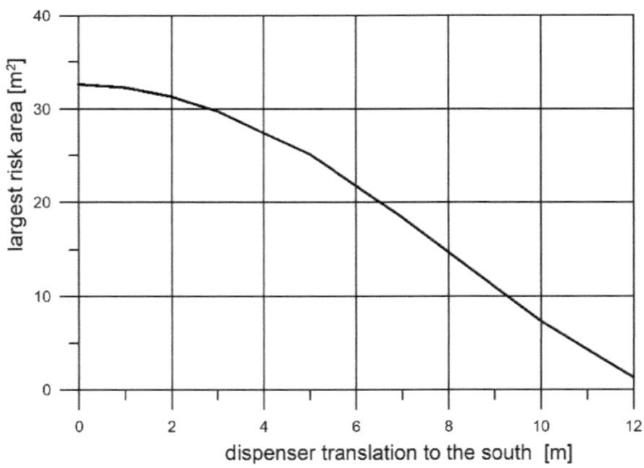

Fig. 9.19 Surface area of the highest risk zone versus dispenser translation

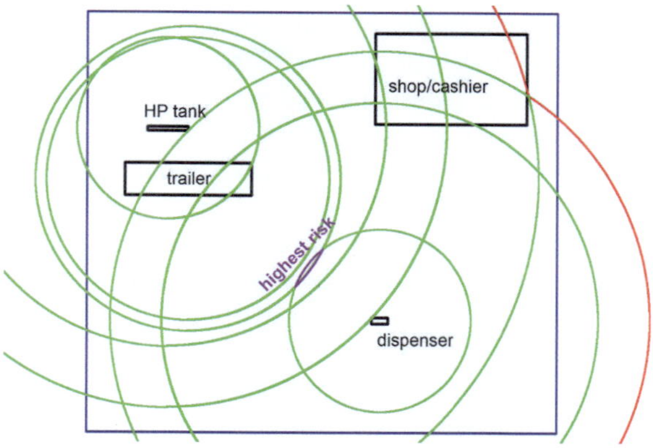

Fig. 9.20 Refueling station risk analysis—Case 4 after modification

9.5 Summary and Conclusions

The results presented in this chapter prove that using hydrogen as a fuel for vehicles is related to hazards that derive from the properties of this gas. The main concern is a potential leakage and uncontrolled release of hydrogen, which create a possibility of fire or explosion. The amount of the released hydrogen and its parameters (mostly pressure) in different structures directly affect the level of hazard for people and the environment.

The risk analysis presented here and in Chap. 6 was conducted for a refueling station with three structures assumed to be the main sources of the risk: a dispenser, a trailer, and an HP tank. The hazard zones calculated for the structures have larger ranges in case of a fire than in case of an explosion. Also, the death probability is higher for a fire. This is true for both types of damage: a puncture and a rupture. The largest zone with a range of around 30 m exists for a rupture of a trailer. This is due to the significant amount of hydrogen stored in one tank. If only punctures are considered, then the largest zone of 27 m exists for the HP tank due to the highest pressure. In case of explosion, the largest hazard zone exists again for the trailer, and it has a radius of 14 m. The zones for the dispenser and the HP tank are similar with radii of 9 m because these two structures contain almost an identical amount of hydrogen. The death zones related to a puncture followed by a phenomena of a moving hydrogen cloud have a range of 2, 3, and 4 m for a trailer, an HP tank and a dispenser, respectively.

The risk analysis described here included also the determination of the location and the size of the highest risk area. This part of the analysis involves potential hazards from multiple sources. In the presented work, three main sources of a hydrogen leak are considered: a trailer, a high-pressure tank representing the storage section, and a dispenser. Different layouts of a station with these three structures have different locations and size of the highest risk area. The methodology shown here allows to calculate the location and the size.

The improvement of the operational safety in hydrogen refueling stations requires actions undertaken at different stages of existence of the stations, mostly during the design phase and during the operation time. According to the results provided here, the arrangement of the structures in a station should minimize the surface of the areas with the highest risk. Also, the domino effect should be avoided meaning that a failure of one structure must not cause failures of other, nearby structures due to spreading heat flux or overpressure wave. Additionally, the design stage should take into consideration the influence of the objects located in the direct vicinity of the station since they may become the sources of the hydrogen ignition (Rusin et al. 2024a, b).

Another aspect is safety systems. Hydrogen stations must be equipped with hydrogen detection systems with an alarming capability. A detection of a leak must be followed by signals received by the maintenance crew and in extreme cases by first responders. Safety systems must encompass a complete hydrogen processing path including transportation trailers, low and high-pressure tanks, compressors, pipelines, heat exchangers, and dispensers.

Safety is greatly improved by preparation of the hydrogen users. Training should involve not only refueling station personnel and service employees but also the drivers refilling their cars. Driving school preparation should be expanded to include at least basic information that emphasizes the hazards related to hydrogen properties, which makes this gas more dangerous than hydrocarbon fuels.

References

Baraldi D, Papanikolaou E, Heitsch M, Moretto P, Cant RS, Roekaerts D, Dorofeev S, Kotchourko A, Middha P, Tchouvelev AV, Ledin S, Wen J, Venetsanos A, Molkov VV (2011) Prioritisation of research and development for modelling the safe production, storage, delivery and use of hydrogen. Joint Research Centre Institute for Energy and Transport

Eisenberg NA, Lynch CJ, Breeding RJ (1975) Vulnerability model. A simulation system for assessing damage resulting from marine spills. U.S. Coast Guard, SA/A-015 245

Graves R (2000) Qualitative risk assessment. PM Netw 14(10):61–66

Groth K, Hecht E, Reynolds JT, Blaylock ML, Carrier EE (2017) Methodology for assessing the safety of hydrogen systems. HyRAM 1.1 technical reference manual. SAND2017-2998

Li Z, Pan X, Ma J (2011) Quantitative risk assessment on 2010 Expo hydrogen station. Int J Hydrogen Energy 36(6):4079–4086

Matthijsen AJCM, Kooi ES (2006) Safety distances for hydrogen filling stations. Fuel Cells Bull 11:12–16

Nakayama J, Suzuki Y, Owada S, Shiota IY, Noguchi K, Miyake A (2022) Qualitative risk analysis of the overhead hydrogen piping at the conceptual process design stage. Int J Hydrogen Energy 47(22):11725–11738

Pan X, Li Z, Zhang C, Lv H, Liu S, Ma J (2016) Safety study of a wind–solar hybrid renewable hydrogen refuelling station in China. Int J Hydrogen Energy 41(30):13315–13321

Rusin A, Stolecka-Antczak K, Kapusta K, Rogoziński K, Rusin K (2021) Analysis of the effects of failure of a gas pipeline caused by a mechanical damage. Energies 14(22):1–21

Rusin A, Stolecka-Antczak K, Kosman W, Rusin K (2024a) The impact of the configuration of a hydrogen refueling station on risk level. Energies 17:5504

Rusin A, Stolecka-Antczak K, Kosman W, Rusin K (2024b) Influence of driver error on the level of hydrogen refuelling station risk. Int J Hydrogen Energy 73:73–85

Suzuki T, Shiota K, Izato Y, Komori M, Sato K, Takai Y, Ninomiya T, Miyake A (2021) Quantitative risk assessment using a Japanese hydrogen refueling station model. Int J Hydrogen Energy 46:8329–8343

Tian Y, Zhang X, Shan M, Qi M, Shu C-M, Li B, Liu Y (2024) Methodology for optimally designing firewalls in hydrogen refueling stations. Int J Hydrogen Energy 49(Part D):1196–1209

Timmers PGJ, Stam G (2017) Risk based safety distances for hydrogen refueling stations. In: International conference of hydrogen safety, Hamburg, Germany

Witkowski A, Rusin A, Majkut M, Stolecka K (2017) Comprehensive analysis of hydrogen compression and pipeline transportation from thermodynamics and safety aspects. Energy 141:2508–2518

Yoo BH, Wilailak S, Bae SH, Gye HR, Lee CJ (2021) Comparative risk assessment of liquefied and gaseous hydrogen refueling stations. Int J Hydrogen Energy 46:35511–35524

Chapter 10
Barrier Management for Hydrogen Refueling Stations

Alessandro Campari, Antonio Javier Nakhal Akel, Abhishek Subedi, and Nicola Paltrinieri

Abstract This chapter explores the critical role of safety barrier management in hydrogen refueling stations. Using Acoustic Emission Testing (AET) for Type IV composite overwrapped pressure vessels (COPVs) as a case study, the chapter presents the integration of Human Reliability Analysis (HRA) and Bayesian Network (BN) modeling in continuous evaluation and support of safety barriers. The study identifies tasks such as sensor performance evaluation and test completion as critical points sensitive to human errors and performance-shaping factors. These findings highlight the need for robust operator training, intuitive interfaces, and adaptive environmental controls to enhance barrier effectiveness. Additionally, the study emphasizes the importance of maintaining barrier independence and robustness, as well as fostering a proactive safety culture within organizations. The integration of digital technologies such as machine learning and real-time data analytics is proposed to improve defect detection and predictive insights, providing a pathway for advancing hydrogen safety management. By aligning these methodologies with established safety theories, the chapter offers a comprehensive framework for improving safety barrier management in hydrogen storage systems and other high-risk industrial applications.

A. Campari · A. Subedi · N. Paltrinieri (✉)
Department of Mechanical and Industrial Engineering, Norwegian University of Science and Technology (NTNU), Trondheim, Norway
e-mail: nicola.paltrinieri@ntnu.no

A. J. N. Akel
Department of Mechanical and Aerospace Engineering, Sapienza University of Rome, Rome, Italy

Department of Engineering and Science, Universitas Mercatorum, Rome, Italy

Y. Liu et al. (eds.), *Safety Measures for Hydrogen Fueling Stations*, Risk, Reliability and Safety Engineering, https://doi.org/10.1007/978-981-96-9412-9_10

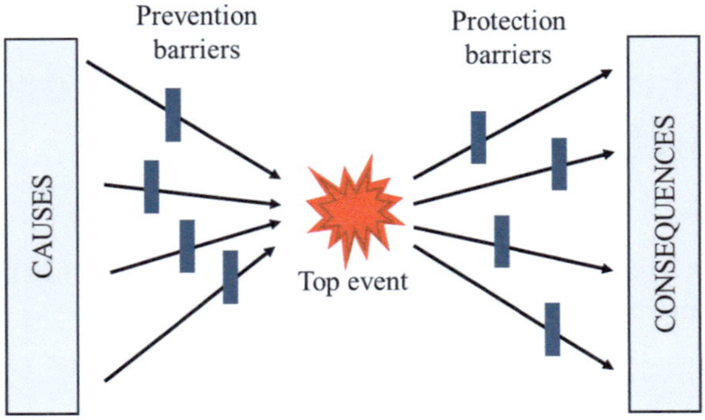

Fig. 10.1 Representation of safety barriers in bow-tie diagram

10.1 Introduction

Safety barriers are physical and/or non-physical means planned to prevent, control, or mitigate undesired events or accidents. The barriers are related to the implementation of safety functions to prevent, control, or mitigate undesired events or accidents. The recognized classification based on bow-tie model as presented in Fig. 10.1 distinguishes proactive (prevention) and reactive (protection) barriers. Proactive barriers intend to prevent or reduce the probability of hazardous event (top event) whereas reactive barriers reduce the consequences of it.

Safety barriers are also classified as active and passive incorporating physical, technical, operational, and organizational barriers. The acknowledged systematic classification of safety barriers available in the literature are provided by ARAMIS (2004), Hollnagel (2004), Reason (1997), and Sklet (2006). Mostly in all classifications, barriers align with having the following properties: specificity, adequacy, independence, dependability, robustness, and auditability. The concept of safety barriers is employed in risk analysis. Risk analysis involves identifying hazards, likelihood of their happening, and the potential consequences when it happens. Initially, it is essential to identifying potential hazards and evaluating their probabilities and consequences. To mitigate risks, safety barriers are implemented to decrease the likelihood and mitigate the unintended consequences.

10.2 Overview of Barrier Management

The Norwegian Petroleum Safety Authority defines barrier management as a systematic and continuous process that involves the identification, establishment, and maintenance of barriers. DNV (2014) and PSA (2017) establishes barrier management to aspects like improvement of safety, inclusion of safety culture, operational risk management, and organizational learning. In general, barrier management can be understood as the integration of several tasks which aim to achieve the required barrier function for risk mitigation.

Barrier management starts from the identification of the appropriate barriers and then analyzes their roles in preventing and mitigating accidents. The barrier analysis methods are a mix of qualitative and quantitative approaches. Table 10.1 summarizes different barrier analysis methods.

The principles of barrier management include systematically integrating barrier management with risk management to enhance safety measures, maintaining the independence and diversity of safety barriers to prevent simultaneous failures, encouraging interdisciplinary collaboration to leverage expertise from various fields, and streamlining communication to ensure effective risk identification, assessment, and management. These principles collectively aim to achieve robust risk mitigation and continuous safety improvement.

However, safety barriers can be unreliable and might have intermittences of failures. "Swiss Cheese Metaphor" by Reason (1990) illustrates system failures by comparing safety barriers to slices of Swiss cheese. The holes in the cheese symbolize the flaws in each barrier, through which hazards can pass through, potentially causing accidents. Consequently, in the rapidly advancing domain of hydrogen technologies,

Table 10.1 Summary of different barrier analysis methods

Barrier analysis method	Description	Application stage	Type
Hazard-barrier matrices	Identify all hazards and the barriers	Initial risk assessment	Qualitative
Safety barrier diagram	Illustrate event sequences and barriers. It can usually be converted into a logic tree	Design and analysis	
Bow-tie diagram	Depicts the relationship between hazardous events, causes, consequences, and barriers	Risk management	
Energy Flow/ Barrier Analysis (EFBA)	Identify paths from energy sources to vulnerable assets and to assess whether these paths have adequate barriers	Design and operation	
Barrier and Operational Risk Analysis (BORA)	Analyze proactive and reactive barriers in operational phase	Operation	Semi-quantitative
Layer of Protection Analysis (LOPA)	Decide if existing safety barriers are enough or additional barriers are required	Design and operation	
Probabilistic models	Computationally efficient approach with random variables and probability distributions. E.g.: fault tree analysis, event tree analysis, and Bayesian network	All stages	Quantitative
State-transition models	Model stochastic process to assess barrier performance	All stages	

it is essential to perpetually monitoring, evaluating, and updating safety barriers to address new risks. For effective management and implementation of barriers in hydrogen systems, it is imperative to adhere to established standards. The standards as presented in Table 10.2 provide comprehensive guidelines for the effective management through proper implementation, monitoring, and maintenance of safety barriers, ensuring their reliability and robustness.

Table 10.2 Comparative overview of different standards for hydrogen safety barriers

Standard	Name	Application and scope
ISO 22734	Hydrogen generators using water electrolysis—industrial, commercial, and residential applications	Hydrogen generators: requirements of operating conditions, risk management, mechanical equipment, electrical equipment, control systems, ion transport medium, and protection of service personnel, and warning signs, test methods and operation/handling document
NFPA 2	Hydrogen technologies code	Hydrogen's stationary, portable, and vehicular infrastructure application: general hydrogen requirements, explosion protection, GH2 and LH2 fueling facilities, fuel cell power systems, hydrogen generation systems, combustion applications, laboratory operations, parking and repair garage, and industrial hydrogen trucks
CAN/BNQ 1784	Canadian hydrogen installation code	Gaseous and liquid hydrogen applications: general gaseous and liquid hydrogen installation requirements, hydrogen equipment, electrical equipment, residential installations, and portable equipment
HSE IPG 2009	Health and safety executive, installation permitting guidance for hydrogen and fuel cell stationary applications	Hydrogen and fuel cell system: hazards associated with fuel cell installation and operation, general and higher level requirements for design, manufacturing, installation, regulatory approval considerations, and operational/maintenance considerations, system specific and site considerations for hydrogen generation, containment, piping, explosion prevention and protection, hydrogen sensing, fire precautions, and interconnectivity of generators to local power distribution networks
Global	Property loss prevention data sheets: hydrogen	Liquefied and gaseous hydrogen storage, and systems for dispensing, end-use equipment, and laboratories: loss prevention recommendations, release prevention, limiting, and mitigation safeguards along with damage limiting and emergency exhaust ventilation

10.3 Continuous Evaluation and Support of Safety Barriers

Ensuring the safety of industrial systems involving hazardous materials is a critical challenge across sectors, particularly in applications requiring the handling and storage of flammable and high-pressure substances. Effective safety management relies on a robust framework of barriers to prevent, control, and mitigate risks. However, these barriers—whether technical, operational, or organizational—are not infallible and can be undermined by factors such as human error, environmental conditions, and system complexity. This underscores the need for continuous evaluation and support of safety barriers to ensure their reliability and effectiveness in managing risks.

Hydrogen storage systems, particularly Type IV composite overwrapped pressure vessels (COPVs) in hydrogen refueling stations, exemplify the challenges of barrier reliability and performance. These tanks operate at pressures exceeding 300 bar, making them susceptible to failure mechanisms like hydrogen-induced cracking. The inspection of these vessels is a crucial element of safety management, yet it is fraught with complexity. Acoustic Emission Testing (AET), a nondestructive inspection method used to detect structural defects, must contend with high levels of background noise, intricate data interpretation, and the potential for human error in execution. Ensuring that safety barriers—such as procedural guidelines, operator training, and technical safeguards—are effectively supported is essential for mitigating these risks.

In this study, the inspection of a hydrogen storage tank is used as a representative case to illustrate the challenges and solutions in safety barrier management. The methodology integrates barrier management principles with Human Reliability Analysis (HRA) and Bayesian Network (BN) modeling. Barrier management focuses on the systematic implementation and monitoring of safety measures, ensuring that both proactive and reactive barriers are robust and adaptable. HRA identifies potential human errors during AET procedures, quantifying their likelihood and exploring how factors like operator expertise, environmental conditions, and task complexity affect performance. BN modeling provides a probabilistic framework for understanding the dependencies between these factors and their influence on inspection outcomes.

By focusing on hydrogen storage systems, the study highlights the broader implications of effective safety barrier management in complex industrial applications. The findings emphasize the need for ongoing support and evaluation of safety barriers to address evolving risks, minimize human errors, and enhance the reliability of inspection methods. This approach not only strengthens safety practices in hydrogen technologies but also serves as a model for managing safety in other high-risk industries.

10.4 Case Study

This study addresses the inspection of a hydrogen storage tank at a refueling station. Section 10.4.1 provides an overview of the facility's design, with a detailed description of the Type IV composite overwrapped pressure vessel and the pipes and instrumentations connected. Furthermore, Sect. 10.4.2 summarizes the fundamentals of acoustic emission testing (AET) and outlines the step-by-step procedure to perform this type of inspection on the hydrogen pressure vessel.

10.4.1 Hydrogen Storage System in a Refueling Station

Hydrogen refueling stations (HRSs) are categorized as off-site or on-site, depending on where the hydrogen is produced. In this case, an off-site setup is considered, where hydrogen is generated at a separate location and transported to the station via truck. Although all off-site HRSs include a multi-stage compressor, cooling system, dual-pressure dispensers, relief devices, leak detectors, and fire extinguishing systems, this study focuses on the storage system. Pressure tanks operate from 900 to 300 bar (Pratt et al. 2015). Type IV hydrogen storage tanks, typically found in hydrogen refueling stations, feature a polymeric liner wrapped in a structural fiber composite. The liner acts as a protective barrier, preventing hydrogen leaks and protecting the composite fibers from microcracking and debonding. The composite overwrapped pressure vessel (COPV) has a diameter of 600 mm, a length of 20 m, a minimum wall thickness of 15 mm, and is made from carbon fiber-reinforced polymer. The pipe connected to the storage system is equipped with various components for pressure monitoring and control, including a pressure gauge (PG1) and pressure transducer (PT1), along with a solenoid valve (VS1) to regulate interactions between the compressor, storage, and dispenser. A pressure relief valve (PRV1) ensures the tank does not exceed its maximum allowable pressure, venting when the internal pressure exceeds 110% of the design value. Compressed inert gas purges the hydrogen tank and associated pipeline during the inspection activity. The solenoid valve (VS2) is opened to connect the tank with the bottles of inert gas. Additionally, isolation valves are installed on the dispenser (VL1), compressor (VL2), storage (VL3 and VL4), and inert gas pipes (VL5 and VL6) to guarantee flow control and ensure safe operations (Castiglia and Giardina 2013). Figure 10.2 illustrates the schematic of the hydrogen storage system, indicating the ID of each component.

10.4.2 Acoustic Emission Testing

Type IV tanks require routine inspections to maintain their structural integrity and prevent releases of high-pressure hydrogen in the environment. As per API 510

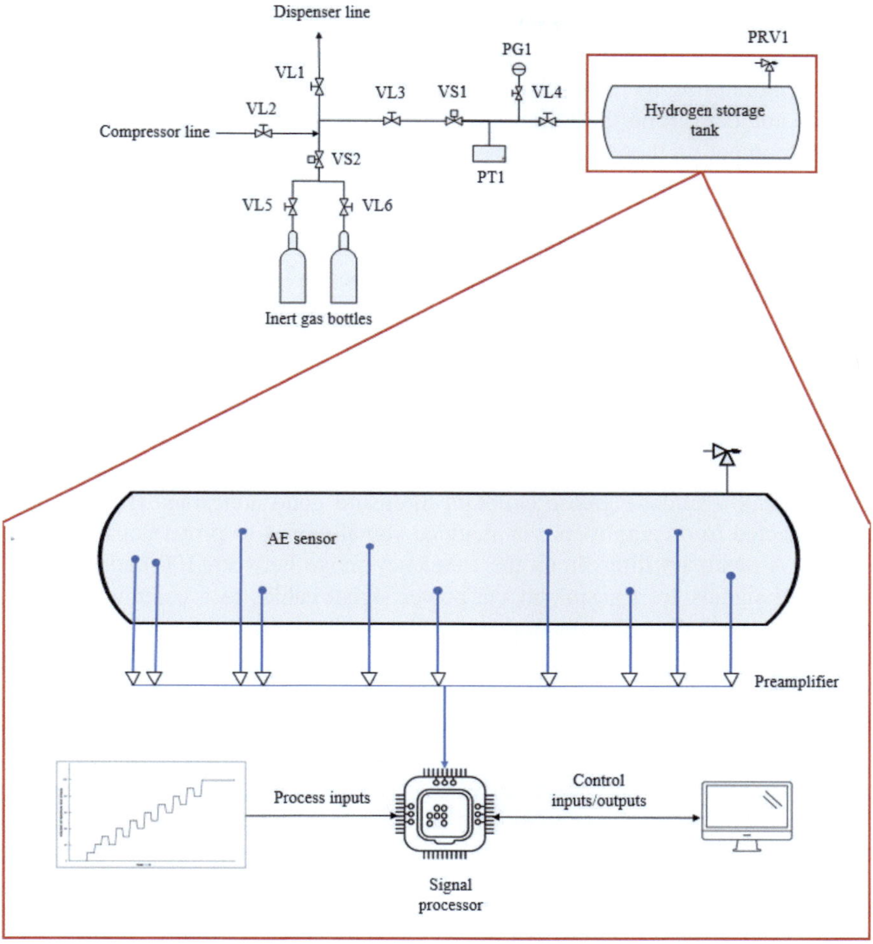

Fig. 10.2 Schematic of the storage system and the AET equipment connected to the hydrogen tank

(American Petroleum Institute 2022), internal and external examinations of pressure vessels for flammable substances should be conducted once every five years. However, inspection intervals can be optimized through the risk-based inspection (RBI) approach in accordance with the guidelines in API RP 580 (American Petroleum Institute 2016) and API RP 581 (American Petroleum Institute 2019). Acoustic emission testing (AET) is a nondestructive testing method that can detect damage in composite overwrapped pressure vessels. Along with the typical resin cracking, delamination, bond failure, fiber debonding, and pullout, hydrogen-induced cracking can also compromise the integrity of such equipment. AET is based on detecting the acoustic emissions generated when the material undergoes stress, particularly near its yield strength, where it transitions from elastic to plastic behavior. Cracks and macroscopic flaws amplify the acoustic waves, with larger amplitudes corresponding to faster crack growth. Conversely, minor defects or flaws subjected to insufficient loads do not emit significant acoustic signals. This inspection method involves monitoring the acoustic waves produced during load cycles at increasing pressure and converting these waves into electrical signals for recording and analysis. By comparing these waves' arrival times and attenuations, trained inspectors can identify structural defects using planar and zonal techniques (Dahmene et al. 2015). One of the key advantages of AET is its ability to inspect the entire tank by placing sensors in areas most susceptible to damage, while causing limited disruption to normal operations. However, while AET is effective for locating potential flaws, additional inspection methods, such as ultrasonic testing, may be needed to size and characterize any defects identified (American Society of Mechanical Engineers 2019).

The ASTM E1067 standard (American Society of Mechanical Engineers 2018) outlines the minimum requirements for acoustic emission testing of fiber-reinforced plastic tanks. High-frequency piezoelectric sensors are positioned on the tank's surface using a silicone grease couplant to ensure good adhesion. These sensors are connected to preamplifiers via shielded signal cables to protect against interference. A bandpass filter limits the frequency range between 100 and 200 kHz. Amplified signals are transmitted via power signal cables to a central processor, which analyzes various parameters such as hits, peak amplitude, signal strength, and duration. Additionally, the central processor receives data from process monitoring equipment (such as the pressure sensors). Operators can use the display to monitor real-time process conditions, control inputs and outputs, activate automatic valves, and review stored data for further analysis.

The AET procedure involves several distinct steps. Initially, the tank must be disconnected from the compressor and the dispensers. It is emptied by opening the pressure relief device to vent the stored hydrogen gas. This step is crucial, as performing the test without emptying the tank could create ignitable mixtures in the case of unintended release. Once the tank is empty, AET sensors are mounted on the external surface of the tank. The sensors' peak amplitude, attenuation, and background noise are checked to ensure proper functionality, as outlined in ASTM E1067. Upon confirming that the sensors are correctly installed and functioning, the tank is connected to an inert gas supply and pressurized at 30% of the design pressure. During this phase, background noise and acoustic emissions are recorded while maintaining this pressure for a specified duration. Subsequently, the pressure relief valve is opened to reduce the pressure by 10%, and acoustic emissions are recorded also at this lower pressure. The pressure is increased by 20%, and the system is held at this pressure while continuing to monitor background noise and AEs. This cycle is repeated until reaching 110% of the tank's design pressure, with AEs being recorded at each stage. Once the test is completed, the tank is vented, disconnected from the inert gas supply, and reconnected to the compressor and dispensers. The collected data will be analyzed to detect structural issues within the pressure vessel. Table 10.3 summarizes the sub-tasks comprising the AET procedure for H_2 storage tanks. The component IDs refer to Fig. 10.2.

Table 10.3 List of tasks for conducting AETs on a hydrogen storage tank

Task	ID	Sub-task
Tank isolation and draining	1.1	Isolate the tank from the dispenser by closing the isolation valve VL1
	1.2	Isolate the tank from the compressor by closing the isolation valve VL2
	1.3	Empty the tank by opening the pressure relief valve PRV1
	1.4	Empty the storage line by opening the solenoid valve VS1
	1.5	Close the PRV1 when the tank has reached the atmospheric pressure
Sensors' performance check	2.1	Measure and record the attenuation of the acoustic sensors
	2.2	Mount the acoustic sensors on the tank's exterior
	2.3	Check and record the peak amplitude of each acoustic sensor
	2.4	Conduct an automatic test on all the sensors
	2.5	Measure and record the background noise
Tank pressurization	3.1	Connect the inert gas line by opening the isolation valves VL5 and VL6
	3.2	Connect the tank with the inert gas line by opening the solenoid valve VS2
	3.3	Pressurize the tank at a constant pressurization rate, checking the pressure transducer
	3.4	Monitor the background noise and interrupt the test if it reaches a critical value
	3.5	Isolate the tank from the inert gas line by closing solenoid valve VS2 when the pressure is increased by 20%
	3.6	Maintain a constant pressure for four minutes
	3.7	Record the background noise for future analysis
Tank depressurization	4.1	Reduce the pressure of the tank by opening the pressure relief valve PRV1
	4.2	Isolate the tank by closing the pressure relief valve PRV1 when the pressure is reduced by 10%
	4.3	Maintain a constant pressure for four minutes
Test completion	5.1	Carry out an automatic performance test on each sensor
	5.2	Disconnect the tank from the insert gas line by closing the isolation valves VL5 and VL6

(continued)

Table 10.3 (continued)

Task	ID	Sub-task
	5.3	Empty the tank by opening the pressure relief valve PRV1
	5.4	Close the PRV1 when the tank has reached the atmospheric pressure
	5.5	Connect the tank with the compressor by opening the isolation valve VL2
	5.6	Connect the tank with the dispenser by opening the isolation valve VL1

Several technical challenges can affect the accuracy of AETs, creating uncertainty about the actual condition of the hydrogen storage tank. High-pressure environments often generate significant background noise, disturbing the acoustic emissions associated with geometrical defects and structural damages and making it difficult to distinguish them from environmental noise. This can result in failure to identify premonitory signs of degradation. Mechanical vibrations and electronic interference can also cover the AEs, increasing the likelihood of incorrect diagnoses. Lastly, composite materials have multiple sources of acoustic emissions, making it difficult to process and interpret even accurate AE data.

10.4.3 Methodology

The methodology for evaluating human errors during the operational inspection of hydrogen storage tanks involves integrating BN into HRA. The approach is twofold. Firstly, the Petro-HRA methodology was employed to identify the operational task to inspect the industrial equipment and to quantify the likelihood of inappropriate actions during the inspection of high-pressure storage systems via acoustic emission testing. This step provides a systematic analysis of potential human errors that could occur during the inspection process, ensuring that each phase of the inspection is meticulously assessed for risk factors.

On the other hand, the BN approach was utilized to investigate the conditional dependencies among human errors and various Performance-Shaping Factors (PSFs). By modeling these relationships, the BN approach allows for a deeper understanding of how different PSFs—such as environmental conditions, worker experience, and procedural complexities—affect the probability of human error. This probabilistic analysis enables the identification of key factors that significantly influence human performance during the inspection process.

10.4.3.1 Bayesian Network Analysis

A Bayesian Network (BN) is a probabilistic graphical model characterized by nodes connected by edges to express the conditional dependencies between variables in a specific domain (Amin et al. 2021; Sivia and Skilling 2006). Each node corresponds to a random variable, and each edge represents the conditional probability associated with the connected random variables. A BN is a directed acyclic graph, meaning no loops or self-connections are allowed in the network (Sattari et al. 2021; Zhou et al. 2022). Each edge has an orientation and connects a parent node to a child node. The analysis of a BN can be divided into six steps:

1. Definition of the analysis: This step involves defining the system variables, the boundaries of the analysis, and any technical or process data useful for identifying the nodes.
2. Modeling the network: This involves gathering the necessary information to model the network. It includes defining the timeline, the number of steps to perform the analysis, and the variables to be translated into nodes.
3. Define the initial probabilities: After establishing the timeline and nodes, the conditional probability functions for each node must be defined. For example, the $\mathrm{Pr_c}(X\,|Y)$ is the marginal probability of Y given X which can be expressed as

$$\mathrm{Pr_c}(X\,|Y) = \frac{\mathrm{Pr}(Y|X) + \mathrm{Pr}(X)}{\mathrm{Pr}(Y)} \tag{10.1}$$

   where $\mathrm{Pr}(X)$ and $\mathrm{Pr}(Y)$ are the prior probabilities of X and Y, respectively.
4. Performing the inference: This step estimates the probability distribution function of unknown states given some known observations and initial probability distribution. It requires defining the types of inference to be performed.
5. Analyzing the network: Identify which variables have the most significant impact on the output by ensuring that the model accurately represents the defined scenario. This helps in understanding the critical factors in your network and exploring the predictions.

10.4.3.2 Human Reliability Analysis

Human Reliability Analysis (HRA) is a systematic method to evaluate the risk of human errors by identifying possible incorrect actions, estimating their probabilities, and determining the factors and systems that influence the likelihood of such errors (Joice et al. 1998; Swain 1990). This study utilizes the Petro-HRA approach, which was originally designed for the oil and gas sector but is flexible enough to be applied to other industries, including the hydrogen sector. Given that hydrogen and natural gas share similar hazards—particularly their high flammability and explosivity—equipment designed for natural gas transport and storage can often be adapted for hydrogen applications (Laumann et al. 2014). Furthermore, both gases require similar safety measures, routine inspections, and compliance with industry standards.

The Petro-HRA methodology involves the following steps:

1. Scenario definition—Establishes the scope of the analysis; in this case, the focus is on the inspection of a pressurized hydrogen storage tank using acoustic emission testing. This includes preparing for the test, executing it, collecting data, and restoring the facility's normal operations.
2. Qualitative data collection—Gathers relevant information from inspection protocols, facility documentation, and discussions with operators and experts. This step clarifies key aspects such as the location, environmental conditions, procedural activities, duration, involved equipment, implemented safety barriers, and the roles and responsibilities of the inspection team.
3. Task analysis—Breaks down the inspection process into detailed tasks and sub-tasks, addressing both physical actions and intellectual reasoning. The information is structured systematically using a hierarchical approach.
4. Human error identification—Pinpoints possible errors in task execution, examines their consequences, evaluates potential recovery actions, and identifies critical performance-shaping factors (PSFs) that influence the success or failure of the AET.
5. Human error modeling—Maps the relationships between errors, actions, performance shaping factors, and the socio-technical systems involved. This step uses an action event tree to define success and failure conditions for each part of the inspection procedure.
6. Human error quantification—Calculates the probability of human error for each task by combining a base probability adjusted with nine performance shaping factors, as specified in Eq. (10.2):

$$\begin{aligned} \mathrm{HEP} = \mathrm{NHEP} \cdot \mathrm{PSF_t} \cdot \mathrm{PSF_{ts}} \cdot \mathrm{PSF_{tc}} \cdot \mathrm{PSF_{ot}} \\ \cdot \mathrm{PSF_p} \cdot \mathrm{PSF_{hmi}} \cdot \mathrm{PSF_{as}} \cdot \mathrm{PSF_{tw}} \cdot \mathrm{PSF_e} \end{aligned} \tag{10.2}$$

NHEP indicates the probability of nominal human error. It is assumed that a generic task has a likelihood of being incorrectly performed of 1%. This base value is corrected with nine performance shaping factors. $\mathrm{PSF_t}$ accounts for the effect of time, evaluating if there is enough time to complete an intellectual or physical action. $\mathrm{PSF_{ts}}$ considers the effect of the threat stress and discomfort conditions that could impact the successful task completion. $\mathrm{PSF_{tc}}$ indicates the complexity of the task and depends on the number of actions required, their order, correlation, and inherent complexity, along with the possibility of unexpected conditions. $\mathrm{PSF_{ot}}$ accounts for the level of expertise of the operators carrying out the task; if they have already experienced similar situations in the past or have received systematic training, the likelihood of correct and timely action will dramatically increase. $\mathrm{PSF_p}$ accounts for the quality and level of details of written procedures that should guide the operators when performing their tasks. $\mathrm{PSF_{hmi}}$ indicates the Human–Machine Interface (HMI), i.e., the quality of the hardware and software used to carry out the actions. $\mathrm{PSF_{as}}$ indicates the attitude to safety and management support, i.e., if the management prioritizes safety and supports precautionary actions of the operators. $\mathrm{PSF_{tw}}$ accounts for teamwork, good communication, and trust between the members. PSF_e indicates the physical working environment, i.e., if the working site is accessible, the devices are ergonomic, and the environmental conditions are favorable (Joice et al. 1998; Swain 1990).

10.4.4 Evaluation Results

The human error probabilities are calculated by selecting and combining the PSFs for each task. While the time, threat stress, task complexity, human–machine interface, procedure, and teamwork depend on the task considered, others apply equally to all the actions. All the tasks are performed by the same inspection team, which has a particular experience and training. The attitude to safety depends on the company's safety culture and managerial choices. Finally, the physical working environment depends on the location of the HRS, the meteorological conditions at the time of inspection, the presence of background noise or other sources of disturbance, etc. Therefore, four sets of PSFs are considered. The first set assumes optimal conditions during the inspection: the management supports the team and shows a high degree of mindfulness about safety, the working environment does not affect the success of the inspection, and the operators have extensive experience and knowledge to perform the AET correctly. The second set considers the case in which the management does not prioritize and support the inspection activities, i.e., shows a limited attitude to safety. The third assumes that the AET is conducted under favorable environmental conditions, i.e., with excessive background noise, low temperatures compromising the sensor's sensitivity, or a tank located in a hard-to-reach place. The fourth considers the effect of limited experience and training of the inspectors, i.e., if the AET is conducted by specialized personnel. Table 10.4 reports the tasks to conduct an AET and the human error probabilities for each scenario.

Therefore, it is possible to compute the prior probabilities of each operational task by using the human error probabilities defined in Table 10.4. These prior probabilities for each task and sub-task of the AET test have been defined as Boolean values, either "property" or "not property." The value "property" refers to the task or sub-task being developed correctly and completely. On the other hand, the value "not property" denotes the task or sub-task being developed inappropriately or incompletely. This Boolean representation allows for a clear and concise assessment of each task's performance status, facilitating the analysis of operational efficiency and identifying potential areas for improvement. By utilizing the HEPs, it is possible to systematically evaluate the likelihood of successful completion for each stage of the AET process. Figure 10.3 illustrates the BN analysis of the tasks and sub-tasks involved in performing the AET test. This visual representation helps in understanding the dependencies and interactions between various elements of the process. The BN framework provides a structured approach to model the probabilistic relationships among the tasks, allowing for a comprehensive risk assessment. Therefore,

Table 10.4 Human error probabilities obtained from Petro-HRA for four sets of PSHs

N.	Task	Optimal conditions	Limited attitude to safety	Adverse working environment	Limited operators' training
1.1	Close the isolation valve VL1 in the dispenser line	5×10^{-5}	10^{-3}	5×10^{-4}	2.5×10^{-3}
1.2	Close the isolation valve VL2 on the compressor line	5×10^{-5}	10^{-3}	5×10^{-4}	2.5×10^{-3}
1.3	Open the pressure relief valve PRV1	5×10^{-5}	10^{-3}	5×10^{-4}	2.5×10^{-3}
1.4	Open the solenoid valve VS1 on the storage line	5×10^{-5}	10^{-3}	5×10^{-4}	2.5×10^{-3}
1.5	Close the pressure relief valve PRV1 when the tank is completely empty	5×10^{-4}	10^{-2}	5×10^{-3}	2.5×10^{-2}
2.1	Characterize the attenuation of the acoustic emission sensors	2.5×10^{-4}	5×10^{-3}	2.5×10^{-3}	1.25×10^{-2}
2.2	Mount the acoustic emission sensors on the tank surface	2.5×10^{-3}	5×10^{-2}	2.5×10^{-2}	1.25×10^{-1}
2.3	Verify the peak amplitude of each acoustic emission sensor	2.5×10^{-4}	5×10^{-3}	2.5×10^{-3}	1.25×10^{-2}
2.4	Perform a global automatic sensor test	2.5×10^{-4}	5×10^{-3}	2.5×10^{-3}	1.25×10^{-2}
2.5	Verify the background noise	5×10^{-3}	0.1	5×10^{-2}	2.5×10^{-1}
3.1	Open the isolation valves VL5 and VL6 on the inert gas line	2.5×10^{-4}	5×10^{-3}	2.5×10^{-3}	1.25×10^{-2}
3.2	Open the solenoid valve VS2 on the inert gas line	2.5×10^{-4}	5×10^{-3}	2.5×10^{-3}	1.25×10^{-2}
3.3	Check the pressure transducer to keep a pressurization rate of 3.45 MPa/h	2.5×10^{-3}	5×10^{-2}	2.5×10^{-2}	1.25×10^{-1}
3.4	Monitor the background noise and stop the pressurization if it is excessive	10^{-2}	2×10^{-1}	10^{-1}	5×10^{-1}
3.5	Close solenoid valve VS2 when the pressure is increased by 20%	2.5×10^{-3}	5×10^{-2}	2.5×10^{-2}	1.25×10^{-1}
3.6	Hold this pressure level for 4 min	2.5×10^{-4}	5×10^{-3}	2.5×10^{-3}	1.25×10^{-2}
3.7	Record the level of background noise	2.5×10^{-4}	5×10^{-3}	2.5×10^{-3}	1.25×10^{-2}
4.1	Open pressure relief valve PRV1	2.5×10^{-4}	5×10^{-3}	2.5×10^{-3}	1.25×10^{-2}

(continued)

Table 10.4 (continued)

N.	Task	Optimal conditions	Limited attitude to safety	Adverse working environment	Limited operators' training
4.2	Close pressure relief valve PRV1 when the pressure is reduced by 10%	2.5×10^{-3}	5×10^{-2}	2.5×10^{-2}	1.25×10^{-1}
4.3	Hold this pressure level for 4 min	2.5×10^{-4}	5×10^{-3}	2.5×10^{-3}	1.25×10^{-2}
5.1	Perform a global automatic sensor test	2.5×10^{-4}	5×10^{-3}	2.5×10^{-3}	1.25×10^{-2}
5.2	Close isolation valves VL5 and VL6 on the inert gas line	5×10^{-5}	10^{-3}	5×10^{-4}	2.5×10^{-3}
5.3	Open pressure relief valve PRV1	5×10^{-5}	10^{-3}	5×10^{-4}	2.5×10^{-3}
5.4	Close pressure relief valve PRV1 when the tank is completely empty	5×10^{-4}	10^{-2}	5×10^{-3}	2.5×10^{-2}
5.5	Open isolation valve VL2 on the compressor line	5×10^{-5}	10^{-3}	5×10^{-4}	2.5×10^{-3}
5.6	Open isolation valve VL1 on the dispenser line	5×10^{-5}	10^{-3}	5×10^{-4}	2.5×10^{-3}

the BN analysis depicted in Fig. 10.3 also highlights critical factors in the AET where interventions may be necessary to enhance the accuracy and reliability of the AET.

Figure 10.3 shows the 96.60% probability of successfully inspecting a hydrogen tank using AET, assuming optimal working conditions, a strong safety commitment, and proper operator training. This success relies on five key steps: Tank isolation and draining; Sensor's performance evaluation; Tank pressurization; Tank depressurization; and Test completion. By integrating HEPs with BN analysis, this approach enhances the safety and efficiency of hydrogen tank inspections, providing actionable insights for reducing errors and improving overall system reliability.

The relatively high probability of correctly and successfully completing the test is attributed to the HEPs associated with the sub-tasks. Table 10.5 highlights that the likelihood of successful tank depressurization and test completion are the lowest, mainly due to the simplicity of these actions. These straightforward steps have well-defined procedures, resulting in minimal error rates. In contrast, these stages are more critical to performing a correct AET. The tasks involved in these phases require a higher level of skill and understanding, and the human–machine interface used during these phases can be challenging to interpret. This complexity increases the likelihood of errors, making these steps the most vulnerable in the AET process.

On the other hand, Fig. 10.4 presents the result of tests on sensitivity and strength of influence for each task and sub-task of the AET test. The thickness of the arrows shows the strength of influence. Then, the task associated with the sensor performance

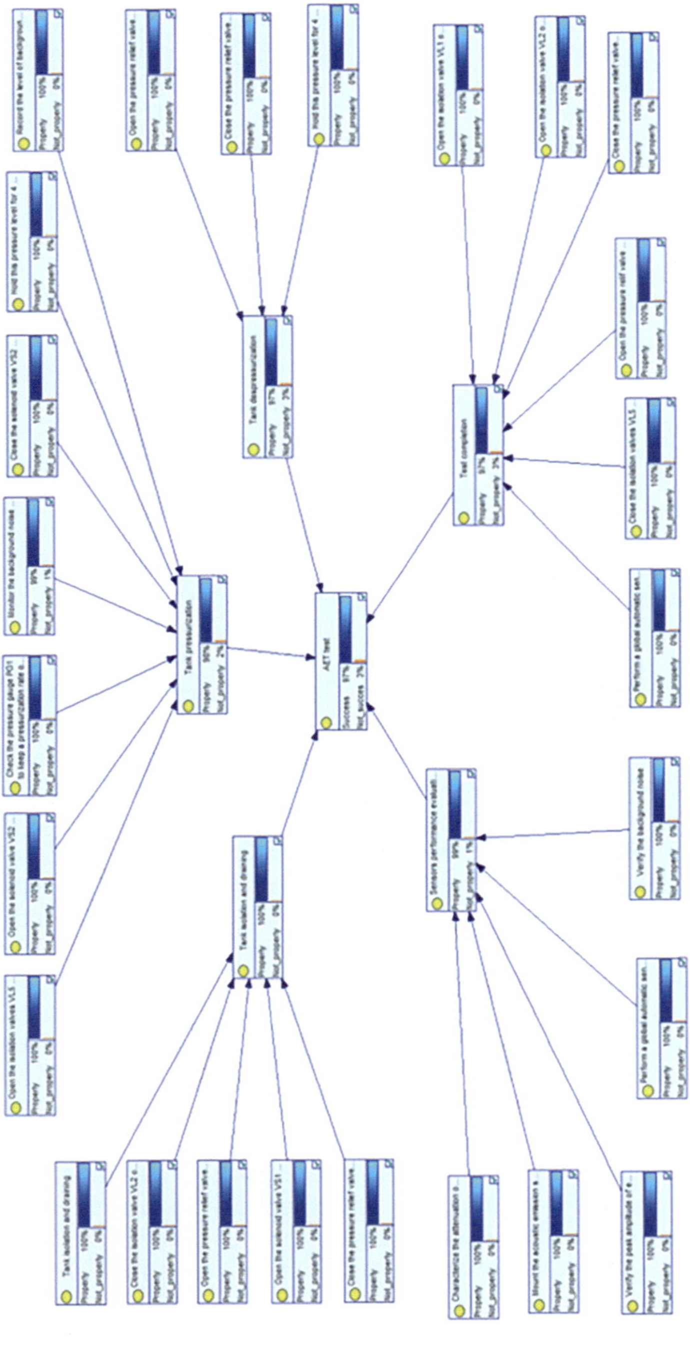

Fig. 10.3 BN model analysis by showing the behavior of each task by assigning the prior probabilities based on the human error

Table 10.5 Probability of success of each task in the AET test

Main task	Success probability (%)
Tank isolation and draining	99.93
Sensor's performance evaluation	99.13
Tank pressurization	97.62
Tank depressurization	97.31
Test completion	97.24

evaluation, tank completion, tank depressurization, tank pressurization, tank isolation, and draining has the highest influence to impact the overall system performance. Indeed, the analysis depicted that the main activities of the AET are highly interconnected and influence the entire system. Additionally, there are no other notable strong influences within the network (Table 10.6). The unique strong influence that could be interesting to present is between pairs of tasks like opening the pressure relief valve PRV1 related with the main activity tank depressurization highlighting key dependencies and interactions within the system. On the other hand, the colored elements in Fig. 10.4 represent the degree of sensitivity. The result shows that the AET test in the network has been highly sensitive to the changes in Test completion and Sensor's performance evaluation. Therefore, this analysis reveals that any changes in these main tasks significantly influence the outcomes of the AET test within the BN. This means that any variations in the parameters or probabilities associated with the Test completion and Sensor's performance evaluation, or their sub-tasks significantly affect the beliefs or probabilities associated with the outcome of the AET test. Then if the probabilities of success for Test completion and Sensor's performance evaluation decrease, it may lead to a decrease in the probability of passing the AET test and vice versa.

Therefore, the BN analysis reveals valuable insights into the complex system's dependencies and sensitivities. The thickness of the arrows indicates the strength of influence between nodes, with the main task of the AET emerging as critical factors impacting overall system performance. Moreover, the sensitivity analysis highlights the critical role of main tasks (including test completion and sensor performance evaluation) and their associated sub-tasks, because any change in these factors can significantly influence the outcomes of the AET test. This underscores the need of precise management and careful assessment of these parameters to ensure reliable test results.

This integrated approach, combining HEPs and BN analysis, offers a robust methodology for assessing and enhancing the operational safety and efficiency of hydrogen tank inspections. It emphasizes the importance of detailed, data-driven analysis in developing effective risk management practices in complex technical environments.

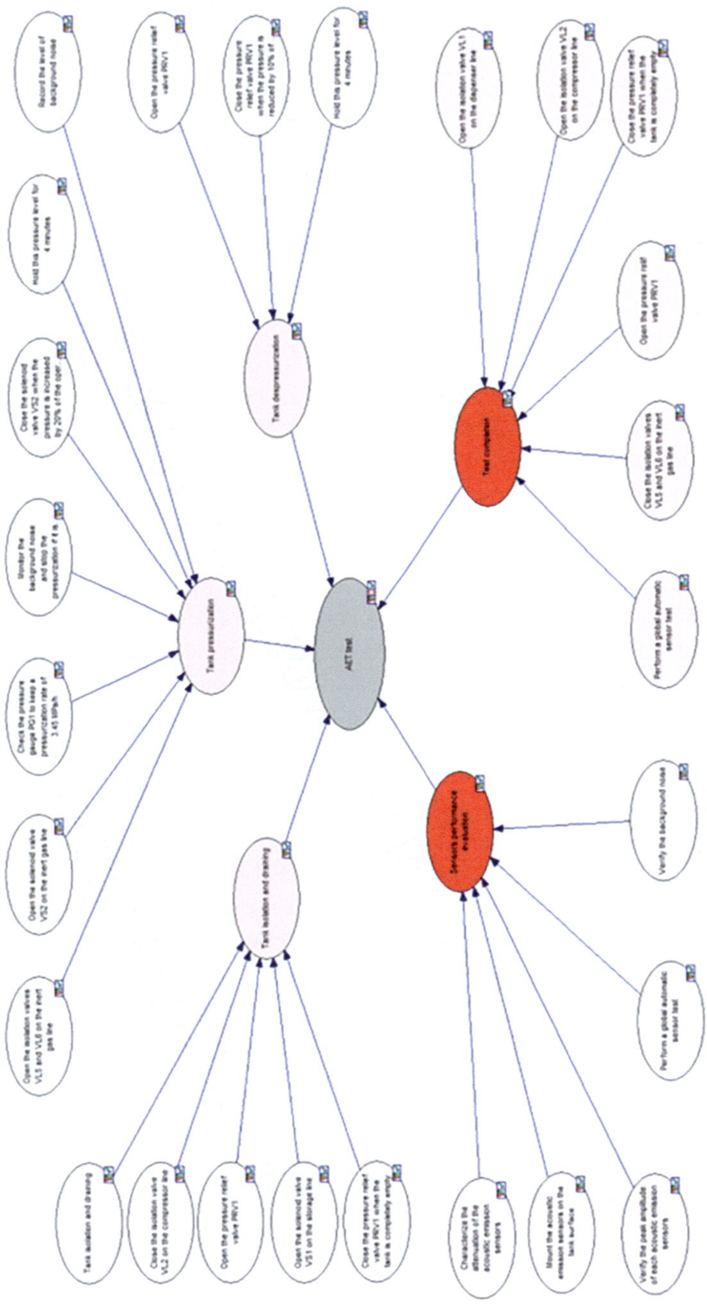

Fig. 10.4 Sensitivity analysis of the BN

Table 10.6 Influence values of the fifth highest node of the BN

Parent node	Child node	Average
Test completion	AET test	0.125
Tank pressurization	AET test	0.125
Tank isolation and draining	AET test	0.125
Tank depressurization	AET test	0.125
Sensor performance evaluation	AET test	0.125

10.5 Discussions

The results of the case study provide critical insights into the effectiveness of safety barrier management in hydrogen storage systems, with specific emphasis on the challenges posed by high-pressure environments and complex inspection processes. Building on the foundational theories by Sklet (2006), Reason (1997), and Hollnagel (Hollnagel 2004; Stefana et al. 2024), the findings extend the understanding of how safety barriers operate within the dynamic and high-risk domain of hydrogen technologies.

One key aspect worth further exploration is the interplay between human factors and technological barriers. The results indicate that tasks such as sensor performance evaluation and test completion exhibit higher sensitivity to human error, underscoring the need for robust training programs and intuitive HMI. Hollnagel's emphasis on dependability and auditability (Hollnagel 2004) is particularly relevant here, as it suggests that regular audits and feedback loops could enhance the alignment between operator actions and system requirements.

Moreover, the study highlights the role of environmental conditions, which act as external performance-shaping factors influencing the reliability of both human operators and technical barriers. Adverse conditions, such as excessive noise or extreme temperatures, were shown to degrade the effectiveness of AET. This aligns with the principle of robustness as discussed by Sklet (2006), which advocates barriers that can withstand variable operational conditions. A recommendation emerging from this finding is the incorporation of environmental monitoring systems and adaptive barrier designs that can compensate for external disturbances.

Another significant area for discussion is the potential for integrating digital technologies to enhance barrier management. Advanced digital tools, such as real-time data analytics, artificial intelligence, and predictive maintenance algorithms, could further strengthen the proactive barriers in hydrogen storage systems. For example, integrating AET data with machine learning models could improve defect detection accuracy and provide predictive insights into potential failure points. This aligns with the properties of specificity and independence emphasized by the ARAMIS project (ARAMIS 2004), enabling targeted and independent decision-making.

The organizational dimension also warrants attention. The scenarios analyzed in this study reveal the significant impact of safety culture on the overall effectiveness of barrier management. As Reason's framework illustrates (Reason 1997), the prioritization of safety at the managerial level cascades through the organization, influencing operator behavior and the integrity of safety processes. The results suggest that promoting a proactive safety culture, including transparent communication and clear accountability structures, could enhance the implementation and maintenance of safety barriers.

Lastly, the findings also raise broader implications for regulatory and standardization efforts in hydrogen technologies. While existing standards such as ISO 22734 and NFPA 2 (ISO 2019; NFPA 2023) provide comprehensive guidelines for hydrogen safety, the study demonstrates the need for more dynamic and adaptive regulatory frameworks. These frameworks should incorporate emerging insights from human reliability and probabilistic analyses to ensure that standards remain relevant to the evolving nature of hydrogen technologies.

This contribution reinforces the theoretical underpinnings of safety barrier management but also provides practical insights into addressing the unique challenges of hydrogen systems. By integrating human, technological, and organizational dimensions, the findings offer a holistic approach to enhancing the safety and reliability of high-risk industrial applications. Future research could explore the use of advanced digital tools and adaptive standards to further refine safety barrier management practices.

10.6 Conclusions

This chapter highlights the critical role of effective safety barrier management in hydrogen storage systems, addressing challenges posed by technical, human, and environmental factors. Using HRA and BN modeling, the study identifies tasks such as sensor performance evaluation and test completion as critical points of sensitivity, emphasizing the need for robust operator training, intuitive interfaces, and adaptive environmental controls.

The results demonstrate the importance of barrier independence and robustness, with proactive measures like training and predictive maintenance complementing reactive safeguards such as real-time monitoring. Organizational factors, particularly safety culture and managerial support, are shown to significantly influence barrier effectiveness.

The study advocates for leveraging digital technologies, including machine learning and data analytics, to enhance barrier specificity and adaptability, enabling better defect detection and predictive insights. By integrating technical, human, and organizational dimensions, this approach advances hydrogen safety management while offering a transferable framework for other high-risk systems. Future research should explore the potential of digital tools and refined regulatory approaches to further strengthen safety barriers.

References

American Petroleum Institute (2016) API RP 580—risk-based inspection

American Petroleum Institute (2019) API RP 581—risk-based inspection methodology

American Petroleum Institute (2022) API 510—pressure vessel inspection code: maintenance, inspection, rating, repair, and alteration

American Society of Mechanical Engineers (2018) E1067/E1067M-18—standard practice for acoustic emission examination of fiber glass reinforced plastic resin (FRP) tanks/vessels. https://doi.org/10.1520/E1067-18.2

American Society of Mechanical Engineers (2019) ASME BPVC-V—nondestructive examination. In: ASME boiler and pressure vessel code

Amin MT, Khan F, Ahmed S, Imtiaz S (2021) A data-driven Bayesian network learning method for process fault diagnosis. Process Saf Environ Prot 150:110–122. https://doi.org/10.1016/j.psep.2021.04.004

ARAMIS (2004) Accidental risk assessment methodology for industries—user guide. http://aramis.jrc.it

Castiglia F, Giardina M (2013) Analysis of operator human errors in hydrogen refuelling stations: comparison between human rate assessment techniques. Int J Hydrogen Energy 38(2):1166–1176. https://doi.org/10.1016/j.ijhydene.2012.10.092

Dahmene F, Yaacoubi F, El Mountassir M, Langlois C, Bardoux O (2015) Towards efficient acoustic emission testing of COPV, without Felicity ratio criterion, during hydrogen-filling. Int J Hydrogen Energy 41(2):1359–1368. https://doi.org/10.1016/j.ijhydene.2015.11.065

DNV (2014) Good practices: barrier management in operation for the rig industry

Global FM (2021) Property loss prevention data sheets

Hollnagel E (2004) Barriers and accident prevention. Routledge. https://doi.org/10.4324/9781315261737

HSE (2009) Installation permitting guidance for hydrogen and fuel cell stationary applications: UK version

ISO (2019) ISO 22734—hydrogen generators using water electrolysis—industrial, commercial, and residential applications

Joice P, Hanna GB, Cuschieri A (1998) Errors enacted during endoscopic surgery—a human reliability analysis. Appl Ergon 29(6):409–414. https://doi.org/10.1016/S0003-6870(98)00016-7

Laumann K, Øien K, Taylor C, Boring RL, Rasmussen M (2014) Analysis of human actions as barriers in major accidents in the petroleum industry, applicability of human reliability analysis methods (Petro-HRA). In: PSAM 2014—probabilistic safety assessment and management, Honolulu

NFPA (2023) NFPA-2 hydrogen technologies code

Pratt J, Terlip D, Ainscough C, Kurtz J, Elgowainy A (2015) H2FIRST reference station design task project

PSA (2017) Principles for barrier management in the petroleum industry

Reason J (1990) Human error

Reason J (1997) Managing the risks of organizational accidents. Ashgate Publishing Limited

Sattari F, Macciotta R, Kurian D, Lefsrud L (2021) Application of Bayesian network and artificial intelligence to reduce accident/incident rates in oil & gas companies. Saf Sci 133. https://doi.org/10.1016/j.ssci.2020.104981

Sivia D, Skilling J (2006) Data analysis. A Bayesian tutorial, 2nd edn. Oxford University Press

Sklet S (2006) Safety barriers: definition, classification, and performance. J Loss Prev Process Ind 19(5):494–506. https://doi.org/10.1016/j.jlp.2005.12.004

Stefana E, Ramos M, Paltrinieri N (2024) Machine learning-based literature review on the concept of safety barriers against hazardous events. J Loss Prev Process Ind 92:105470. https://doi.org/10.1016/J.JLP.2024.105470

Swain AD (1990) Human reliability analysis: need, status, trends and limitations. Reliab Eng Syst Saf 29(3):301–313. https://doi.org/10.1016/0951-8320(90)90013-D

Zhou Y, Li X, Yuen KF (2022) Holistic risk assessment of container shipping service based on Bayesian network modelling. Reliab Eng Syst Saf 220. https://doi.org/10.1016/j.ress.2021.108305

Chapter 11
Technical Safety Barriers and Integrity Assessment

Yiliu Liu and Svetlana Erofeeva

Abstract The objectives of this chapter are to present the technical safety barriers that are used in hydrogen fueling stations and introduce methods for evaluating the performance of these barrier systems. Design principles and operational modes of technical safety barriers are discussed, followed by the introduction of failure modes. Dangerous undetected failures are the main contributor of unavailability of technical safety barriers. The chapter provides a framework of reliability and integrity assessment of such barriers in hydrogen fueling stations. Since the applications of relatively new and existing reliability information are not sufficient, reliability prediction is necessary. This chapter then employs approximation formulas and the Markov methods for integrity assessment, with case studies on pressure relief valve systems using single barrier components and 1-out-of-2 voting structures. The impact of common cause failures is also considered in the case studies.

11.1 Introduction

11.1.1 Definitions Related to Technical Safety Barriers

The concept of safety barriers has obtained considerable attention in industrial safety management as an effective approach to reducing risks and avoiding accidents (Liu 2020; Yuan et al. 2022; Liu 2025). Safety barriers can be classified based on their modality or format, and technical safety barriers belong to entitative barriers (Liu 2020), referring to the equipment using various technologies to perform barrier functions. These technologies include instrumentation, electrical and electronic control technologies, and digital technologies.

Y. Liu (✉) · S. Erofeeva
Department of Mechanical and Industrial Engineering, Norwegian University of Science and Technology (NTNU), Trondheim, Norway
e-mail: yiliu.liu@ntnu.no

Y. Liu et al. (eds.), *Safety Measures for Hydrogen Fueling Stations*, Risk, Reliability and Safety Engineering, https://doi.org/10.1007/978-981-96-9412-9_11

Technical safety barriers are also called safety-critical systems in the context of high-risk industries such as nuclear power (NEA 2003). According to international standard for process safety IEC 61511 (2016), safety-critical systems leveraging instrumentation technologies are regarded as safety-instrumented systems (SISs). In IEC 61508 (2010), these systems are referred to as Electrical/Electronic/Programmable Electronic (E/E/PE) safety-related systems.

In this book, failure of a safety-critical system or technical safety barrier may result in a harm or at least a higher risk to the asset it is protecting, or equipment under control (EUC), as defined in IEC 61508 (2010). However, in the absence of a hazard, the failure of such a system does not influence the functioning of EUC. In other words, the barrier is not directly related to the essential function of the EUC. This distinction separates safety-critical systems from safety-related systems, which serve protective functions but are also integral to EUC operations.

The safety realized by a safety-critical system or technical safety barrier is called *functional safety* in IEC 61508 (2010).

11.1.2 Common Technical Safety Barriers in Hydrogen Fueling Stations

Technical safety barriers are critical and have been installed in HRSs. They can serve as preventive barriers to avoid the occurrences of hazardous events or as mitigative barriers to alleviate the consequences of these events when they occur.

Considering the scenario where leaked hydrogen gas accumulates to the ignition threshold as a hazardous event in a gaseous hydrogen refueling station, examples of technical safety barriers for preventing its occurrence include the following (e.g., see Erofeeva 2024):

- **Pressure relief valves and burst discs**: These components are installed to prevent over-pressurization in hydrogen tanks, pipelines, and other associated equipment. When pressure exceeds a predetermined limit, the relief valves open, releasing gas to prevent damage or rupture of the storage tank and other containers. Burst discs rupture when the system reaches critical pressure, offering an additional layer of protection.
- **Gas detectors and alarms**: A variety of sensors are deployed to detect hydrogen leaks before they can reach hazardous concentrations. These sensors should be placed near possible leak points, such as fuel dispensers, flanges, process equipment, and valves. In hydrogen applications, catalytic detectors using platinum or palladium as catalysts can be employed. Additionally, acoustic sensors are used to detect the ultrasonic waves (typically above 25 kHz) emitted by gas leaks. However, their reliability can be affected by background noise. Other possible detection methods include igniting small gas volumes and making them detectable via flame or heat sensors. Electrochemical sensors, semiconductor oxides, thermal

conductivity sensors, and mass spectrometers are also useful in HRSs. In practice, these detection systems work together with alarms to immediately alert personnel in the event of a leak.

- **Fault diagnosis and degradation monitoring**: Advanced fault diagnosis sensors and software can be employed to monitor the health and operational status of critical components in real time. For example, monitoring the wear and tear of valves, pumps, and dispensers can provide early warnings of equipment failure, enabling preventive maintenance and thus reducing the likelihood of gas leakage.
- **Ventilation system**: Adequate ventilation is very important in preventing the accumulation of hydrogen gas, especially in confined or semi-enclosed areas. Since hydrogen is much lighter than air, it tends to rise and disperse quickly in open spaces. However, in areas with limited ventilation, hydrogen can easily accumulate to the dangerous level. Ventilation systems are therefore installed to continuously circulate air and ensure that any leaked hydrogen is quickly dispersed, reducing the likelihood of the gas reaching a flammable concentration.

On the other hand, examples of technical safety barriers for mitigating consequences of such a hazardous event include the following (see Verfondern et al. 2021):

- **Fire detectors**: Fire detection systems are critical for identifying the presence of a fire, alerting personnel, and initiating emergent control measures. These systems can reduce the risk of fire spreading to nearby equipment or exposing individuals to danger. Fire detectors should be installed in areas where hydrogen may leak, spill, or accumulate, particularly around hydrogen dispensers, storage vessels, and primary safety functions. These systems need to provide prompt and reliable signals indicating the presence, location, and size of a hydrogen flame, which is often invisible to the naked eye.
- **Emergency shutdown (ESD) system**: An ESD is installed to prevent abnormal situations from escalating into a severe accident. The ESD continuously monitors key parameters such as gas concentrations within the HRS. When the system detects a deviation from safe operating limits, it triggers an automatic shutdown of critical processes. For example, if a hydrogen leak is detected in the piping system, the ESD will immediately close the supply valves to stop the flow of hydrogen gas.
- **Fire suppression and protection system**: Fire protection can be active and passive. Active fire protection is always technical safety barriers, including water sprays, foam systems, and gas-based suppression designed to quickly extinguish fires and cool down critical components. On the other hand, passive fire protection is not realized by technical systems, but realized by those structures,vessels, and other critical elements that have sufficient fire resistance to delay the spread of fire and allowing more time for response.

Other safety systems, such as emergency power systems, emergency lighting, and communication systems, are essential components in an HRS considering that an accident like fire or explosion has occurred. They also fall under the category of technical safety barriers.

In the remainder of this chapter, we will present the operational modes and failures of technical safety barriers, as well as how to model and assess the performance of these barriers, with simply illustrative case studies in hydrogen fueling stations.

11.2 Design and Operation of Technical Safety Barriers

11.2.1 Design of Technical Safety Barriers

In the literature, researchers have proposed generic requirements for the design of technical safety barriers (e.g., see Hollnagel 2004; Johansen and Rausand 2015), such as

- **Adequacy**: This requires that a safety barrier should be able to reduce the risk to the asset to an acceptable or at least to the expected level. If the barrier is found inadequate, more barriers need to be introduced. In the paper by Liu (2020), effectiveness is taken as a measure for evaluating how well the requirement of adequacy is satisfied.
- **Specificity**: A technical safety barrier should be designed specifically for a particular hazard. For instance, acoustic sensors can be used in the sensing subsystems of technical barriers in an HRS since they are designed with the capability to reveal very small amount of gas leakage, which pose significant risks in hydrogen applications.
- **Independence**: This means that each safety barrier should be designed and operated without dependence on other barriers or the asset it protects. With independent design, the failure or hazard on the asset does not affect the performance or function of the barrier.
- **Auditability**: This means that the design, installation, and maintenance of a technical barrier should be tracked and verified according to the established safety standards and regulations.

11.2.2 Structures of Technical Safety Barriers

As the exemplar pressure safety valve system illustrates Fig. 11.1, a technical safety barrier is always consisting of three subsystems: sensing subsystem, logic solver or controlling subsystem, and actuating subsystem:

- **Sensing subsystem**: It involves one or more sensors that monitor the condition of the asset, with the aim to detect any abnormal situations. In the example shown in Fig. 11.1, this subsystem includes three pressure transmitters (PT1, PT2, and

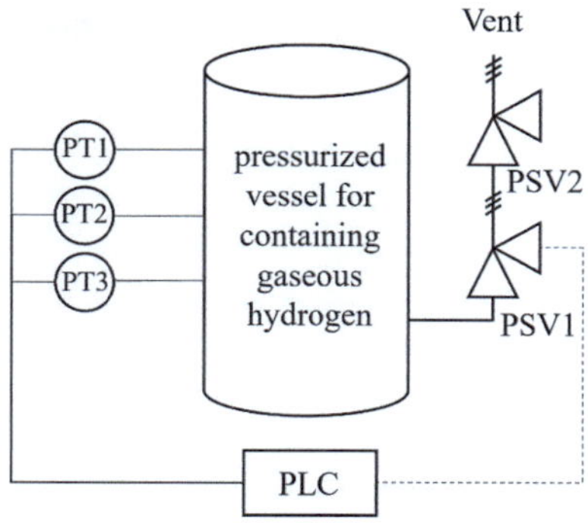

Fig. 11.1 Illustration of a pressure safety valve system

PT3). The sending subsystem serves as the input element of a technical safety barrier, collecting information and sending signals to the controlling subsystem.

- **Controlling subsystem**: It includes at least one controller, such as a programmable logic controller (PLC) or computer, for analyzing the information sent by the sensing subsystem. In the example shown in Fig. 11.1, the controlling subsystem is a safety PLC. This subsystem can be regarded as the decision-maker of the technical barrier to determine necessary safety actions, and it will send commands to the actuating subsystem when needed.
- **Actuating subsystem**: It includes one or more physically actuating components to execute the safety actions requested by the controlling subsystem. In the example shown in Fig. 11.1, the actuating subsystem comprises two pressure safety valves (PSV1 and PSV2) that can open to release excess gas when they receive signals of opening.

It can be found that the sensing and actuating subsystems in the example include more than one component. This redundancy in design aims to enhance the reliability and availability of the technical safety barrier. In this context, the barrier component that can perform the barrier function of a subsystem is referred to as a *channel*. When a barrier subsystem contains n channels, it is functioning as long as at least k channels are functioning, and this configuration is regarded as a k-out-of-n ($k$oo$n$) voting structure. In the example, the sensing subsystem has a 2oo3 voting structure, while the actuating subsystem has a 1oo2 voting structure.

It is worth noting that systems consisting of only one or two of these subsystems can also function as technical safety barriers. For example, a standalone sensing system may trigger alarms to warn operators to evacuate. While such systems require human intervention to achieve the safety function, they are still relevant and included as study objects in this chapter.

11.2.3 Operational Modes of Technical Safety Barriers

Technical safety barriers have particular operational modes compared with those production systems. A hazardous event in the HRS can generate a *demand* to the associated safety barrier, namely a requirement to activate the barrier function. A demand is also referred to as a *deviation*. If a technical safety barrier is functioning well, it can respond to this demand to eliminate or at least mitigate the potential harm led by the event. Otherwise, if the technical safety barrier is faulty or unavailable, the event may harm the asset if the barrier is the ultimate barrier or bypass this barrier and trigger response of the other barriers.

Considering potential demands, technical safety barriers generally operate in one of two modes:

- **Continuous mode**: A technical barrier operates continuously, actively manages any deviations, and interrupts an accidental scenario.
- **Demanded mode**: A technical barrier remains in a dormant state when there is no demand and is only activated by a specific demand to perform its barrier function.

Within a technical safety barrier, its subsystems may operate in different modes. In the example shown in Fig. 11.1, the sensing- and controlling subsystems operate in the continuous mode, continuously collecting data and performing calculations. In contrast, the two pressure safety valves operate in the demanded mode because they activate only when the pressure in the vessel becomes excessively high, opening the pipeline for ventilation.

The demanded mode can be further classified into the *low-demand mode* and the *high-demand mode* based on the frequency of demands. Liu (2014) has highlighted that the classification considers the likelihood of a safety barrier failure resulting in an accident or being bypassed, particularly since such failures are typically detected only during regular proof tests.

11.3 Integrity Assessment of Technical Safety Barriers

Chapters 4 and 9 of this book have presented the framework of risk analysis and management where barriers perform their functions. In this chapter, we will introduce the detailed tasks in technical barrier specification and barrier integrity assessment.

11.3.1 Framework of Assessment

The main tasks in technical barrier assessment are illustrated in Fig. 11.2.

The detailed tasks in the six steps include the following:

Fig. 11.2 The main steps in technical barrier assessment

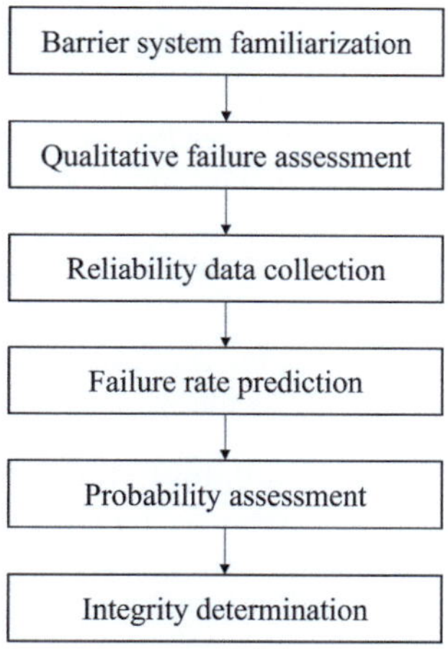

Step 1—Barrier system familiarization: This step involves specifying the required barrier functions and the integrity requirements of these barrier functions. It is also necessary to clarify the physical components and subsystems that are used to realize the barrier functions.

Step 2—Qualitative failure assessment: The tasks in this step include identifying the primary failure modes of the elements within the studied barrier system, as well as their causes and impacts on barrier functions. For the critical failure modes, a further quantitative assessment is needed. This will be discussed in more detail in the next section.

Step 3—Reliability data collection: To perform a quantitative assessment for critical failure modes, it is necessary to collect the relevant failure rate data of the barrier components. Such data can be found in the existing databases and reliability data handbooks. In the cases where sufficient data is available, the reliability and integrity of the equipment can be more accurately estimated. For example, OREDA (2009) is a database for reliability prediction in the oil and gas applications, and MIL-HDBK-217F (1991) is for electronic components.

However, for new technologies and novel applications, such as the equipment used in HRSs, there may be limited accident records available to support quantitative analysis. In such cases, failure rate prediction is necessary.

Step 4—Failure rate prediction: In this step, failure rate data from similar equipment in existing databases is used as a reference to calculate the failure rate of the studied item. For example, the failure modes and failure rates of PSVs have been well recorded in the offshore oil and gas databases with the experiences of decades. When such equipment is used as an HRS, its performance can be similar, but it will be influenced by risk influencing factors (RIFs) that can be different in this new application. RIFs include factors such as design changes, the physical properties of hydrogen gas, environmental stresses, and maintenance requirements.

Several established models for failure rate prediction are available, such as the Brissaud's approach (Brissaud et al. 2010), Rahimi's approach (Rahimi and Rausand 2013), artificial neural networks (Kumar 2016), and Bayesian belief networks (Zhang et al. 2016). The common principle behind these methods is to calculate the failure rate of the studied object based on the contributions of RIFs to the rates of specific failure modes and the differences between the reference and the new conditions. Due to page limitations, further details on failure rate prediction methods are not provided in this chapter. Readers can refer to Kumar (2016) for an overview of failure rate prediction methods.

Step 5—Probability assessment: The next step is to rely on the failure rate prediction result to assess the long-term availability and unavailability of barrier components, subsystems, and the overall barrier system. It is expected that the assessment results can be compared with experimental results when resources and conditions for experiments are available. The associated methods will be introduced in later sections of this chapter.

Step 6—Integrity determination: Based on the results of the probability assessment, the integrity of the technical safety barrier can be determined, taking into account its operational mode. This will also be discussed in the following sections.

11.3.2 Failure Analysis of Technical Safety Barriers

For the technical barriers in the demanded mode, it is needed to conduct tests and inspections to examine the status of these systems even when they are in the dormant state. There are two main types of tests used for technical safety barriers:

- **Diagnostic test**: It is also known as a diagnosis, depending on a built-in diagnostic module or feature to identify specific types of failures of barrier components. Diagnostic tests can be conducted with very high frequency to identify failures shortly after they occur.
- **Proof test**: Such a test is conducted to confirm that a barrier component is able to perform its barrier function as intended. Proof tests are scheduled regularly, occurring at relatively long fixed intervals.

Considering these two types of tests, failures of a technical safety barrier can be categorized as

- **Detected failure**: A failure that can be identified immediately after it occurs or in a short time, e.g., by a diagnostic test.
- **Undetected failure**: A failure that remains unfound until a complete examination in a proof test.

On the other hand, failures of a technical safety barrier can be classified based on their severities of consequences:

- **Dangerous (D) failure**: A failure that compromises or disables the barrier function, and thus may lead to a severe harm on the EUC.
- **Safe (S) failure**: A failure that does not stop barrier function and does not lead to serious harm to the EUC, and is thus relatively safe.

Based on these two classifications for the failures, there are four distinct types of potential failures in a technical safety barrier: *dangerous detected* (DD) failure, *dangerous undetected* (DU) failure, *safe detected* (SD) failure, and *safe undetected* (SU). When evaluating the integrity of technical safety barriers, greater emphasis is placed on DD- and DU-failures, with DU-failures being the main contributor to system unavailability.

As mentioned in Sect. 11.2, redundant structures are always adapted in technical safety barriers, where *common-cause failures* (CCFs) may occur. A CCF means that two or more components fail due to a shared underlying factor. For example, the hydrogen embrittlement can be a shared cause for many metal barrier components used in an HRS, leading to their simultaneous failure within a short time frame. CCFs, especially DU-CCFs, can significantly compromise the effectiveness of redundant safety barrier structures that are specifically designed to enhance fault tolerance.

11.3.3 Safety Integrity Metrics

Probability of failure on demand (PFD) and *average frequency of failures* (PFH, the abbreviation of the old name of "probability of having a dangerous failure per hour") are the two commonly used integrity metrics of technical safety barriers. PFD is the suitable metric for barriers in the low-demand mode, while PFH is for the barriers operating in the high-demand or continuous mode.

In the calculation of PFD, the following assumptions are always needed:

- Proof tests are performed with a constant interval, denoted as τ.
- In a proof test, all faults can be found, and the following maintenance is perfect, meaning that a barrier can be restored to an "as-good-as-new" state.
- The times to DD-failures and DU-failures follow the exponential distribution, with the rates of λ_{DD} and λ_{DU}, respectively.
- At most one failure occurs on a barrier component simultaneously.

Based on the above assumption, the proportion of the DD-failure rate to the total rate of D-failures is termed as *diagnostic coverage* (DC):

$$\mathrm{DC} = \frac{\lambda_{\mathrm{DD}}}{\lambda_{\mathrm{D}}} = 1 - \frac{\lambda_{\mathrm{DU}}}{\lambda_{\mathrm{D}}} \tag{11.1}$$

Given that a barrier component is failed, it is not available until the failure is fixed. PFD(t) denotes the probability that a barrier component is not available by time t, and $\mathrm{PFD}_{\mathrm{avg}}$ is the average value of PFD(t) within a test interval.

In this section, we consider that DD-failures can be immediately identified and repaired in a short time; their impact on the unavailability of a barrier is relatively less. The influence of DD-failures can thus be skipped in simple analysis.

On the other hand, with a DU-failure, the barrier is unavailable until the next proof test. Given that time 0 is the moment that a proof test is completed, or the item is put into operation, $\mathrm{PFD}_{\mathrm{avg}}$ can be calculated as

$$\mathrm{PFD}_{\mathrm{avg}} = \frac{1}{\tau}\int_0^{\tau} F(t)\mathrm{d}t = 1 - \frac{1}{\tau}\int_0^{\tau} R(t)\mathrm{d}t = \frac{1}{\tau}\int_0^{\tau} \mathrm{PFD}(t)\mathrm{d}t \tag{11.2}$$

where $F(t)$ is the failure probability by time t, and $R(t)$ is the survival probability.

Given the renewal assumption, Eq. (11.2) can be applied for any interval. It can be noted that both item reliability (reflected by $F(t)$) and the operational strategy (proof test interval τ) can influence the value of $\mathrm{PFD}_{\mathrm{avg}}$.

With the exponential distribution assumption, Eq. (11.2) can be written as

$$\mathrm{PFD}_{\mathrm{avg}} = 1 - \frac{1}{\tau}\int_0^{\tau} \mathrm{e}^{-\lambda_{\mathrm{DU}}t}\mathrm{d}t = 1 - \frac{1}{\lambda_{\mathrm{DU}}\tau}\left(1 - \mathrm{e}^{-\lambda_{\mathrm{DU}}\tau}\right) \tag{11.3}$$

Then, $\mathrm{e}^{-\lambda_{\mathrm{DU}}\tau}$ can be replaced by its Taylor series as

$$\begin{aligned}\mathrm{PFD}_{\mathrm{avg}} &= 1 - \frac{1}{\lambda_{\mathrm{DU}}\tau}\left[\lambda_{\mathrm{DU}}\tau - \frac{(\lambda_{\mathrm{DU}}\tau)^2}{2} + \frac{(\lambda_{\mathrm{DU}}\tau)^3}{3!} - \cdots\right] \\ &= \frac{\lambda_{\mathrm{DU}}\tau}{2} - \frac{(\lambda_{\mathrm{DU}}\tau)^2}{6} + \cdots \end{aligned} \tag{11.4}$$

Since most barrier components are highly reliable, namely $\lambda_{\mathrm{DU}}\tau$ is a small positive number (< 0.1), the following approximation is valid:

$$\mathrm{PFD}_{\mathrm{avg}} \approx \frac{\lambda_{\mathrm{DU}}\tau}{2} \tag{11.5}$$

For a 1oo2 voting structure, its survival probability $R_{\mathrm{S}}(t)$ by time t is

$$R_{\mathrm{S}}(t) = 1 - [1 - R_i(t)]^2 = 2\mathrm{e}^{-\lambda_{\mathrm{DU}}t} - \mathrm{e}^{-2\lambda_{\mathrm{DU}}t}$$

Based on the similar consideration in Eq. (11.4), $\mathrm{PFD}_{\mathrm{avg}}$ of this 1oo2 voting structure can be approximated as

$$\mathrm{PFD}_{\mathrm{avg}}^{1\mathrm{oo}2} = \frac{1}{\tau}\int_0^{\tau} F(t)\mathrm{d}t \approx \frac{(\lambda\tau)^2}{3} \tag{11.6}$$

For a *koon* voting structure with independent and identical barrier components, it is functioning as long as at least k of its n components are functioning. A generic formula for $\mathrm{PFD}_{\mathrm{avg}}$ can be obtained as

$$\mathrm{PFD}_{\mathrm{avg}}^{koon} \approx \binom{n}{n-k+1} \frac{(\lambda_{\mathrm{DU}}\tau)^{n-k+1}}{n-k+2} \tag{11.7}$$

Based on this, $\mathrm{PFD}_{\mathrm{avg}}$ of some commonly used barrier subsystem configurations can be found in Table 11.1.

Considering CCFs, the beta-factor model proposed by Fleming (1975) is the most widely used method for CCF analysis and is suggested by IEC 61508 (2010). In such a model, symbol β is used to denote the conditional probability of a CCF on all the n components when a failure has been observed on the specific component, namely

$$\beta = \Pr(\mathrm{CCF}|\text{Failure of component}) \tag{11.8}$$

Under the exponential distribution assumption, a barrier component can be regarded as a series structure vulnerable to both independent failure and CCF, and the overall failure rate of this item is the sum of the rate of independent failure (I), and rate of CCF; therefore, the overall failure rate is

$$\lambda = \lambda^{\mathrm{I}} + \lambda^{\mathrm{CCF}} \tag{11.9}$$

The beta-factor, β, can be calculated as

$$\beta = \frac{\lambda^{\mathrm{CCF}}}{\lambda} = \frac{\lambda^{\mathrm{CCF}}}{\lambda^{\mathrm{I}} + \lambda^{\mathrm{CCF}}} \tag{11.10}$$

For more discussions on CCFs and numerical formulas for CCF calculations, readers can refer to the textbook of safety-critical systems by Rausand (2014).

A technical safety barrier fails if any of its subsystems fail. Assuming that the three subsystems of a barrier system are independent and highly reliable, the probability that more than one subsystem fails at the same time is very low. Thus, the average unavailability of the entire barrier system can be determined as

Table 11.1 Approximation formulas of *koon* voting structures of barriers

k/*n*	1	2	3	4
1	$\frac{\lambda_{\mathrm{DU}}\tau}{2}$	$\frac{(\lambda_{\mathrm{DU}}\tau)^2}{3}$	$\frac{(\lambda_{\mathrm{DU}}\tau)^3}{4}$	$\frac{(\lambda_{\mathrm{DU}}\tau)^4}{5}$
2	–	$\lambda_{\mathrm{DU}}\tau$	$(\lambda_{\mathrm{DU}}\tau)^2$	$(\lambda_{\mathrm{DU}}\tau)^3$
3	–	–	$\lambda_{\mathrm{DU}}\tau$	$2(\lambda_{\mathrm{DU}}\tau)^2$
4	–	–	–	$2\lambda_{\mathrm{DU}}\tau$

$$\mathrm{PFD}_{\mathrm{avg}}^{(\mathrm{BS})} \approx \mathrm{PFD}_{\mathrm{avg}}^{(\mathrm{SE})} + \mathrm{PFD}_{\mathrm{avg}}^{(\mathrm{CO})} + \mathrm{PFD}_{\mathrm{avg}}^{(\mathrm{AC})} \tag{11.11}$$

where $\mathrm{PFD}_{\mathrm{avg}}^{(\mathrm{SE})}$, $\mathrm{PFD}_{\mathrm{avg}}^{(\mathrm{CO})}$, $\mathrm{PFD}_{\mathrm{avg}}^{(\mathrm{AC})}$ are the average unavailability of the sensor-, controlling, and actuating subsystem, respectively.

$\mathrm{PFD}_{\mathrm{avg}}$ is actually evaluating the proportion of downtime of a safety barrier in the long term, and if no demand occurs during the downtime of the technical safety barrier, there is no accident. Since the barrier is operated in the low-demand mode, the very low frequency of demand means very low accident probability, and thus the demand has no or very little influence on the value of $\mathrm{PFD}_{\mathrm{avg}}$.

Given that a safety barrier is in the continuous or high-demand mode, when the barrier is unavailable, it is more likely that an accident will occur on the EUC, while the possibility of no protection (of barrier) but no accident is low. The value of $\mathrm{PFD}_{\mathrm{avg}}$ is thus influenced by the demand frequency. In such cases, PFH should be used, which is calculated as

$$\mathrm{PFH} = \frac{F(t)}{t} = \frac{E[N(0,t)]}{t} \tag{11.12}$$

where $E(N(0,t))$ is the expected number of failures occurring in the period of $(0,t)$. Proof test interval τ is excluded from Eq. (11.12), meaning that it is not an influence factor of PFH.

In terms of value, PFH is actually the same with the *rate of occurrence of failures* (ROCOF), as the unconditional failure rate of an item at any given time. For more explanations of ROCOF, readers can refer to Rausand et al. (2020). Let $N(t)$ represent the cumulative number of failures that have occurred up to time t, with its expected value denoted as $E[N(t)]$, and the derivative of $E[N(t)]$ is ROCOF by time t:

$$\mathrm{ROCOF}(t) = \frac{\mathrm{d}}{\mathrm{d}t}E[N(t)] = \lim_{\Delta t \to 0} \frac{E[N(t+\Delta t) - N(t)]}{\Delta t} \tag{11.13}$$

Since Δt is a very short interval, we can assume that at most one failure occurs within this interval. Consequently, $\mathrm{ROCOF}(t)$ is approximated to the probability that the item has a failure in the interval of length Δt.

Within a certain time interval $(0, t_1)$, the average failure frequency of a barrier component operated in a high-demand mode is given by

$$\mathrm{PFH} = \mathrm{ROCOF}(0, t_1) = \frac{E[N(0, t_1)]}{t_1} = \frac{1}{t_1}\int_0^{t_1} \mathrm{ROCOF}(t)\mathrm{d}t \tag{11.14}$$

The next work is integrity determination, as Step 6 outlined in Sect. 11.3.1. IEC 61508 (2010) introduces *safety integrity levels* (SILs) to evaluate barrier integrity and facilitate risk management and communication. SIL normally includes four discrete levels. The highest degree of safety integrity is SIL 4, while the lowest is SIL 1.

Table 11.2 Determination of SIL by PFD and PFH (adopted from IEC 61508)

SIL	PFD_{avg}	PFH
4	$\geq 10^{-5}$ to $< 10^{-4}$	$\geq 10^{-9}$ to $< 10^{-8}$
3	$\geq 10^{-4}$ to $< 10^{-3}$	$\geq 10^{-8}$ to $< 10^{-7}$
2	$\geq 10^{-3}$ to $< 10^{-2}$	$\geq 10^{-7}$ to $< 10^{-6}$
1	$\geq 10^{-2}$ to $< 10^{-1}$	$\geq 10^{-6}$ to $< 10^{-5}$

SIL is determined based on the calculation result of PFD_{avg} or PFH, as shown in Table 11.2 according to IEC 61508.

11.4 Modeling Approaches in Integrity Assessment

This section will provide an overview on the system modeling approaches for analyzing performance of technical safety barriers and calculating PFD_{avg} or PFH. Due to the page limitation, we just briefly present the basics of the most commonly used modeling approaches. Readers who have interests in barrier modeling can reference the textbook by Rausand (2014) and the book chapter for safety barrier performance modeling and assessment (Zhang and Liu 2022).

11.4.1 Fault Tree and Reliability Block Diagram

Among all the modeling approaches for technical safety barriers, fault tree analysis (FTA) and reliability block diagram (RBD) are the most important tools in system reliability analysis and evaluation.

FTA is a failure-oriented approach to describe how a system failure occurs due to different causes, for example, component failures. This method is based on a top-down logic diagram like a tree. Meanwhile, RBD is a success-oriented approach to describe how a specified system function is fulfilled with the connections of subfunctions. RBD is a kind of network with functional blocks (rectangles) and connections (lines) between blocks.

An RBD can be transferred to a fault tree, and vice versa. The parallel structure in an RBD has a similar role with the AND gate in a fault tree for system reliability analysis, whereas a series structure in an RBD is similar to an OR-gate. Fault tree and RBD models have similarities in foundational principles, and both models are effective in the modeling of large-scale systems and structural analysis, and they support both qualitatively causal analysis and quantitative performance and evaluation.

Since these two methods are based on the Boolean algorithm, it is a bit difficult to apply them for degradation modeling or analysis for systems with more complex behaviors.

Both FTA and RBD are recommended in standard IEC 61508 (2010) for technical barrier analysis. In addition, readers can learn information about FTA in the standard IEC 61025 (2006), and more details about RBD in the standard IEC 61078 (2016).

11.4.2 Bayesian Belief Network

A Bayesian network, or Bayesian belief network (BBN), is a graphical model of causal relationships, and it can be used to evaluate technical safety barriers when there are more risk influencing factors. A BBN consists of two parts: A directed acyclic graph (DAG) with a finite set of nodes and a set of directed edges, and conditional probability tables (CPTs) associated with nodes. In a BBN, a node can be a parent node to several child nodes and be a child of multiple parent nodes simultaneously. The CPT of a node illustrates the probability distribution of an event, conditional on the states of its parent nodes.

A fault tree model can be accurately translated into a BBN without any loss of details. Quantitative analysis of BBN is based on the Bayesian probability theory.

Some existing studies, such as Cai et al. (2013) and Wu et al. (2016), have successfully used BBN for technical system modeling and reliability analysis. However, it should be noted that the barrier analysis based on BBN needs more workloads in data collection and establishment of CPTs when the number of nodes increases.

11.4.3 Markov Method

When dynamic behaviors, such as degradation of barrier components and tests and maintenances, have critical impacts on barrier performance, state transition models are more effective than the structure-based models, such as RBD. One of the most widely used state transition models is the Markov method, especially the continuous time Markov chain (CTMC).

The Markov method relies on a Markov diagram describing the transitions between discrete states of an item. With CTMC and the exponential distribution assumption (or Markov property), it is possible to conduct numerical analysis on the sojourn probability of a barrier system in a certain state in the long term, as the foundation of calculation of PFD_{avg}.

The Markov method is both intuitive and highly flexible, making it a popular choice for system reliability analysis. However, as the system scale grows and more states are incorporated, the size of the model increases rapidly, which can lead to significant computational complexity.

IEC 61508 (2010) has included the Markov method in the integrity analysis of technical safety barriers. This method will also be applied throughout the remainder of this chapter to analyze the performance of technical barriers. For a deeper understanding of the mathematical principles, readers can refer to the book by Rausand et al. (2020), which provides comprehensive insights into the theory and application of Markov models in reliability analysis.

11.4.4 Petri Net

Petri Net (PN) is another state transition model that includes both static and dynamic elements. The static elements consist of places, transitions, and directed arcs, which are used for structural analysis of the system, while dynamic elements, represented by tokens, are used to analyze the distribution and movement of resources within the system.

Unlike the Markov method, PN modeling does not require the identification of all potential states of a system in advance. This makes it particularly useful for analyzing complex behaviors. PN is also highly effective for modeling parallel events and conditions, such as the implementation of different testing and maintenance strategies for technical safety barriers. However, it is important to note that PN modeling requires more specialized knowledge and training.

IEC 62551 (2013) specifies the terminology of PN, and it is worth noting that PN has seen several extensions over recent decades to accommodate more complex system behaviors. IEC 61508 (2010) incorporates a reliability block diagram-driven PN (RBD-PN) approach for analyzing the performance of technical safety barriers. This hybrid approach combines the structured reliability block diagram with the dynamic modeling capabilities of PN to offer a more comprehensive tool for assessing barrier effectiveness.

The models discussed above can be applied in various scenarios for barrier assessment. The choice of the appropriate modeling approach can be guided by the framework shown in Fig. 11.3.

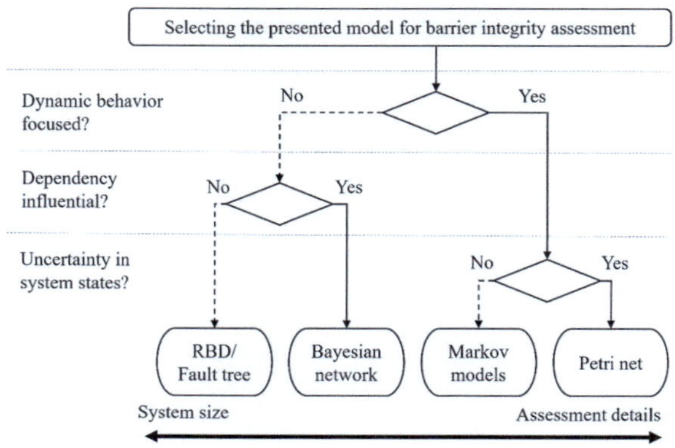

Fig. 11.3 Modeling approach selection (based on the figure in the thesis of Zhang (2018))

11.5 Case Studies

In this section, we consider the pressure safety valve (PSV) system used in the high-pressure gaseous hydrogen storage system as the case study for probability assessment and integrity determination. When the gas pressure exceeds the specified value, it will automatically open to discharge the gas to prevent the pressure from exceeding the system safety value. When the pressure returns to normal, the valve will automatically close and prevent the gas from continuing to be discharged. An illustration of such a safety barrier system is illustrated in Fig. 11.1.

11.5.1 Markov Modeling

In this case study, we focus on the modeling and assessment of the actuating subsystem. Firstly, we consider only one valve is used as a 1oo1 subsystem, and only one failure mode that the valve cannot be opened when it is required. This is a DU-failure. Therefore, the 1oo1 barrier can have two possible states as shown in Table 11.3.

We can build a simple Markov diagram for this barrier, shown in Fig. 11.4 to illustrate the two states and the transitions between them. The process behind this model can be regarded as a CTMC since the barrier performance is time-dependent.

In Fig. 11.4, the circles represent system states: state 0 is the functioning state, and state 1 is the DU-faulty state.

Similar to what we have done in Sect. 11.3.3, the following assumptions are needed for Markov analysis:

- Time-to-failure is exponentially distributed, namely the DU-failure rate keeps constant.
- Failures occur independently, and their severities are same.

Table 11.3 States of a 1oo1 barrier only with DU-failure

State	Short name	State description
0	W	The barrier is functioning (working)
1	DU	The barrier has a DU fault

Fig. 11.4 A simple Markov diagram for a 1oo1 barrier subsystem

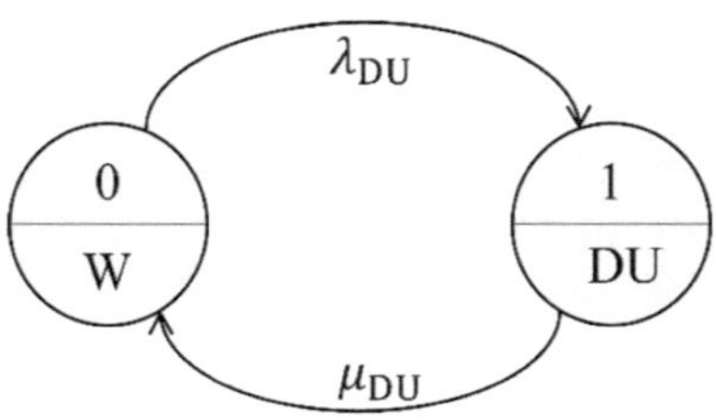

- Proof tests are performed regularly with constant intervals.
- After a proof test, the barrier component can be restored to the "as-good-as-new" condition.
- The demand frequency/rate is assumed to be constant, implying that the period between demands follows an exponential distribution.

The transition from state 0 to state 1 is a DU-failure with a rate of λ_{DU}, and the transition from state 1 to state 0 is the restoration of the system with a rate of μ_{DU}. Given the proof test interval is τ, μ_{DU} can be calculated by the reverse of the mean downtime (MDT):

$$\mu = \frac{1}{\text{MDT}} \approx \frac{1}{\frac{\tau}{2} + \text{MRT}} \quad (11.15)$$

The approximation in Eq. (11.15) is valid since the barrier component has a constant DU-failure rate; when it is found to be failed in the proof test, its time of failure is uniformly distributed over the test interval $(0, \tau]$, and thus its MDT is the expected unavailable time before test $\tau/2$ plus the mean repair time (MRT). For barrier integrity assessment, PFD_{avg} in this case is calculated as the sojourn probability in state 0.

Then, we consider DD-failure and S-failure. In the normal situation, the PSV is closed. Leakage may occur if there are problems on the seal, and such a failure can be detected immediately when diagnosis modules are in place. Thus, it is a DD-failure. On the other hand, when the PSV is open, the over-pressurized gas can be released. After the pressure in the tank comes back to the safe range, the PSV is expected to be closed again. If the valve fails to close, the failure can result in gas loss. Given that ventilation system is working well, this failure is relatively safe, and here we regard it as a SU-failure of the PSV. Based on the considerations above, the states of this safety barrier are listed in Table 11.4.

A Markov diagram can be built for this barrier considering diverse failures, as illustrated in Fig. 11.5.

In this model, the transition from state 1 to state 0 has a rate as the converse of mean time to repair (MTTR), namely $\mu_{DD} = 1/\text{MTTR}$. This value is much smaller than μ_{DU}, since MTTR is often several hours, much shorter than a proof test interval. The transition rate from state 2 to state 0, μ_{SU}, can be regarded equal to μ_{DU}, assuming that the SU-failure can also be revealed in the regular proof tests.

Table 11.4 States of a 1oo1 barrier with dangerous and safe-failures

State	Short name	State description
0	W	The barrier is functioning
1	DD	The barrier has a DD fault
2	SU	The barrier has a SU fault
3	DU	The barrier has a DU fault

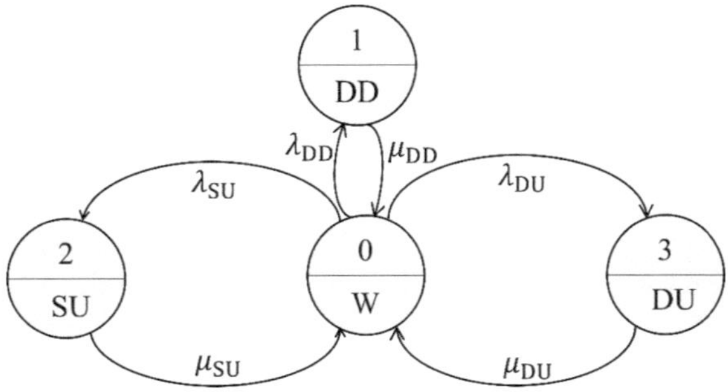

Fig. 11.5 Markov model for a 1oo1 barrier with DD-, SU-, and DU-failures

Next, we can consider a redundant structure of two PSVs as illustrated in Fig. 11.1. When only DU-failures are taken into account for barrier availability analysis, this barrier subsystem has three states, as listed in Table 11.5.

The associated Markov diagram is shown in Fig. 11.6.

The subsystem can transfer from state 0 to state 1, with a rate of $2\lambda_{DU}$, because any of the two functioning barrier components may fail. Then, from state 1 to state 2, the transition rate is λ_{DU} since only one barrier component is still working. Since a proof test is expected to reveal all DU-faults, if it is assumed that the maintenance resource is enough for fixing these failed components at the same time, both transition rates from state 2 to state 0 and from state 1 to state 1 are μ_{DU}.

However, the model above ignores the impact of CCF. When CCF is in evaluation, the Markov diagram can be modified based on the beta-factor model as shown in Fig. 11.7.

It can be found that there are two outgoing transitions from state 0: One is for CCF, with the rate of $\beta\lambda_{DU}$, and the other is for individual DU-failure on one component, with the rate of $2(1-\beta)\lambda_{DU}$ considering that each of two barrier components may

Table 11.5 States of a 1oo2 barrier subsystem only with DU-failure

State	Short name	State description
0	2W	2 barrier components are functioning
1	1W1DU	1 barrier component is functioning, and 1 barrier component is the DU-failed state
1	DU	The barrier has a DU fault

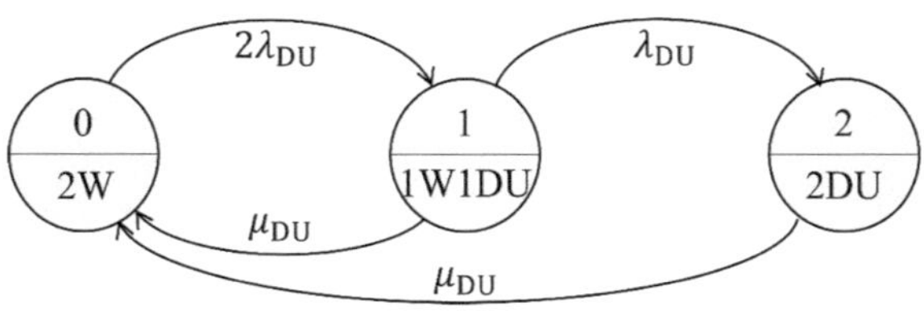

Fig. 11.6 Markov model for a 1oo2 barrier only with independent DU-failures

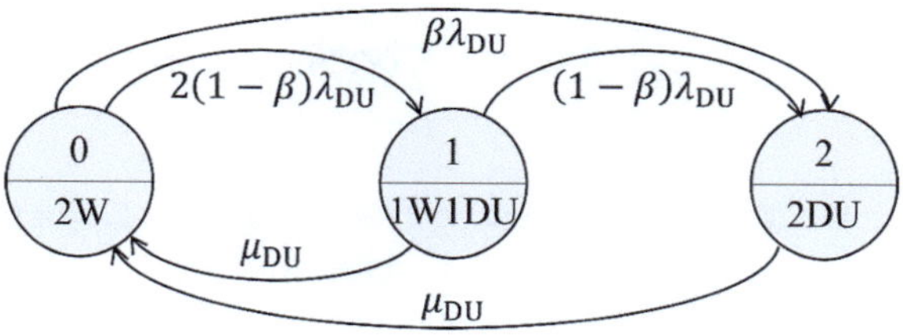

Fig. 11.7 Markov model for a 1oo2 barrier only with CCF and independent DU-failures

fail. In this model, we also can assume that the maintenance resource is sufficient, and no matter one or two faults are revealed in a proof test, they can be fixed at the same time. Thus, the transition rates from state 1 and state 1 to state 0 are μ_{DU}.

11.5.2 Numerical Analysis

Following the Markov models, we can use D with losing generality to denote the set of the states where the barrier is unavailable. The $\mathrm{PFD}_{\mathrm{avg}}$ can then be calculated based on the percentage of time that the barrier is sojourning in all the unavailable states in a long term, namely

$$\mathrm{PFD}_{\mathrm{avg}} = \sum_{i \in D} p_i \tag{11.16}$$

where p_i is the steady-state probability in state i.

Markov model-based numerical analysis is always based on the assumption that the state transitions are in a stochastic process following the Markov property. This implies that the future states are only dependent on the present state and not influenced by how the process arrived in the present state. Such a property also means that all transitions in the CTMC are exponentially distributed in terms of transition time.

Consider a CTMC $\{X(t), t \geq 0\}$ with state space $\Omega = \{0, 1, 2, \ldots, n\}$. The transition probability from state i to state j by time t is

$$p_{ij}(t) = \Pr\left[X(t) = j | X(0) = i\right] \tag{11.17}$$

For a system with n states, we can set a transition matrix as

$$\boldsymbol{P}(t) = \begin{bmatrix} p_{00}(t) & p_{01}(t) & \cdots & p_{0n}(t) \\ p_{10}(t) & p_{11}(t) & & p_{1n}(t) \\ \vdots & & \ddots & \vdots \\ p_{n0}(t) & p_{n1}(t) & \cdots & p_{nn}(t) \end{bmatrix} \tag{11.18}$$

When a process is in state i at time 0, it must be in a certain state within the state space Ω at time t, implying

$$\sum_{j=0}^{N} p_{ij}(t) = 1 \quad \text{for all } i \in \Omega \tag{11.19}$$

Let α_i denote the leaving rate of the process from state i, and p_{ij} the probability that the process transits from state i to state j at a certain time. The total leaving rate from state i is

$$\alpha_i = \sum_{\substack{j=0 \\ j \neq i}}^{n} \alpha_{ij} \tag{11.20}$$

α_{ij} can be organized into a transition rate matrix $\mathcal{A}$ for all possible states:

$$\mathcal{A} = \begin{pmatrix} \alpha_{00} & \alpha_{01} & \cdots & \alpha_{0n} \\ \alpha_{10} & \alpha_{11} & & \alpha_{1n} \\ \vdots & & \ddots & \vdots \\ \alpha_{n0} & \alpha_{n1} & \cdots & \alpha_{nn} \end{pmatrix} \tag{11.21}$$

Then, the following notation for the diagonal elements (Rausand et al. 2020):

$$\alpha_{ii} = -\alpha_i = -\sum_{\substack{j=0 \\ j \neq i}}^{n} \alpha_{ij} \tag{11.22}$$

We skip the deduction process due to the page limitation, but come to the Kolmogorov forward equation of a CTMC; its complete expression can be written as

$$\left[p_0(t), p_1(t), \ldots, p_n(t)\right] \cdot \begin{pmatrix} \alpha_{00} & \alpha_{01} & \cdots & \alpha_{0n} \\ \alpha_{10} & \alpha_{11} & & \alpha_{1n} \\ \vdots & & \ddots & \vdots \\ \alpha_{n0} & \alpha_{n1} & \cdots & \alpha_{nn} \end{pmatrix} = \left[\dot{p}_0(t), \dot{p}_1(t), \ldots, \dot{p}_n(t)\right] \tag{11.23}$$

For the details in obtaining the Kolmogorov forward equation, readers can refer to the textbook for reliability analysis (Rausand et al. 2020).

In the long term, the sojourn probability of the Markov process in each state can approach to a constant (p_i), and thus $\dot{p}_i(t) = 0$ when $t \to \infty$. Equation (11.23) can be then written as

$$\left[p_0, p_1, \ldots, p_n\right] \cdot \begin{pmatrix} \alpha_{00} & \alpha_{01} & \ldots & \alpha_{0n} \\ \alpha_{10} & \alpha_{11} & & \alpha_{1n} \\ \vdots & & \ddots & \vdots \\ \alpha_{n0} & \alpha_{n1} & \cdots & \alpha_{nn} \end{pmatrix} = [0, 0, \ldots, 0] \tag{11.24}$$

Also based on Eq. (11.19), in the long term

$$\sum_{i=0}^{n} p_i = 1 \tag{11.25}$$

It is possible to obtain n linear algebraic equations based on Eq. (11.13). It is possible to $(n-1)$ of the equations along with Eq. (11.25) to calculate the steady state probabilities, $p_0, p_1, \ldots, p_n$.

Now we can numerically analyze the Markov models built in this chapter. For a 1oo1 barrier with the simplest Markov model in Fig. 11.4, the transition rate matrix is

$$\mathcal{A} = \begin{pmatrix} -\lambda_{\mathrm{DU}} & \lambda_{\mathrm{DU}} \\ \mu_{\mathrm{DU}} & -\mu_{\mathrm{DU}} \end{pmatrix}$$

Then, we can have

$$-\lambda_{\mathrm{DU}} p_0 + \mu_{\mathrm{DU}} p_1 = 0$$

$$p_0 + p_1 = 0$$

Resolving the two equations, we can obtain that $\mathrm{PFD_{avg}} = p_1$ in this case, and thus $\mathrm{PFD_{avg}} = \frac{\lambda_{\mathrm{DU}}}{\lambda_{\mathrm{DU}}+\mu_{\mathrm{DU}}}$.

Using the reference data in the PDS data handbook, λ_{DU} of a PSV can be 2.2×10^{-6} per hour for the oil and gas application, and for the new hydrogen applications, we can use the reliability prediction method mentioned in Sect. 11.3.1 to roughly predict the DU-failure rate as 2.5×10^{-6} per hour. Given that the proof test interval is half year ($\approx$ 4380 h) and mean repair time is 8 h, $\mathrm{PFD_{avg}}$ can be calculated as

$$\mathrm{PFD_{avg}} = \frac{2.5 \times 10^{-6}}{2.5 \times 10^{-6} + \frac{1}{8+4380/2}} = 5.4650 \times 10^{-3}$$

For the 1oo2 barrier subsystem with the Markov model in Fig. 11.6, it becomes unavailable when it falls into state 2, meaning that $\mathrm{PFD_{avg}} = p_2$. The corresponding transition rate matrix of this Markov model can be written as

$$\mathcal{A} = \begin{bmatrix} -2\lambda_{DU} & 2\lambda_{DU} & 0 \\ \mu_{DU} & -(\mu_{DU} + \lambda_{DU}) & \lambda_{DU} \\ \mu_{DU} & 0 & -\mu_{DU} \end{bmatrix}$$

The steady-state probabilities can then be determined based on the following three equations:

$$-2\lambda_{DU}p_0 + \mu_{DU}p_1 + \mu_{DU}p_2 = 0$$

$$\lambda_{DU}p_1 - \mu_{DU}p_2 = 0$$

$$p_0 + p_1 + p_2 = 1$$

p_2 can be obtained by solving these equations:

$$p_2 = \frac{2\lambda_{DU}^2}{2\lambda_{DU}^2 + 3\lambda_{DU}\mu_{DU} + \mu_{DU}^2} \approx \frac{2\lambda_{DU}^2}{\mu_{DU}^2}$$

We have assumed that both barrier components fail following the exponential distribution with the same rate, and the failures have an equal probability of occurring at any time within a proof test interval. When the subsystem is revealed as DU-faulty in a proof test, it is reasonable to consider that the two DU-failures are evenly distributed in the period: One DU-failure occurs after $\tau/3$ from the last test or restoration, and the other DU-failure occurs by $2\tau/3$. Therefore, the mean downtime for the group is calculated as $\text{MDT} = \left(1 - \frac{2\tau}{3}\right) + \text{MTTR} = \frac{\tau}{3} + \text{MTTR} = 1/\mu_{DU}$.

When the proof test interval is much longer than MTTR, PFD_{avg} can be approximated as

$$\text{PFD}_{\text{avg}} = p_2 \approx \frac{2(\lambda_{DU}\tau)^2}{9}$$

This is lower than the result of the approximation formula in Eq. (11.15) since the latter ignores the situation where one DU-faulty component can be found in a proof test. Using the same data of the 1oo1 barrier example, PFD_{avg} can be calculated as 2.6445×10^{-5}, corresponding to SIL 4.

The Markov model in Fig. 11.6 also can be used for a 2oo2 barrier subsystem, while the PFD_{avg} should be calculated as $p_1 + p_2$.

When CCFs are considered, using the beta-model, the corresponding transition rate matrix of the Markov model in Fig. 11.7 is

$$\mathcal{A} = \begin{bmatrix} -(2-\beta)\lambda_{DU} & 2(1-\beta)\lambda_{DU} & \beta\lambda_{DU} \\ \mu_{DU} & -\mu_{DU} - (1-\beta)\lambda_{DU} & (1-\beta)\lambda_{DU} \\ \mu_{DU} & 0 & -\mu_{DU} \end{bmatrix}$$

Similar to the previous examples, p_2 can then be calculated as

$$p_2 \approx \frac{\beta\lambda_{DU}}{\mu_{DU} + 2(1-\beta)\lambda_{DU}{}^2 + \beta\lambda_{DU}} \approx \frac{\beta\lambda_{DU}}{\mu_{DU}}$$

When the value of β is not too small, for example $\beta \geq 0.01$, the occurrence probability of a CCF within a proof test is much higher than the probability of two sequential independent DU-failures, and it can be assumed that the revealed faulty state of a 1oo2 barrier subsystem is due to CCF. Therefore, the mean downtime can be approximated as $\tau/2$. For example, given that $\beta = 0.05$, PFD_{avg} of such a barrier subsystem can be calculated as 5.4750×10^{-4}, falling in SIL 3. It can be found that the effectiveness of redundancy is greatly compromised by the CCF, even though the integrity is still higher than the 1oo1 voting structure.

11.6 Discussion and Conclusions

This book chapter briefly introduces technical safety barriers that can be installed in hydrogen fueling stations, as well as the framework and selected methods in technical safety barrier assessment. In the case study, the failure to open PSVs under the overpressure of the hydrogen containers is regarded as the DU-failure of a technical safety barrier, and its impact on the long-term barrier availability is evaluated. The Markov method is included in this chapter for numerical analysis.

Generally, in the design, installation, and operation of technical safety barriers, there are two types of questions that need to be answered: Are the exiting or specified safety barriers sufficient for controlling the risk based on the current state information, and how should the testing and maintenance activities be scheduled to ensure the required integrity level of the barrier. Considering the simple approximation formula of PFD, Eq. (11.5), the first question is related to the value of λ_{DU}, while the second question is related to τ.

In practice, the availability of technical safety barriers relies on numerous factors, such as barrier system structures, types and severities of hazardous events, and the chronic degradations of barrier components due to mechanisms of corrosion, wear-out, etc. These factors challenge the sufficiency of failure rate-based methods with binary system state assumptions in this chapter. The Ph.D. thesis by Zhang (2021) can provide more insights into barrier degradation modeling and assessment.

For low-demand safety barriers, it is sometimes costly and unreasonable to conduct full proof test periodically for simulating the whole trip of barrier protection actions. For example, a full shutdown of a valve may interrupt the normal operation and make damages on the valve and pipeline. Therefore, it is also meaning to introduce partial proof tests that are used to examine some functions of safety barriers. Wu et al. (2018) have studied the impacts of different testing strategies on technical barrier integrity.

With the development of digital technologies, the digital twin approach has been proved useful in the design and operational phases of safety barriers and can support barrier health management. More information about the digitalization and digital twin approach for technical safety barriers can be found in Khan et al. (2022).

References

Brissaud F, Charpentier D, Fouladirad M, Barros A, Bérenguer C (2010) Failure rate evaluation with influencing factors. J Loss Prev Process Ind 23(2):187–193

Cai B, Liu Y, Liu Z, Tian X, Zhang Y, Ji R (2013) Application of Bayesian networks in quantitative risk assessment of subsea blowout preventer operations. Risk Anal 33(7):1293–1311

Erofeeva S (2024) Technical safety barrier assessment of liquid hydrogen storage system. Norwegian University of Science and Technology, Trondheim, Norway

Fleming KN (1975) A reliability model for common mode failures in redundant safety systems

Hollnagel E (2004) Barriers and accident prevention. Ashgate

IEC 61025 (2006) Fault tree analysis (FTA). International Electrotechnical Commission, Geneva, Switzerland

IEC 61078 (2016) Reliability block diagram. International Electrotechnical Commission, Geneva, Switzerland

IEC 61508 (2010) Functional safety of electrical/electronic/programmable electronic safety-related systems. International Electrotechnical Commission, Geneva, Switzerland

IEC 61511 (2016) Functional safety—safety instrumented systems for the process industry sector. International Electrotechnical Commission, Geneva, Switzerland

IEC 62551 (2013) Analysis techniques for dependability—petri net techniques. International Electrotechnical Commission, Geneva, Switzerland

Johansen IL, Rausand M (2015) Barrier management in the offshore oil and gas industry. J Loss Prev Process Ind 34:49–55

Khan F, Pasman HJ, Yang M (eds) (2022) Methods to assess and manage process safety in digitalized process system. Academic Press

Kumar A (2016) Prediction of failure rates for subsea equipment. Norwegian University of Science and Technology, Trondheim, Norway

Liu Y (2014) Discrimination of low- and high-demand modes of safety-instrumented systems based on probability of failure on demand adaptability. Proc Inst Mech Eng Part O J Risk Reliab 228(4):409–418. https://doi.org/10.1177/1748006X14523263

Liu Y (2020) Safety barriers: research advances and new thoughts on theory, engineering and management. J Loss Prev Process Ind 67:104260. https://doi.org/10.1016/j.jlp.2020.104260

MIL-HDBK-217F (1991) Reliability prediction of electronic equipment. U.S. Department of Defense, Washington, DC

NEA (2003) Engineered barrier systems (EBS) in the context of the entire safety case. Nuclear Energy Agency

OREDA (2009) Offshore reliability data, 5th edn. O. Participants

Rahimi M, Rausand M (2013) Prediction of failure rates for new subsea systems: a practical approach and an illustrative example. Proc Inst Mech Eng Part O J Risk Reliab 227(6):629–640

Rausand M (2014) Reliability of safety-critical systems: theory and applications. Wiley

Rausand M, Barros A, Hoyland A (2020) System reliability theory: models, statistical methods, and applications, 3rd edn. Wiley-Blackwell

Verfondern K, Cirrone D, Molkov V, Makarov D, Coldrick S, Ren Z, Wen J, Proust C, Friedrich A, Jordan T (2021) Handbook of hydrogen safety: chapter on LH2 safety. In: Pre-normative research for safe use of liquid hydrogen (PRESLHY)

Wu S, Zhang L, Zheng W, Liu Y, Lundteigen MA (2016) A DBN-based risk assessment model for prediction and diagnosis of offshore drilling incidents. J Nat Gas Sci Eng 34:139–158
Wu S, Zhang L, Barros A, Zheng W, Liu Y (2018) Performance analysis for subsea blind shear ram preventers subject to testing strategies. Reliab Eng Syst Saf 169:281–298
Yuan S, Yang M, Reniers G, Chen C, Wu J (2022) Safety barriers in the chemical process industries: a state-of-the-art review on their classification, assessment, and management. Saf Sci 148:105647. https://doi.org/10.1016/j.ssci.2021.105647
Zhang J (2018) Contribution to reliability and availability analysis of novel subsea technologies—methods and approaches to apply in the early design phase. Norwegian University of Science and Technology, Trondheim
Zhang A (2021) Prognostics and health management of safety-instrumented systems: approaches of degradation modeling and decision-making. Norwegian University of Science and Technology, Trondheim. https://hdl.handle.net/11250/2726046
Zhang A, Liu Y (2022) Performance evaluation of digitalized safety barriers. In: Khan F, Pasman HJ, Yang M (eds) Methods to assess and manage process safety in digitalized process system. Academic Press
Zhang J, Liu Y, Lundteigen MA, Bouillaut LU (2016) Using Bayesian networks to quantify the reliability of a subsea system in the early design. In: 2016 European safety and reliability conference, Glasgow, Scotland
Barrier Engineering: Models and Methods for Technical Safety

Chapter 12
Accident Preparedness and Emergency Response for Hydrogen Refueling Stations

Dimitrios Tzioutzios, Yiliu Liu, and Nicola Paltrinieri

Abstract This chapter explores the academic discussion surrounding emergency preparedness for hydrogen system accidents, with a focus on hydrogen refueling stations (HRS). It synthesizes recent research and offers practical recommendations to address key challenges and gaps in accident response strategies. The discourse presented emphasizes the importance of a comprehensive, proactive approach to safety, outlining specific measures to manage incidents involving hydrogen systems. A key takeaway is the value of a capabilities-based approach to preparedness, which provides a structured framework for identifying and addressing potential risks. By prioritizing risk identification, capability development, and collaborative planning, stakeholders can strengthen their resilience to hydrogen system accidents, minimizing their impact on people, property, and the environment. The chapter also highlights the iterative nature of emergency preparedness, emphasizing continuous improvement and adaptability to evolving risks and technologies. The active involvement of all relevant parties, including first responders and facility operators, is crucial to maintaining effective emergency management systems. The discussion concludes with future research directions, including the integration of advanced technologies into emergency response frameworks, as well as the exploration of simulation-based training, multi-agency coordination, and public communication strategies for enhancing preparedness and response to HRS accidents.

D. Tzioutzios (✉) · Y. Liu · N. Paltrinieri
Department of Mechanical and Industrial Engineering, Norwegian University of Science and Technology (NTNU), Trondheim, Norway
e-mail: dimitrios.tzioutzios@ntnu.no

Y. Liu et al. (eds.), *Safety Measures for Hydrogen Fueling Stations*, Risk, Reliability and Safety Engineering, https://doi.org/10.1007/978-981-96-9412-9_12

12.1 Introduction

12.1.1 Emergency Planning for Accidents

Going beyond technical measures and safety design aspects for accident prevention and risk mitigation (see the discussions about safety barriers in Chaps. 10 and 11), a holistic approach to accident risk management ought to also address organizational and coordination elements for emergency response. In this context, a precautionary approach is essential, requiring the development of contingency strategies for worst-case scenarios—irrespective of their rarity—when dealing with technological accidents (Krausmann and Necci 2021). This is especially pertinent to the emerging hydrogen technologies, where risk-based assessment methods may be hindered by limited statistical data, leading to ambiguity and potentially unacceptable levels of risk (Dadashzadeh et al. 2018; Genovese et al. 2023). Such considerations are fundamental in establishing a reliable and adaptive framework for hydrogen safety engineering.

Hydrogen refueling stations (HRS) are considered to involve hazardous materials (hazmat) due to the inherent risks posed by hydrogen. Hydrogen is classified as a dangerous substance because it can cause harm to people due to its chemical properties or the way it is used. Like other fuels such as petrol, liquefied petroleum gas (LPG), and liquefied natural gas (LNG), hydrogen presents risks because of its flammability, potential for explosive reactions, and the high pressure at which it is stored and transported in HRS. Several industrialized countries have specific regulations to address these hazards. For instance, in the United Kingdom (UK), the Dangerous Substances and Explosive Atmospheres Regulations (DSEAR) regard hydrogen systems like HRS as falling within the scope of hazardous substances due to the dangers they present (DSEAR 2002). Nowadays, the establishment of mixed fuel stations further complicates incidents at such installations, as the combination of different fuels may introduce additional chemical risks (Nakayama et al. 2016; Sun and Li 2021). While accidents at HRS typically do not escalate to the scale of catastrophic disasters (unless they are involved in conjoint hazard scenarios), they still present potentially life-threatening and destructive hazards. Therefore, proactive emergency planning and response strategies tailored to the specific risks of HRS are essential to mitigate these hazards effectively.

Incidents involving hazardous materials pose significant risks to life and property, often resulting in fire or explosion. According to the Federal Emergency Management Agency (FEMA 2021) of the United States (US), such events require complex strategies and are influenced by various factors, such as weather conditions, while also demanding specialized resources for effective management. In response to these challenges, emergency organizations need to develop safety goals, refine internal structures, coordinate resources, and, at times, establish new frameworks. Ultimately, emergency planning aims to create a robust response system to address hazards before they occur (Perry and Lindell 2006).

The response process includes multiple interconnected activities (FEMA 2019). Key steps involve recognizing the incident, isolating the area through perimeters and safety zones, notifying relevant agencies, identifying the hazardous substances, and conducting rescue and recovery operations (Karagiannis 2024). Other critical activities include firefighting, spill control, and leak containment. Spill control methods vary based on the release type, using techniques such as diking, absorption, and dilution for land spills, booming and damming for waterway releases, and ventilating or blanketing for airborne hazards. Containment, on the other hand, focuses on restricting materials to their original container using measures like valve shutoff, patching, and plugging breaches. Finally, protective actions should be taken in hazmat emergencies to protect the public if there is a potential for harm (Karagiannis 2024). Protective actions depend on the scale of the incident and can include evacuation or shelter-in-place measures. Evacuation involves relocating people away from the hazard to increase safety, while shelter-in-place requires people to stay indoors to avoid exposure.

Building upon general hazardous materials response procedures, recent research has focused on adapting and expanding these strategies to address the unique challenges posed by hydrogen-based technologies and their associated emergencies. Several studies have emphasized the need to enhance accident preparedness and emergency response strategies tailored to hydrogen systems (Li et al. 2021; Nakayama et al. 2016; Sun and Li 2021; Tretsiakova-McNally et al. 2017). Prominent international organizations have acknowledged the importance of streamlining these processes to support first responders (such as fire, police, and emergency medical services), facility operators, emergency planners, and decision-makers in mitigating associated risks.

Academic literature has traditionally concentrated on the technical aspects, including inherent safety design and engineering barriers, to prevent and mitigate accident risks and their consequences (Li et al. 2021). Few studies have explored non-technical issues related to the response to accidents at HRS. In more detail, Weiner et al. (2013) reviewed eight web-based tools for hydrogen safety knowledge, highlighting their potential to complement traditional resources such as conferences, journals, and magazines. However, these tools require regular updates and maintenance to ensure accuracy and responsiveness to user feedback. Tretsiakova-McNally et al. (2017) introduced an innovative training strategy that combines e-learning with a virtual reality pedagogical approach to help emergency responders understand the complexities of emerging hydrogen fuel cell vehicles (HFCVs). This training package has been promoted through initiatives by various local authorities.

Further research by Li et al. (2021) explored firefighters' perceptions of HFCVs, revealing concerns about economic costs and a predominant focus on rescue-oriented safety over fire-oriented safety. Key themes and areas for improvement were identified, leading to recommendations for addressing these concerns. Additionally, a study by Sun and Li (2021) on hazard analysis and emergency strategies for hydrogen vehicle accidents highlighted key risks such as hydrogen releases and tank ruptures. They proposed practical strategies, including rapid assessment methods, firefighting tactics, and rescue operations, to address these challenges and enhance emergency response capabilities. Nonetheless, a holistic approach to emergency response planning for hydrogen systems has yet to be explored.

In this context, this chapter contributes to the academic discourse on emergency preparedness and response to accidents involving hydrogen systems, with a particular focus on HRS. It synthesizes recent research findings with practical strategies to address identified gaps and challenges in accident preparedness and response for hydrogen systems. Following a clarification of the concepts of accident preparedness and emergency response, the chapter discusses specific measures for accidents involving hydrogen systems. The subsequent section explores key challenges in emergency response planning. Finally, the chapter concludes with a summary of the main points.

12.1.2 Definitions of Accident Preparedness and Emergency Response

Drawing upon Disaster Risk Management (DRM) literature, we can disambiguate the concepts of planning for emergency preparedness and response for incidents involving hazmat. FEMA (2012) defines emergency response and incident planning as "*planning associated with an actual or potential incident, likely under emergency conditions, that involves developing procedures for responding to actual or projected effects.*" The terms "*preparedness,*" "*emergency response,*" and "*incident planning,*" as used throughout this chapter, are explained below.

Preparedness involves developing the knowledge and capacities necessary to anticipate, respond to, and recover from disasters (UNDRR 2016). It is rooted in a thorough analysis of disaster risks and effective integration with early warning systems. Key activities include contingency planning, stockpiling resources, establishing coordination mechanisms, organizing evacuation plans, disseminating public information, and conducting training and drills. These efforts require formal institutional, legal, and budgetary support. Preparedness plans set out pre-arranged actions to address specific potential hazards or disaster scenarios. For instance, preparedness measures for HRS would entail:

- Developing emergency response plans specific to hydrogen-related incidents.
- Equipping first responders with specialized training and equipment for hydrogen fires.
- Installing and maintaining early warning systems and gas detectors.
- Conducting regular drills and simulations to test emergency procedures.

Emergency response encompasses the immediate actions taken before, during, or shortly after a disaster to save lives, minimize health impacts, ensure safety, and meet essential needs (UNDRR 2016). It focuses on short-term goals and relies on pre-established preparedness measures to enhance effectiveness and timeliness.

Response activities typically involve the collaboration of public and private emergency services, community sectors, and volunteers. Specialized emergency agencies, such as civil protection authorities, police, and fire services, play a central role. Examples of emergency response actions include:

- Activating emergency response teams and evacuating the area.
- Containing the hydrogen leak or fire using appropriate techniques.
- Providing medical assistance to any injured individuals.
- Implementing traffic control measures around the affected area.

Incident planning refers to the process of developing and implementing plans to address an actual or potential emergency, typically under crisis conditions (FEMA 2012). As Karagiannis and Synolakis (2017) explain, this involves continuously evaluating the situation, determining necessary actions, and developing a coordinated plan through four key stages:

1. *Situation Analysis*: Gather and organize critical incident data, conduct hazard analysis, and develop running estimates to understand evolving conditions.
2. *Incident Assessment*: Identify disaster-generated needs, evaluate available resources, and determine additional capability requirements.
3. *Course of Action Development and Analysis*: Create and assess alternative response strategies, outlining tasks, responsibilities, timelines, and potential challenges.
4. *Incident Plan Development*: Present response options for approval, finalize the selected plan, and document it based on complexity and urgency.

This process is iterative, as the situation evolves and new plans are formulated based on updated information (Karagiannis and Synolakis 2017). In contrast to (emergency/accident) preparedness, which is done proactively before an incident, incident planning is reactive and happens when a hazard is imminent or has already occurred. Incident planning involves the coordination of various groups across different administrative levels, with their roles and locations determined by existing regulations, the structure of the emergency response system, and the specific needs of the incident (Haddow et al. 2020). In the case of a hazmat release, senior members of first response agencies typically take charge of incident planning from an Incident Command Post (ICP). At the ICP, incident action planning is conducted, focusing on managing the incident by identifying objectives, coordinating resources, developing courses of action, and assigning tasks (Karagiannis and Synolakis 2017).

For example, in the event of a hydrogen system accident, incident planning would be initiated when a hydrogen leak or fire is detected. The incident plan would involve identifying the location of the hazard, containing the leak, and evacuating the premises to ensure the safety of personnel. Tasks would be assigned to fire and emergency response teams to extinguish the fire, isolate the area, and provide medical assistance to any affected individuals.

12.2 Accident Preparedness

12.2.1 Emergency Response Plan

In many industrialized countries, facilities handling hazardous materials are required to draft basic on-site emergency response plans to address minor releases and outline the initial response actions to prevent accident escalation (Karagiannis 2024). Within the facility grounds, the operator is typically responsible for managing incidents until the arrival of first response teams. Additionally, local and regional authorities develop off-site emergency plans to address hazardous materials incidents that could impact people, property, or the environment beyond the facility's boundaries.

In essence, HRS safety does not diverge from the abovementioned approach for hazmat accidents. A case in point is the UK's DSEAR, which stipulates that employers at HRS—among other types of refueling stations—must carry out risk assessments and implement measures to eliminate or reduce risks, which may include developing procedures and maintaining and installing appropriate equipment (DSEAR 2002). Going beyond technical safety measures, employers are also required to provide equipment and *procedures to manage incidents involving fire or explosions* and to identify hazard zones where explosive atmospheres might occur during specific activities. Thus, provisions for a rudimentary accident preparedness and emergency response plan for HRS are not an uncommon prerequisite for their establishment and operation.

Nonetheless, as discussed in the introductory section of this chapter, planning for emergency preparedness and response is a process involving numerous stakeholders across administrative tiers. Establishing robust emergency plans is essential for public authorities and operators to address accidents effectively. These plans aim to localize accidents, contain them when possible, and minimize harm to health, the environment, and property (OECD 2023). A critical first step involves identifying hazardous installations and assessing potentially affected populations, environments, and assets, including cross-border considerations. Such on-site and off-site emergency plans should detail technical and organizational procedures to mitigate accident impacts comprehensively.

Emergency planning is expected to account for various complicating factors, such as transport incidents, extreme weather, natural hazards, loss of utilities, communication breakdowns, synergistic effects of multiple substances, domino effects, and malicious acts (OECD 2023). Identifying potential scenarios, assessing environmental and health consequences, and evaluating risks are integral to the planning process. Also, emergency plans should allow for flexible responses to a spectrum of situations, from minor incidents to worst-case scenarios, while addressing practical challenges in their implementation.

Several factors ought to be considered during planning and preparation to avoid failure. Limited information availability at the time of an accident necessitates clear, adaptable strategies. Regular training and drills are vital to ensure preparedness across all elements of the plan. Potential failures in communication systems and the impact of stress on individuals during emergencies highlight the importance of simplicity and clarity in the plan to reduce complexity and optimize response efficacy (OECD 2023).

Particularly for hydrogen-related incidents, an effective emergency action plan should provide comprehensive guidelines for managing various scenarios, including hydrogen leaks, spills, fires, and transportation emergencies (Pacific Northwest National Laboratory 2024). It should outline the actions required by initial responders, specify the location of emergency response equipment in the hydrogen facility, and detail appropriate fire suppression measures. Clear communication protocols should be established and maintained throughout the emergency, ensuring seamless coordination. The plan should also address first aid and medical response procedures and the establishment of site security. Finally, it should include explicit criteria for determining when public authorities need to be alerted to respond to incidents with potential off-site impacts, identify the appropriate contacts, and outline the procedure for summoning external assistance.

In anticipation of broader impacts, the plan should account for potential media coverage and include protocols for salvage and restoration operations following the incident (Pacific Northwest National Laboratory 2024). Incident Command Posts should be established with a pre-designated chain of authority to streamline decision-making and coordination. Additionally, the plan should include an inventory of hazardous materials and their locations, alongside clearly defined evacuation routes and staging areas for personnel not directly involved in the emergency response (OECD 2023).

12.2.2 Identifying Target Areas and Populations

Local-level emergency planning is crucial for engaging stakeholders, building trust, and fostering acceptance. Clearly identifying areas and populations at risk, including those exposed to thermal radiation or structural hazards, is fundamental (OECD 2023). Particular attention should be given to critical infrastructure, environmentally sensitive zones, and locations with vulnerable populations, such as hospitals, schools, or nursing homes. Emergency plans should include processes for direct communication with these groups, offering specific instructions during incidents. Lastly, risk assessments can benefit greatly from classifying areas with similar adverse effects into zones. This zoning facilitates targeted planning for warnings, evacuations, and other response measures, ensuring a systematic approach to minimize the consequences of chemical accidents.

12.2.3 Communication for Emergency Response Planning

Emergency response planning should include opportunities for public input to address community concerns and integrate their feedback into the planning process (OECD 2023). Media representatives should also be involved in the development and promotion of the plan, particularly regarding public warning methods and information dissemination. Emergency plans are needed to guide the public on appropriate responses during an incident, covering warnings, sheltering in place, evacuation procedures, and emergency contacts. Communication strategies should be tailored to the specific communities at risk, ensuring clarity and effectiveness (Tzioutzios et al. 2024).

Effective emergency warning systems are essential to alert the public (OECD 2023). These systems should deliver timely, reliable warnings through mechanisms such as sirens, automated phone messages, public address systems, mobile notifications, or smartphone apps. Public awareness about the purpose and use of these systems should be raised, and regular testing should be conducted to ensure functionality and understanding.

Public alert and warning systems are vital for delivering emergency information via mobile phones, radio, television, and other channels, helping save lives during hazardous incidents (Karagiannis 2024). Authorities ought to ensure systems are in place to inform the public during and after accidents. Communication during the crisis should be timely, factual, simple, and continuous, covering off-site effects, risks of further impacts, public actions, and follow-up details (OECD 2023). For this purpose, effective public and media communication requires preparation, including templates, training, and long-term relationship building. Utilizing social and traditional media effectively is essential for maintaining trust and ensuring accurate dissemination of emergency information.

12.2.4 Review and Test Emergency Response Plans

On-site and off-site emergency plans have to be regularly tested and reviewed, incorporating both tabletop and field exercises, in order to keep them updated with changes in risks, nearby developments, response technologies, lessons from previous incidents, and personnel changes (OECD 2023). Testing and exercises build the response personnel's competency and confidence in managing real emergencies. After each test, a systematic review process should identify lessons learned and implement necessary improvements. Regular testing should include defining clear goals and objectives using the SMART criteria (Specific, Measurable, Attainable, Realistic, Timely), setting the scope of exercises, and ensuring participation from various stakeholders, including internal and external parties, industry, public authorities, and the public (OECD 2023). Cross-border coordination may also be necessary in certain cases to exchange knowledge and practices, while exercises should foster a learning environment without any blame attribution. Finally, training methods should be carefully chosen, and evaluation should prioritize openness, transparency, and incorporating improvements based on extracted lessons.

12.2.5 Resource Availability

Preparedness planning requires a thorough assessment of existing skills, equipment, and resources compared to those needed for various accident scenarios, including worst-case events (OECD 2023). This evaluation identifies gaps and informs the acquisition of additional capabilities. Responsible parties should ensure that human, financial, and material resources, including communication tools and personal protective equipment, are readily available for immediate deployment.

Transparency with stakeholders about available resources, limitations, and expected public behaviors during emergencies is essential (OECD 2023). Emergency plans should ensure the availability of critical supplies, information, and data collection mechanisms, such as hazmat detection tools. Plans should also incorporate resilience by including backup communication systems, alternative command centers, and provisions for responder relief and recovery, such as feeding stations and sanitary facilities.

12.2.6 Emergency Response Training

Training for hazmat incident response is critical due to its complexity, requiring scientific knowledge, technical skills, and specialized equipment (Karagiannis 2024). Emergency response exercises are mandatory in many industrialized countries and aim to test systems under realistic conditions. Equipment for hazardous materials response includes personal protective equipment (PPE), detection devices, firefighting tools, and spill control materials. Facilities handling hazardous substances may also need fixed response systems like fire suppression equipment and spill containment measures.

All personnel involved in emergency response should receive training aligned with the off-site emergency plan. Training should cover personal risks, protective measures, properties of hazardous substances, contamination hazards, and first aid procedures (OECD 2023). Exercises help assess and improve emergency management systems while increasing responder competency (Karagiannis and Synolakis 2017; Van Niekerk et al. 2015). Particularly for hydrogen-related incidents, first responder training should sufficiently cover topics on incident command, controlling access, staying upwind, techniques for recognizing hydrogen leakages and flames, eliminating ignition sources, and safe firefighting practices (Pacific Northwest National Laboratory 2024).

Furthermore, hydrogen safety training is essential for all employees handling hydrogen or related materials (Pacific Northwest National Laboratory 2024). Therefore, this applies to employees at HRS. Cultivating a safety culture for the HRS operation is crucial. Training plans should address potential job hazards, include periodic retraining, and be updated when new equipment or information is introduced. Employees need to understand hydrogen properties, safety requirements, equipment maintenance, leak and spill procedures, and emergency response policies. Ongoing reminders, such as safety bulletins, labels, and signs, should reinforce critical information. Staff should be encouraged to report safety concerns regarding equipment or systems.

12.2.7 Operators' Role in Accident Preparedness

The management of hazardous installations is responsible for preparing and implementing on-site emergency plans in collaboration with employees (OECD 2023). All staff, including subcontractors, should understand the plan's provisions and know how to respond during emergencies, such as limiting hazardous substance releases or evacuating to designated assembly points. Procedures for raising alarms should be clear to ensure swift action in the event of an accident.

Furthermore, facility operators need to ensure key personnel are familiar with the capabilities and response plans of fire authorities and other emergency responders. Visitors should be briefed on emergency actions and generally accompanied by facility staff. Management is strongly suggested to maintain sufficient emergency supplies, monitor items with limited shelf lives, and develop plans to source additional materials quickly, such as firefighting foam or absorbent materials (OECD 2023). If a facility lacks expertise in emergency planning, it should seek assistance from trade associations, consultants, neighboring installations, suppliers, or public authorities. Lastly, facilities in the same area, such as ports or industrial parks, should coordinate on-site emergency plans to ensure consistency, establish mutual aid agreements, and mitigate risks of domino effects.

12.2.8 Public Authorities' Role in Accident Preparedness

Public authorities should ensure the development, testing, implementation, and regular updating of off-site and on-site emergency plans for hazardous installations. Responsibility for these tasks varies by country, with off-site plan development often assigned to local officials, committees, or regional/national authorities, depending on applicable laws and policies. It is important that decision-making responsibilities are always clearly defined in such plans.

In addition, central authorities should provide guidance and support to local authorities, fostering motivation to build effective emergency preparedness and response capabilities. At the local level, authorities ought to ensure that on-site and off-site plans are well-coordinated and align with central principles to ensure consistent and effective implementation.

12.3 Emergency Response

12.3.1 Alarm Systems Response

Effective alert and warning systems are crucial for responding to chemical accidents or imminent threats (OECD 2023). These systems should immediately notify response personnel, triggering the emergency plan and informing management of the situation. Public warning systems should be activated during accidents to notify potentially affected individuals. Designated spokespersons from industry and public authorities should possess the necessary information, skills, authority, and credibility to communicate effectively. This ensures timely and accurate information reaches the public, enhancing preparedness and safety during emergencies.

Particularly, it is essential that hydrogen-related facilities have systems to detect abnormal conditions, malfunctions, and potential failures, supported by alarm systems with distinct audible and visible signals for each emergency type (Pacific Northwest National Laboratory 2024). Regular drills ensure employees recognize alarms and respond appropriately. Signal meanings should be clearly posted in operational areas, and alternative notification methods, such as public address systems, should be implemented if the abovementioned alarm systems are unavailable.

12.3.2 Hydrogen Incident Response

Responding to hydrogen incidents requires specialized training and equipment for fire protection and suppression. Therefore, only trained firefighters should attempt to manage hydrogen fires, as extinguishing them without halting the hydrogen supply poses risks of re-ignition and explosion (Pacific Northwest National Laboratory 2024). Personnel working around hydrogen should understand the unique properties of hydrogen fires and follow proper procedures, focusing on evacuation, assisting others, and alerting first responders rather than attempting firefighting.

Hydrogen's wide flammability range and low ignition energy mean that any leak or release is likely to ignite. Hydrogen fires are challenging to detect due to their colorless, odorless, nearly invisible flames, low radiant heat, and lack of smoke. Tools like thermal imaging cameras or flame detectors should be readily available to verify the presence of a hydrogen flame (Pacific Northwest National Laboratory 2024). In their absence, personnel can cautiously approach suspected leaks, observing for thermal waves or using methods like placing combustible objects into the flame to detect it safely. While hydrogen fires produce no smoke, burning nearby materials can generate harmful smoke, posing inhalation risks. These fires also transmit heat through radiation and convection, potentially damaging or igniting surrounding objects.

Best practices for handling gaseous and liquid hydrogen fires may have slight differences, but they always prioritize safety and preventing explosions (Pacific Northwest National Laboratory 2024). For *gaseous hydrogen fires*, the fire should not be extinguished until the hydrogen supply is shut off, as unburned hydrogen may cause an explosion. Adjacent equipment should be cooled with water sprays. Firefighters should avoid extinguishing fires involving hydrogen cylinders unless they are in open or well-ventilated areas free from combustibles and ignition sources. Burning cylinders should not be removed; instead, keep them and the surrounding cylinders cool using water. If multiple cylinders are involved in the fire, it is important to maintain a safe distance to prevent injury from potentially flying debris.

For *liquid hydrogen fires*, the fire should generally be allowed to burn under control until the hydrogen flow is stopped by closing block or isolation valves remotely from a safe distance (Pacific Northwest National Laboratory 2024). As liquid hydrogen vaporizes, it produces hydrogen gas, which poses a significant risk of creating flammable or explosive mixtures in the surrounding area. Water sprays can be used to extinguish secondary fires and prevent spreading but should not be directed at pressure-relief devices to avoid ice formation, which could impair function. Cooling hydrogen-containing equipment is vital. While carbon dioxide may be used, dry chemicals are preferable as they make hydrogen flames visible. It should be noted that extensive exposure to liquid hydrogen, its cold vapor, or associated chilled equipment can result in severe tissue damage through direct physical contact. Training programs for sites using or storing liquid hydrogen should include comprehensive education on cold injury risks, preventive measures, and emergency response procedures to ensure personnel are fully prepared to handle potential hazards associated with liquid hydrogen safely and effectively (Pacific Northwest National Laboratory 2024). Both scenarios require careful adherence to safety protocols, proper cooling measures, and maintaining safe distances to minimize risks and ensure controlled fire suppression.

12.3.3 Safety Zones and Evacuation

Upon arrival at an accident site, emergency responders should establish safety zones—*hot*, *warm*, and *cold*[1]—aligned with the external and internal emergency

[1] *Hot zone*: Highest risk area, closest to accident source; life-threatening; restricted to skilled emergency responders with top-level protection. *Warm zone*: Possible contamination; patient treatment and decontamination; respiratory protection required; ambulance personnel can assist. *Cold zone*: No special protection needed; incident command center; outer barrier of accident scene.

plans (OECD 2023). These zones limit the spread of hazardous substances and designate areas where responders can operate safely without special protective equipment. This system organizes the response effectively at the scene.

Moreover, deciding between evacuation and shelter-in-place is a critical and complex aspect of emergency management for hazmat incidents, as Karagiannis (2024) noted. While evacuation removes individuals from the danger zone, it is logistically challenging, may expose evacuees to risks such as traffic accidents or thermal and chemical effects, and requires mass care services. Conversely, shelter-in-place may fail if structures are too close to the hazard or not designed to withstand explosions or toxic clouds. Its effectiveness depends on the building's ability to shield against specific threats and the threat's duration.

Evacuation considerations might be relevant in the case of an explosion threat at an HRS accident, in which scenario nearby residents might be subject to an overpressure blast or launched projectiles. The situation might be very complicated, considering the potential compromise of structural integrity (e.g., due to a strong earthquake) or the presence of other hazardous material releases involved (e.g., a cloud of toxic fumes from the combustion of fuel at a multi-fuel station).

12.3.4 Clean-Up and Recovery

Post-accident clean-up (for hazmat involved in the incident or produced during the response) and recovery should be considered from the early stages of incident planning and throughout the emergency response (OECD 2023). During the transition from emergency response to recovery, all involved parties should collaborate and share information to ensure safety and support environmental and property restoration. This is relevant for incidents at multi-fuel stations possibly involving toxic or contaminating materials alongside hydrogen. The emergency response phase concludes when the incident no longer poses an acute risk to human health, the environment, or significant property damage. This decision should be made by the incident commander in consultation with all relevant response organizations, ensuring a safe and coordinated shift to recovery efforts.

12.3.5 Documenting Processes and Reporting

During an emergency response, organizations should document decisions and actions taken to assess the effectiveness of the intervention, improve future emergency plans, contribute to investigations, and derive lessons for future response activities (OECD 2023). Additionally, following any incident or near-miss, the facility should conduct a thorough investigation to identify lessons and prevent similar occurrences in the future (Pacific Northwest National Laboratory 2024). Investigations should be carried out by subject matter experts using a "no-blame" approach, focusing on determining root causes, developing corrective actions, and creating documentation for internal and external use. Incident reports should be widely shared to promote organizational learning. Lastly, additional special regional, national, or local reporting requirements may apply depending on the local regulatory framework. For instance, in the USA, incidents resulting in fatalities or hospitalizations of three or more individuals must be reported to the Occupational Safety and Health Administration (OSHA) within 48 h, detailing the accident's circumstances, fatalities, and injury extent, with further reports as necessary (Pacific Northwest National Laboratory 2024).

12.3.6 Operators' Role in Emergency Response

In the event of a hydrogen-related emergency, facility operators or designated officials should immediately activate the on-site emergency plan. Key actions include promptly alerting local emergency response authorities if the incident threatens health, the environment, or property and shutting off the hydrogen source if safe to do so (Pacific Northwest National Laboratory 2024). Affected areas should be secured by restricting access, posting signs, and stationing personnel upwind to warn others. Good ventilation in confined spaces is critical to prevent hydrogen from mixing with oxidizers, which could lead to deflagration or detonation. On-site personnel may have various duties during an emergency, such as isolating hazardous material flow, evacuating the facility while assisting others, warning others of risks, and calling for emergency response.

A designated responsible person should provide emergency responders with essential information, including a facility overview, hazardous material inventories and locations, facility safety equipment and instrumentation, any actions taken to prevent further escalation, and a copy of the Emergency Response Plan (OECD 2023; Pacific Northwest National Laboratory 2024). It should be emphasized that, once first responders arrive, the senior emergency response official assumes control of the site-specific incident. Collaboration between facility personnel and emergency responders is vital to ensure safety, minimize risks, and effectively manage the situation.

12.3.7 Public Authorities' Role in Emergency Response

When a hazmat accident occurs, response authorities should immediately activate the emergency plan to localize and contain the incident, minimizing harm to health, the environment, and property. Emergency measures should also activate systems to mobilize equipment, specialists, and resources, including regional or national support or assistance from neighboring communities.

The incident commander leads the efforts to prevent human exposure to hazardous substances both on-site and off-site while ensuring the protection of emergency responders from contamination (OECD 2023). Guidance from facility management and other experts should be sought in these situations as needed to optimize response strategies. In cases where the emergency exceeds local capabilities, additional resources and expertise may be requested from external entities, including national or international organizations. Incident planning and decision-making during the response should be adaptable to evolving circumstances and updated information, with continuous monitoring and follow-up to identify all consequences and ensure appropriate actions are taken to safeguard health and the environment. Real-time systems should be available to first responders to provide them with immediate access to critical information, including details about hazardous substances, potential escalation risks, patient projections, and injuries (OECD 2023). This information may include guidance from Safety Data Sheets (SDS) or electronic databases on chemical properties, transformation products, toxicological effects, and strategies for managing fires or spills. Accurate and timely dissemination of such information to responders, medical personnel, and the public is essential to effectively manage the emergency and its aftermath.

In the case of hydrogen-related accidents, initial actions from first responders arriving at the scene should prioritize safety and control: keep unauthorized personnel away, stay upwind, listen for venting gas, watch for thermal waves signaling flames, and eliminate ignition sources (Pacific Northwest National Laboratory 2024). An ICP should be promptly established at a safe distance by the first responders (Pacific Northwest National Laboratory 2024). Coordination will be managed by the incident commander, with support from senior officials of on-site personnel. The protocol suggests that firefighters should allow hydrogen-fed fires to burn out safely while protecting adjacent surfaces.

In this context, first responders need to be familiar with the safety information regarding hydrogen's properties and recommendations for protecting personnel and property during leaks, releases, or fires. They should understand the exclusion zone criteria to protect the public. For instance, awareness of fragment evacuation distances is crucial to minimize injury risks from explosions. Police assistance may be required to enforce protective measures and maintain public safety.

12.4 Discussion

12.4.1 Capabilities-Based Approach for Accident Preparedness

A capabilities-based approach to accident preparedness and emergency response offers valuable insights for managing HRS accidents. These recommendations should not be viewed in isolation but rather as part of a broader emergency management strategy that encompasses mitigation, response, and recovery efforts. As with any emergency, an effective response to hydrogen system accidents requires advanced preparation. Facility operators and regional authorities are expected to engage in deliberate activities designed to build, sustain, and enhance response capabilities (Haddow et al. 2020). Using a capabilities-based approach as a framework, we delineate recommendations for hydrogen accident preparedness into six interconnected steps: (i) *identify and assess risk*, (ii) *estimate capability requirements*, (iii) *build and sustain capabilities*, (iv) *plan to deliver capabilities*, (v) *validate capabilities*, and (vi) *review and update*. While these steps are often described as a linear, circular process, emergency managers typically revisit and refine them in an iterative manner.

Identify and assess risk: Effective preparedness for hydrogen systems accidents begins with comprehensive risk assessments that incorporate both internal and external hazard factors. Safety reports, hazard analyses, and similar evaluations should explicitly identify potential risks specific to hydrogen systems, including equipment failures, leaks, fires, or explosions (Haddow et al. 2020; OECD 2023). At the facility level, assessments should consider vulnerabilities associated with HRS, hydrogen production sites, and transportation routes. Emergency management organizations should collaborate with facility operators to ensure risks are integrated into local, subnational, and national planning frameworks.

A critical step in the assessment process involves evaluating potential hazards using advanced tools and methodologies. For example, consequence analysis helps estimate the spatial extent and severity of endpoint hazards, such as thermal radiation from a hydrogen fire or overpressure from an explosion. Tools such as "HyRAM+" (Ehrhart et al. 2024; Groth and Hecht 2017) can be used to model these scenarios, providing valuable insights into the potential impacts on people, property, and the environment. Such analyses form the basis for informed decision-making and resource allocation.

It is noteworthy that HRS may be situated near other hazardous facilities, particularly in industrial zones, requiring consideration of potential domino effects that could exacerbate accident scenarios. Additionally, HRS, more often than not, being located near high-traffic areas or transportation hubs could amplify the consequences of an accident. In rare conjoint hazard scenarios, such as large-scale natural hazards with a wide area of impact (e.g., earthquakes, floods, or storms), multiple hazmat releases could occur at a single HRS or in multiple installations simultaneously, thus compounding the risk and exponentially complicating emergency response priorities. These conjoint disaster scenarios[2] may disrupt critical infrastructure and hinder emergency response efforts, further escalating the impact of the incident. They may constitute rare scenarios, but ought to be incorporated in the hazard identification step, considering their large-scale impact.

Estimate capability requirements: Accurate risk assessments inform the specific capabilities required to respond to and recover from hydrogen system accidents effectively. With the tools available to them, HRS operators should preliminarily define firefighting tactics as part of an emergency response plan to effectively mitigate

[2] Hazard events that involve the release of hazardous materials triggered by the impact of a natural hazard are called "Natech" (OECD 2023).

accidents at the station (Nakayama et al. 2016). These tactics should be tailored to address specific consequences or HRS configurations, ensuring that appropriate measures are in place to manage a variety of potential scenarios. Additionally, regular training for maintenance personnel is essential to ensure they remain proficient in safe handling practices and emergency procedures, reinforcing the overall safety framework of the station (Genovese et al. 2024).

Moreover, emergency planners should leverage innovations in technology to enhance response capabilities (Haddow et al. 2020; Karagiannis 2024). For instance, Unmanned Ground Vehicles (UGVs) can be deployed in hazardous areas to deliver water or fire suppression agents, minimizing risks to personnel. Similarly, Unmanned Aerial Vehicles (UAVs) offer a cost-effective and efficient way to enhance situational awareness during incidents, providing real-time data on the extent and nature of the hazard.

Build and sustain capabilities: Once the necessary capabilities are identified, the next step is to build and sustain them within budgetary constraints (Haddow et al. 2020). Communities and facilities should prioritize investments in equipment, training, and planning that address high-risk hydrogen system scenarios. Resources should be allocated based on a cost–benefit analysis to mitigate incidents with the greatest potential impacts effectively, yet always adhering to a precautionary principle.

One identified challenge should be noted here. Special hazmat response teams are costly to develop and maintain, and their availability can vary significantly across different regions and jurisdictions (Haddow et al. 2020; Karagiannis 2024). Well-funded urban fire and rescue departments are typically able to sustain full-time hazardous materials response teams. However, in rural or underfunded areas, such resources may be restricted to regional teams or accessible through mutual aid agreements (Sun and Li 2021). Technological advancements have provided new and more cost-effective solutions for emergency response; for instance, while helicopters were once used for aerial reconnaissance, unmanned aircraft systems began to take over this role by the early twenty-first century (Haddow et al. 2020).

Plan to deliver capabilities: Emergency planning for hydrogen system accidents should integrate with broader all-hazards planning at the regional administration level. A hybrid planning approach that combines scenario-based, functional, and capabilities-based methodologies is highly effective (FEMA 2021). This approach begins with developing realistic hydrogen system accident scenarios, such as leaks or explosions at HRS, and analyzing their impacts. Emergency planners can use these scenarios to identify agent- and response-generated demands, informing the development of courses of action and planning assumptions. By creating multiple scenarios, emergency planners can identify common functional requirements, such as fire suppression, evacuation, or public risk communication. Functional planning ensures clarity by defining the roles and responsibilities of various organizations. System-based approaches further enhance preparedness by treating emergency response organizations as interconnected systems of functions, interactions, and support services (Haddow et al. 2020).

Furthermore, collaboration among diverse stakeholders is crucial for successful planning. Facility operators bring expertise on hydrogen properties, equipment, and safety mechanisms and should be involved early in the planning process. This collaboration facilitates resource sharing, enhances information exchange, and ensures a unified response effort during emergencies (Tzioutzios et al. 2024).

Validate capabilities: Advanced modeling tools and satellite imagery are indispensable for improving situational awareness and validating planning assumptions. Hazard modeling provides critical information, such as the boundaries of affected areas, while evacuation modeling estimates clearance times and traffic flow under various scenarios. These tools enable planners to evaluate alternative courses of action and establish clear decision-making triggers for evacuation or shelter-in-place orders (Haddow et al. 2020).

Exercises play a vital role in identifying gaps in plans and testing response capabilities (Karagiannis and Synolakis 2017). Regularly scheduled drills and full-scale exercises focused on hydrogen system scenarios, such as a refueling station explosion or a transportation incident, are essential for realistic preparedness. Such exercises also help evaluate the effectiveness of response strategies, personnel coordination, and resource deployment.

While significant efforts have recently been made to develop comprehensive training materials for HRS operators and first responders, both researchers and practitioners continue to recognize the critical need for enhanced and ongoing professional education. For example, Li et al. (2021) have identified two key categories of information essential for effective training for firefighters dealing with hydrogen-related accidents (particularly with HFCVs). The first is a fundamental level of knowledge that includes an understanding of hydrogen flame behavior, the structure of HRS, hydrogen vehicle design, and associated safety mechanisms. Firefighters are aware of the potential hazards that could impact their safety and understand the type of knowledge required to address these risks effectively. However, a significant gap exists in the accessibility and availability of this knowledge, leaving many uncertain about where or how to obtain it. Thus, the second category pertains to the knowledge of available sources and repositories on related hydrogen safety information. This lack of accessibility underscores the need to consolidate relevant information from various sectors and integrate it into existing training programs. This latter approach would provide firefighters with a foundational level of knowledge in a cost-effective manner, partially mitigating safety concerns and improving their preparedness for HRS incidents.

Advanced training technologies, such as virtual reality (VR) simulations, enhance emergency preparedness for hydrogen refueling stations (HRS). A web-based 3D VR program simulates accident scenarios, offering practical experience in handling equipment and emergency responses (Lee et al. 2010). Additionally, a mixed e-learning and VR approach has been used to train first responders, improving their readiness to manage hydrogen-related incidents (Tretsiakova-McNally et al. 2017). Integrating these advanced tools into training programs offers a hands-on, immersive learning experience, equipping firefighters with the skills and confidence needed to effectively respond to HRS-related emergencies.

Review and update: In many countries, regulations mandate periodic emergency drills for facilities handling hazardous materials (DSEAR 2002). For instance, facilities are often required to test their internal and external emergency plans at least once every three years. However, more frequent and focused drills specific to hydrogen systems are recommended to ensure readiness for these unique challenges.

Finally, emergency plans and capabilities should undergo regular review and updates to ensure they remain effective and aligned with current risks (Haddow et al. 2020). Annual evaluations, as well as revisions following incidents, exercises, or regulatory changes, are essential to maintaining a robust state of preparedness. These updates should account for new technologies, evolving hazards, and lessons learned from real-world events and training exercises.

12.4.2 Key Challenges for Incident Planning

Emergency response to hazmat incidents, particularly those involving hydrogen systems, is a complex and highly coordinated process that requires input from a range of stakeholders, each contributing specialized expertise. As discussed in the previous section, the involvement of multiple agencies, the availability of resources, and the management of on-the-ground personnel all play critical roles in ensuring an effective response. The challenges associated with managing these incidents are further compounded by the technical nature of the hazards involved, including the unique properties of hydrogen. These complexities underscore the importance of a well-structured emergency response framework that can adapt to rapidly evolving situations.

First and foremost, it becomes apparent that emergency response for hydrogen leaks is not solely the responsibility of first responders but relies heavily on on-site facility personnel for the initial response to the incident. In fact, studies often assume that on-site emergency strategies can effectively manage hazards without significant escalation (Sun and Li 2021). However, this assumption seems rather difficult to achieve, based on the findings of Li et al. (2021), even for experienced first responders—let alone for HRS facility staff. Moreover, Karagiannis and Synolakis (2017) demonstrated that the psychological pressure of dealing with potentially life-threatening crises puts under significant stress experienced emergency responders, who may omit critical steps in incident management. They highlighted various challenges in incident planning, including the volatility and uncertainty of situations, as well as operational impediments such as resource shortages and damage to infrastructure. Additionally, the involvement of multiple responding organizations and decision-making centers can complicate coordination and delay responses (Haddow et al. 2020). Finally, the post-disaster environment typically hampers situational awareness, leading sometimes to intuitive decision-making rather than rational, structured approaches. Therefore, emergency planning should take a more precautionary approach to account for this aspect, on top of the special emergency training for HRS operators.

Another major challenge for emergency response to incidents involving hazmat releases is adapting to the specific situation at hand, given the constant uncertainties and need for real-time updates (Karagiannis and Synolakis 2017; Sun and Li 2021). Karagiannis and Synolakis (2017) have shown that situation assessment may suffer from poor assessment of the situation, inaccurate estimations of affected individuals and incident localization, and inadequate hazard assessment. First responders typically face challenges in identifying hazardous substances, determining the affected area, and estimating potential consequences. This is a particular concern with hydrogen releases, as the substance is colorless, odorless, and produces nearly invisible flames. Similarly, the incident assessment phase may be undermined by first responders underestimating resource needs, mobilization time, and logistical requirements or poorly formulating objectives. The course of action development and analysis also presents challenges in terms of coordination such as incomplete plans, unclear roles, or hastily completed analysis. Finally, the least challenging phase, incident plan development, is hindered by slow decision-making, poorly structured plans, and omitted tasks, especially when multiple organizations share authority.

12.4.3 Effective Incident Planning Strategies

Effective incident planning is crucial to ensuring a swift and coordinated response during emergencies. To improve the planning process, several good practices can be adopted. One key recommendation is to embrace an iterative approach to incident planning (Karagiannis and Synolakis 2017). While traditional planning assumes a linear process consisting of situation analysis, requirement identification, objective-setting, and determining a course of action, incident planning is—in practice—an inherently ongoing, iterative, and dynamic process. As new information emerges, emergency response coordinators are advised to continuously revisit and refine their objectives and strategies. Given the rapidly evolving nature of emergency-related demands, incident plans should allow for real-time adaptation. Research revealed that incident planners who employ "running estimates," i.e., constantly reassessing and adjusting objectives and strategies based on continuously updating estimates in response to emerging information on-site, are often able to develop more accurate and timely response plans (Karagiannis and Synolakis 2017). By making early assumptions when information is lacking and adjusting as more details become available, planners can avoid unnecessary delays and ensure a more responsive approach to the evolving situation.

Furthermore, maintaining flexibility is essential in determining whether to formulate a single course of action or to prepare multiple alternatives. Crichton et al. (2005) highlighted that both approaches can be valuable depending on the circumstances of the emergency. Incident planners who could shift between these strategies based on the available time and the urgency of the situation tend to perform better in this regard, according to Karagiannis and Synolakis (2017). Developing multiple response options allows for preparation across a range of possible developments, whereas a single, focused plan may suffice when time constraints are significant or when the operational context is clearer. The ability to adapt and modify planning strategies as conditions change contributes to quicker goal attainment and enhances the overall effectiveness of the response (Karagiannis and Synolakis 2017).

While flexibility in incident planning enhances response effectiveness, the unpredictable and high-pressure nature of emergency situations also places significant physical and cognitive demands on responders. Managing these stressors is crucial to ensuring that well-developed plans can be executed efficiently and safely. In this vein, the high-stress and volatile environments of emergency response can cause fatigue, impairing responders' cognitive and physiological functions and increasing the risk of human error (Butler et al. 2024). To combat fatigue and mitigate these risks, Dawson et al. (2021) recommended practices such as caffeine access, utilization of nap opportunities, and targeted training and education. They emphasized the need for tailored implementation of such strategies across different emergency organizations and advocated for integrating fatigue management into safety systems to enhance responder well-being and operational effectiveness.

In addition, decision-making in emergency situations often involves uncertainty, which can pose significant challenges to the planning process (Karagiannis and Synolakis 2017). Incident planners may be reluctant to act when operating with limited and incomplete information, sometimes opting to delay taking action in favor of attempting to gather more on-site data to reduce uncertainty. However, this approach can delay the emergency response process without significantly improving decision quality. Effective incident planning relies on decision-makers who embrace uncertainty and integrate flexibility into their approach (Karagiannis and Synolakis 2017). By anticipating a range of possible developments, planners can adapt their decisions as situations evolve, ensuring a more responsive and dynamic emergency management process.

Moreover, decentralizing decision-making to field resources while maintaining strategic direction further enhances response efficiency. A decentralized approach enables teams to react swiftly to rapidly developing emergencies without being hindered by a rigid chain of command (House et al. 2014; Power et al. 2024). This approach empowers individuals to make timely decisions while still aligning with overarching objectives. However, successful decentralization requires interagency coordination, ensuring that different organizations understand each other's roles and processes to minimize uncertainty (House et al. 2014). Striking a balance between flexibility and decentralization strengthens decision-making, leading to a more adaptable, coordinated, and effective emergency response (Butler et al. 2024; Power et al. 2024). By embracing these good practices, incident planners can improve their ability to respond effectively and efficiently to emergencies, ensuring a more adaptable, coordinated, and timely response.

The practical implications of effective incident planning emphasize areas where improvements can enhance emergency response outcomes. Good practices, such as iterative planning, flexible strategies, and adaptive decision-making, are particularly valuable in addressing operational challenges, including situational awareness, resource estimation, decision-making speed, and interagency coordination. By refining these areas through targeted training, enhanced coordination, and streamlined processes, emergency managers can bridge gaps and ensure more effective incident preparedness and response.

First, situational awareness seems to be one of the most challenging components of incident planning, primarily due to difficulties in acquiring, cross-checking, and consolidating information during emergencies. When multiple agencies share responsibility and decision-makers are scattered across remote locations, decision-making tends to slow down significantly (House et al. 2014; Karagiannis and Synolakis 2017). A key issue contributing to this challenge is information fragmentation, where data from the emergency site is dispersed across various systems, agencies, and parallel communication channels. This fragmentation often leads to inconsistencies and inefficiencies in decision-making. For instance, Butler et al. (2024) suggested that one way to address this challenge is to co-locate all on-site emergency teams as soon as practicable, fostering shared situational awareness and reducing miscommunication. This approach ensures that responders have a common understanding of the evolving situation, enhancing coordination and response effectiveness.

Similarly, in situations where field information is limited, incident planners may hesitate to develop running estimates to guide the planning process due to their lack of current understanding of potential hazards and risks. While experts are often consulted to provide assessments of hazard impacts, not all planners have ready access to such subject-matter expertise, and, in some cases, experts may be unable to readily offer informed risk judgments (Karagiannis and Synolakis 2017). This is particularly critical in complex scenarios involving hydrogen releases, where expert insights can guide the development of effective response strategies during an emergency. For instance, if mishandled, hydrogen's propensity to ignite or react with ambient substances (e.g., oxygen, chlorine) could rapidly escalate incidents by triggering explosive chain reactions, further complicating containment efforts. Therefore, enhancing emergency managers' knowledge of the hazards and risks relevant to their jurisdictions can significantly strengthen the incident planning process.

Tasks such as identifying emergency-related demands, estimating resource capacities and mobilization, and formulating and evaluating courses of action are particularly challenging aspects of incident planning (Karagiannis and Synolakis 2017). Concurrently, emergency response guidance consistently emphasizes that prior experience significantly influences performance in such incident planning activities (Haddow et al. 2020). Consequently, the effectiveness of incident planning may be enhanced through initiatives that build practical expertise, such as incorporating targeted training into university programs and professional development courses for first responders and emergency managers. For instance, incident planning competencies can be emphasized and cultivated through a series of increasingly complex exercises, in which participants collaborate in small groups to develop operational and support plans for simulated emergency scenarios. Drills structured around hydrogen-release incidents can equip first responders to navigate the specificities and challenges inherent to the planning process, including, for example, the need to anticipate and mitigate invisible vapor cloud hazards, implement appropriate exclusion zones, and coordinate multi-agency response efforts under time pressure and uncertainty.

In addition to information gathering and course of action development and analysis, Karagiannis and Synolakis (2017) underscored the criticality of decision-making speed; by the time certain decisions were taken during emergency response exercises, they were no longer relevant. A key factor contributing to slow decision-making is the involvement of multiple agencies with shared responsibilities, as coordinating between organizations with distinct roles and priorities presents additional challenges (Butler et al. 2024). Given the increasingly multi-agency nature of today's operational environment, emergency operations plans should account for the additional time required for interagency decision-making. This consideration is particularly critical in situations that demand tightly coordinated efforts, such as evacuation planning. Taking as an example conjoint scenarios, such as the need to evacuate a flooded hydrogen refueling station located near residential and commercial zones, rapid coordination is essential among fire services, hazardous materials units, law enforcement, and local government authorities to issue timely evacuation orders, establish exclusion zones, and manage traffic rerouting. Developing a common operational picture beforehand (Wolbers and Boersma 2013) and ensuring clear communication practices (Power et al. 2024) can save valuable time by facilitating joint and parallel planning during emergencies.

Finally, a well-developed common operating picture is not only essential for synchronizing emergency response efforts but also serves as the backbone of effective interoperability. Power et al. (2024) described this interoperability between agencies as a shared socio-technical system built on trust, teamwork, communication, and flexibility—key elements in overcoming coordination challenges. Establishing cohesive goals, fostering trust, and strengthening team identities enhance interoperability and multi-agency collaboration. In this vein, it is essential to cultivate three types of trust within and across emergency response organizations: interpersonal trust, cognitive trust (confidence in others' abilities), and group-based trust (shared among the emergency management community members). Effective interoperability also depends on robust information-sharing, which helps build shared situational awareness and a common operating picture across agencies (House et al. 2014). This foundation ensures better coordination and joint decision-making.

12.5 Conclusions and Research Outlook

This chapter contributed to the academic discourse on emergency preparedness for hydrogen system accidents, with a particular focus on HRS. It synthesized recent research with practical recommendations, addressing key challenges and gaps in accident response strategies. By examining disaster preparedness and response strategies, the chapter outlined specific measures for managing incidents involving HRS, underscoring the importance of adopting a comprehensive, proactive approach to safety. A central takeaway from the discussion was the value of a capabilities-based approach to hydrogen system preparedness, which provides a structured framework for identifying and addressing potential risks. By prioritizing critical areas such as risk identification, capability development, and collaborative planning, stakeholders strengthened their preparedness and resilience to hydrogen system accidents, thus minimizing the impact on people, property, and the environment.

The chapter further emphasized that a capabilities-based approach fosters a culture of continuous improvement, allowing emergency response strategies to evolve in response to emerging risks, technologies, and lessons learned. It presented effective preparedness as an ongoing, dynamic process, benefiting from the iterative development and refinement of response capabilities. This iterative approach ensures that stakeholders remain agile and capable of adapting to changing circumstances. The active engagement of all relevant parties, from first responders to facility operators, is essential for maintaining effective emergency management systems. Continuous refinement of strategies and incorporation of new insights ensure that preparedness efforts remain aligned with the evolving landscape of hydrogen system risks.

Looking ahead, future research could focus on integrating advanced technologies such as artificial intelligence, machine learning, and real-time data analytics into emergency response frameworks for hydrogen systems. Additionally, studies could explore the effectiveness of simulation-based training, multi-agency coordination, and public communication strategies in enhancing preparedness and response to HRS accidents. These research directions have the potential to provide valuable insights that contribute to the continuous improvement of hydrogen safety standards and emergency management practices.

References

Butler P, Flin R, Bearman C, Hayes P, Penney G, McLennan J (2024) Emergency management decision-making in a changing world: 3 key challenges. Aust J Emerg Manage 39(4):23–32. https://doi.org/10.47389/39.4.23

Crichton MT, Lauche K, Flin R (2005) Incident command skills in the management of an oil industry drilling incident: a case study. J Contingencies Crisis Manage 13(3):116–128. https://doi.org/10.1111/j.1468-5973.2005.00466.x

Dadashzadeh M, Kashkarov S, Makarov D, Molkov V (2018) Risk assessment methodology for onboard hydrogen storage. Int J Hydrogen Energy 43(12):6462–6475. https://doi.org/10.1016/j.ijhydene.2018.01.195

Dawson D, Ferguson SA, Vincent GE (2021) Safety implications of fatigue and sleep inertia for emergency services personnel. Sleep Med Rev 55:101386. https://doi.org/10.1016/j.smrv.2020.101386

DSEAR (2002) The Dangerous Substances and Explosive Atmospheres Regulations (DSEAR), Pub. L. No. 2776. Health and Safety Executive (HSE), United Kingdom Statutory Instruments. https://www.legislation.gov.uk/uksi/2002/2776/contents

Ehrhart BD, Sims C, Hecht ES, Schroeder BB, Liu BR, Groth KM, Reynolds JT, Walkup GW (2024) HyRAM+ (Hydrogen plus other alternative fuels risk assessment models) (Version 5.1.1) [Computer software]. Sandia National Laboratories. https://hyram.sandia.gov

FEMA (2012) FEMA incident action planning guide. Federal emergency management agency (FEMA), United States Department of Homeland Security

FEMA (2019) Hazardous materials incidents: guidance for state, local, tribal, territorial, and private sector partners. Federal emergency management agency (FEMA), United States department of homeland security. https://www.fema.gov/sites/default/files/2020-07/hazardous-materials-incidents.pdf

FEMA (2021) National incident management system incident complexity guide: planning, preparedness and training. Federal emergency management agency (FEMA), United States department of homeland security. https://www.fema.gov/sites/default/files/documents/nims-incident-complexity-guide.pdf

Genovese M, Cigolotti V, Jannelli E, Fragiacomo P (2023) Current standards and configurations for the permitting and operation of hydrogen refueling stations. Int J Hydrogen Energy 48(51):19357–19371. https://doi.org/10.1016/j.ijhydene.2023.01.324

Genovese M, Blekhman D, Fragiacomo P (2024) An exploration of safety measures in hydrogen refueling stations: delving into hydrogen equipment and technical performance. Hydrogen 5(1):1. https://doi.org/10.3390/hydrogen5010007

Groth KM, Hecht ES (2017) HyRAM: a methodology and toolkit for quantitative risk assessment of hydrogen systems. Int J Hydrogen Energy 42(11):7485–7493. https://doi.org/10.1016/j.ijhydene.2016.07.002

Haddow GD, Bullock JA, Coppola DP (2020) Introduction to emergency management, 7th edn. Butterworth-Heinemann, ScienceDirect

House A, Power N, Alison L (2014) A systematic review of the potential hurdles of interoperability to the emergency services in major incidents: recommendations for solutions and alternatives. Cogn Technol Work 16(3):319–335. https://doi.org/10.1007/s10111-013-0259-6

Karagiannis GM, Synolakis C (2017) Twenty challenges in incident planning. J Homel Secur Emerg Manage 14(2). https://doi.org/10.1515/jhsem-2016-0061

Karagiannis GM (2024) Natech emergency preparedness and response. In: Oxford research encyclopedia of natural hazard science. Oxford University Press. https://doi.org/10.1093/acrefore/9780199389407.013.532

Krausmann E, Necci A (2021) Thinking the unthinkable: a perspective on Natech risks and Black Swans. Saf Sci 139:105255. https://doi.org/10.1016/j.ssci.2021.105255

Lee Y, Kim J, Kim J, Kim EJ, Kim YG, Moon I (2010) Development of a web-based 3D virtual reality program for hydrogen station. Int J Hydrogen Energy 35(5):2112–2118. https://doi.org/10.1016/j.ijhydene.2009.12.065

Li H, Welsh R, Morris A (2021) Emergency responders perceptions of hydrogen fuel cell vehicle: a qualitative study on the U.K. fire and rescue services. Int J Hydrogen Energy 46(65):32750–32761. https://doi.org/10.1016/j.ijhydene.2021.07.146

Nakayama J, Sakamoto J, Kasai N, Shibutani T, Miyake A (2016) Preliminary hazard identification for qualitative risk assessment on a hybrid gasoline-hydrogen fueling station with an on-site hydrogen production system using organic chemical hydride. Int J Hydrogen Energy 41(18):7518–7525. https://doi.org/10.1016/j.ijhydene.2016.03.143

OECD (2023) OECD guiding principles for chemical accident prevention, preparedness and response—Third Edition (Guidance for industry (Including management and labour), public authorities, communities, and other stakeholders 3rd edn.; Series on chemical accidents, p 180). Organisation for Economic Co-operation and Development (OECD). https://doi.org/10.1787/162756bf-en

Pacific Northwest National Laboratory (2024) H2tools best practices [Knowledge portal for hydrogen systems]. Hydrogen tools. https://h2tools.org/bestpractices/dealing-with-incidents

Perry RW, Lindell MK (2006) Emergency planning, 1st edn. Wiley

Power N, Alcock J, Philpot R, Levine M (2024) The psychology of interoperability: a systematic review of joint working between the UK emergency services. J Occup Organ Psychol 97(1):233–252. https://doi.org/10.1111/joop.12469

Sun K, Li Z (2021) Development of emergency response strategies for typical accidents of hydrogen fuel cell electric vehicles. Int J Hydrogen Energy 46(75):37679–37696. https://doi.org/10.1016/j.ijhydene.2021.02.130

Tretsiakova-McNally S, Maranne E, Verbecke F, Molkov V (2017) Mixed e-learning and virtual reality pedagogical approach for innovative hydrogen safety training of first responders. Int J Hydrogen Energy 42(11):7504–7512. https://doi.org/10.1016/j.ijhydene.2016.03.175

Tzioutzios D, Suarez-Paba MC, Luo X, Cruz AM (2024) Natech risk communication. In: Oxford research encyclopedia of natural hazard science. Oxford University Press. https://doi.org/10.1093/acrefore/9780199389407.013.529

UNDRR (2016) Report of the open-ended intergovernmental expert working group on indicators and terminology relating to disaster risk reduction (A/71/644; p 41). United Nations International Strategy for Disaster Reduction. http://www.preventionweb.net/files/50683_oiewgreportenglish.pdf

Van Niekerk D, Coetzee C, Botha D, Murphree MJ, Fourie K, Roux TL, Wentink G, Kruger L, Shoroma L, Genade K, Meyer S, Annandale E (2015) Planning and executing scenario based simulation exercises: methodological lessons. J Homel Secur Emerg Manage 12(1):193–210. https://doi.org/10.1515/jhsem-2013-0077

Weiner SC, Fassbender LL, Blake C, Aceves SM, Somerday BP, Ruiz A (2013) Web-based resources enhance hydrogen safety knowledge. Int J Hydrogen Energy 38(18):7583–7593. https://doi.org/10.1016/j.ijhydene.2012.07.028

Wolbers J, Boersma K (2013) The common operational picture as collective sense making. J Contingencies Crisis Manage 21(4):186–199. https://doi.org/10.1111/1468-5973.12027

Chapter 13
Cost-Effectiveness Evaluation for Green Hydrogen Refueling Stations

Murat Gökçek

Abstract Achieving global sustainability requires transitioning to low-carbon energy sources, improving energy efficiency, and reducing dependence on fossil fuels. The transportation sector, responsible for over one-third of CO_2 emissions from end-use sectors, plays a pivotal role in this transition. Hydrogen-powered fuel cell electric vehicles (FCEVs) offer a promising solution by providing zero-emission alternatives, especially for long-distance and heavy-load transportation. When powered by green hydrogen produced via renewable energy sources, FCEVs enable a clean energy cycle, aligning with the IEA's Net Zero Scenario targets. However, several challenges hinder widespread adoption, including high hydrogen production costs, underdeveloped refuelling infrastructure, and significant investment requirements. The interdependence between FCEVs and hydrogen refuelling stations (HRFs) is critical, as the expansion of infrastructure fosters greater adoption of hydrogen-fuelled transportation. This chapter examines the current state of FCEVs and HRFs globally, with an emphasis on hydrogen production costs, retail prices, and infrastructure challenges. A case study evaluates the economic feasibility of HRFs powered by hybrid renewable energy systems in regions with high wind and solar potential. Results highlight significant regional variations in hydrogen production costs, underscoring the importance of location and technology selection. Despite challenges, the development of cost-effective hydrogen infrastructure and renewable energy integration can accelerate the global transition to sustainable transportation. This study provides valuable insights for policymakers and investors to optimise project costs and promote hydrogen technologies as a cornerstone of a fossil fuel-free transportation future.

M. Gökçek (✉)
Department of Mechanical Engineering, Faculty of Engineering, Niğde Ömer Halisdemir University, Niğde, Türkiye
e-mail: mgokcek@ohu.edu.tr

Y. Liu et al. (eds.), *Safety Measures for Hydrogen Fueling Stations*, Risk, Reliability and Safety Engineering, https://doi.org/10.1007/978-981-96-9412-9_13

13.1 Introduction

Global sustainability requires reducing greenhouse gas emissions to limit climate change worldwide, ensuring humanity's access to modern and affordable energy services, improving indoor and outdoor air quality, and strengthening energy security while reducing risks associated with energy supply and use. Building a sustainable global energy system requires significant advances in energy efficiency as well as a shift from fossil fuels to low-carbon energy sources, especially renewable energy (Greene et al. 2020). Transportation plays a vital role in the orderly functioning of modern societies, and the key element of this process is energy. Without energy, it is impossible to move anything from one place to another. Today, the majority of transportation systems around the world rely on fossil fuels for their energy needs. In particular, 96% of the energy used in transportation still relies on fossil oil resources. This dependence makes it difficult to achieve global sustainability goals and poses a major challenge in terms of energy security and environmental impacts. Motor vehicles in land, maritime and air transport continue to rely on internal combustion engines, often powered by fossil fuels, and the transport sector has a significant environmental impact, accounting for more than one-third of CO_2 emissions from end-use sectors. Aligning transport with the IEA's Net Zero Scenario requires the implementation of a comprehensive set of policies that promote a transition to low-carbon travel options. It also requires the implementation of operational and technical energy efficiency measures to reduce carbon intensity across all modes of transport. The Net Zero Scenario requires transport sector emissions to fall by around a quarter by 2030, even as transport demand continues to grow. Policies need to encourage a shift to less carbon-intensive travel options such as walking, cycling and public transport, as well as more efficient technologies such as electric vehicles (IEA). Hydrogen is one of a limited number of energy carriers with the potential to meet a large share of global transportation energy demand with near-zero carbon emissions. These energy carriers include electricity, biofuels and synthetic liquid fuels. However, all sustainable energy sources face various challenges, and hydrogen is not exempt. While each alternative has its own advantages, each has to overcome different challenges. However, the likelihood of a successful global transition to sustainable energy will be maximized if a broad portfolio of these promising options is pursued strategically.

Fuel cell electric vehicles (FCEVs) have a critical role in transportation in line with the Net Zero Scenario. FCEVs offer a zero-emission transportation alternative, reducing reliance on fossil fuels. These hydrogen-powered vehicles emit only water vapor during operation and significantly reduce greenhouse gas emissions (Genovese et al. 2023). It stands out as a complementary technology to battery electric vehicles by offering the advantage of high range and efficiency, especially in long-distance transportation and heavy load transportation segments. When using green hydrogen produced from renewable energy sources, fuel cell vehicles make a significant contribution to reducing carbon intensity in the transportation sector by creating a completely clean energy cycle (Gökçek et al. 2024). Thus, FCEVs are considered as a strategic component in both strengthening energy security and ensuring the alignment of transportation with Net Zero targets.

Hydrogen can be produced by electrolysis of water using renewable energy sources, making it a "green" and "clean" energy carrier as soon as it is produced (Acar and Dincer 2019). In this context, networks of hydrogen refueling stations (HRFs) need to be developed to support this new mobility concept. By enabling the supply of hydrogen to FCEVs, these stations form key elements of a critical distribution infrastructure supporting fuel cell (FC) based sustainable mobility (Apostolou and Xydis 2019). However, the successful implementation of this technology has been limited by several challenges that hinder its sustainable development. Major drawbacks include a more limited and inadequate infrastructure compared to refueling infrastructures for petroleum or battery electric vehicles, costly investment requirements, and a lack of defined regulations, norms and standards. FCEVs play an important role in future transportation systems by providing zero-emission transportation solutions. The proliferation of these vehicles is directly related not only to the development of the technology, but also to the availability of appropriate refueling infrastructure to support these vehicles. This illustrates the interdependence between fuel cell vehicles and the installation of HRFs, similar to the "chicken and egg" paradox (Shi et al. 2020). The more widespread the infrastructure becomes; the more fuel cell vehicles will be preferred. In this context, the economically sustainable operation of hydrogen refueling stations becomes a critical issue for investors and policy makers. Expanding the infrastructure will enable the effective use of these tools and the widespread adoption of the technology. However, the installation and operation of stations faces challenges such as high initial costs and the right sizing of system components. Moreover, while the production of hydrogen with renewable energy sources is an important strategy to improve the environmental sustainability of these stations, ensuring cost efficiency is also a major obstacle. This chapter examines the current state of fuel cell vehicles and refueling stations, hydrogen production costs and retail prices at refueling stations around the world. It also discusses a case study that aims to evaluate the design and economic efficiency of hydrogen refueling stations supported by renewable energy sources (especially wind and solar energy). In this context, it is aimed to contribute to the establishment of a fossil fuel-free transportation infrastructure by examining the feasibility of independent hydrogen refueling stations powered by hybrid renewable energy systems in regions with high solar and wind energy potential.

13.2 Current Status of Fuel Cell Vehicles and Hydrogen Refueling Stations

Fuel cell vehicles, which support environmental sustainability by offering zero-emission transportation solutions and have the potential to reduce carbon emissions by reducing dependence on fossil fuels worldwide, and the infrastructure of hydrogen refueling stations that fuel these vehicles are not widespread enough on a global scale and the cost of hydrogen production can be listed as the main obstacles to the adoption of hydrogen-fueled transportation. In this context, in order for hydrogen fueled transportation to become widespread, a structure consisting of three basic components including station infrastructure, vehicle availability and cheap hydrogen production cost needs to be developed. Due to their high energy density, versatile deployment, fast refueling and zero emission characteristics, fuel cell vehicles have long been part of the innovation efforts of global automakers such as Mercedes-Benz, Honda, Toyota, Hyundai, and Mercedes-Benz, and many commercialized models such as Toyota Mirai, Hyundai NEXO, Honda Clarity, Honda CR-V have emerged. According to the 2024 update, the Advanced Fuel Cell Technology Collaborative Program (AFC TCP) data survey reports that the number of operational fuel cell vehicles worldwide reached 88,025 by the end of 2023 (Grube and Rex 2024). South Korea leads the global fuel cell vehicle distribution with 39% market share, followed by China with 23%. The United States and Japan are the third and fourth largest markets for fuel cell vehicles, with 21% and 11% shares respectively. Figure 13.1 shows the distribution of fuel cell vehicles on the roads by country at the end of 2023. Among European countries, Germany has the largest fleet with more than 2300 vehicles. France is the second European country with a fleet of more than 1000 vehicles, while the Netherlands, the United Kingdom, Switzerland, Norway, Norway, Denmark, Poland and Belgium also have fleets of more than 100 vehicles.

FCEVs fall into categories based on passenger carrying capacity, tonnage and other different characteristics. These are passenger cars, minibuses and commercial vehicles, buses, light goods vehicles, medium goods vehicles and heavy goods vehicles. Figure 13.2 shows the shares of different vehicle categories by the end of 2023. In the global fuel cell vehicle fleet, passenger cars are the dominant category, representing 75% of the total. This means that two out of every three FCVs registered are passenger cars. Buses represent the second highest share at 9%, followed by heavy trucks at 8% and medium trucks at 5%. Light commercial vehicles have a very small share.

Fig. 13.1 Country-by-country distribution of fuel cell vehicles

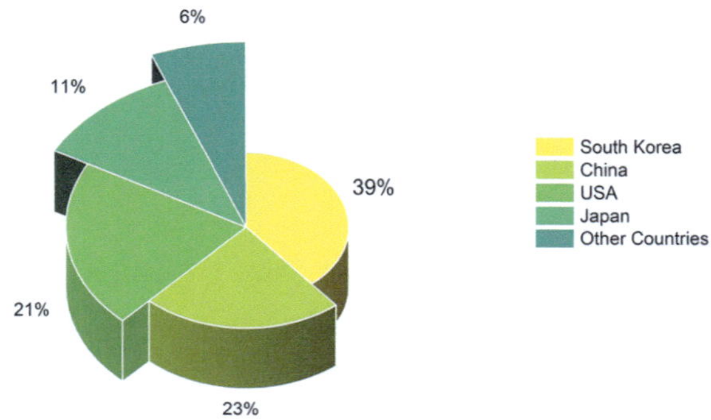

Other Countries: Germany, France, Netherlands, United Kingdom, Switzerland, Norway, Denmark, Poland, Belgium, Canada, Australia, India, Austria, Italy, Spain, Sweden, Iceland, Czech Republic, Latvia, Luxembourg, Costa Rica, Portugal, Slovakia, Estonia, Finland, Brazil, Bulgaria, Lithuania, Romania.

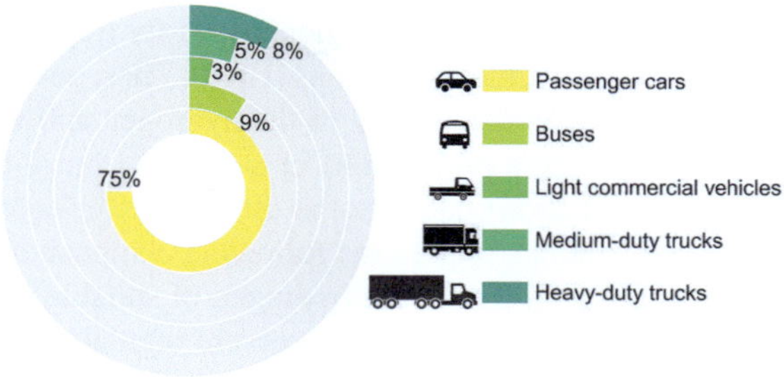

Fig. 13.2 Shares of different fuel cell electric vehicle categories by the end of 2023

An analysis of the regional distribution of fuel cell vehicles shows that the majority of vehicles are concentrated in Asia. Asia is the leader in this field, accounting for around 75% of the global total. North America holds more than 20% of the global market, while Europe has a more limited share of 7%. Australia, on the other hand, has a very small share of 0.3%. At the end of 2023, the number of fuel cell passenger cars was 66,065. Two-thirds of passenger cars are located in Asia. North America hosts more than a quarter of passenger cars, representing a higher proportion compared to the overall distribution. The share of passenger cars in Europe has increased slightly but remains relatively low as a share of the overall population. By the end of 2023, South Korea is the world leader with 33,600 fuel cell passenger cars. This clearly demonstrates South Korea's strong investments in fuel cell vehicle technology and its market size in this field. The US ranks second with 17,900 vehicles, showing that North America is one of the major markets for this technology. Japan ranks third with 7900 vehicles and is another major player in Asia. In Europe, Germany leads the way with 2100 vehicles, but this number is quite low compared to the major markets in Asia and the Americas. Countries such as France, China and the Netherlands have between 1000 and 2000 fuel cell vehicles and are emerging markets for this technology. In other European countries and regions such as Australia, the number of these vehicles is very limited. Overall, South Korea, the US and Japan stand out as the global leaders in this technology, while in Europe the use of fuel cell vehicles remains more limited. When the distribution of fuel cell light commercial vehicles (FC LCVs) by country is analyzed, China has by far the largest number of fuel cell LCVs with 2700 vehicles. Japan ranks second with around 100 vehicles, while the number of fuel cell LCVs in India, Germany, the Netherlands, Switzerland, the United Kingdom, Denmark, Belgium, Costa Rica, Denmark, Belgium, Costa Rica and Canada is quite low and almost at the same level. China appears to be a significant leader in this category. China is the clear leader in the number of fuel cell buses with 6632 thousand fuel cell buses. South Korea is in second place with 651 buses. Germany, Japan, the US and the UK have 178, 149, 100 and 90 fuel cell buses respectively. Other countries lag far behind in the number of fuel cell buses, including Norway, India, France and the Netherlands. In terms of the country-by-country distribution of medium-duty fuel cell trucks, China leads the way with 3900 vehicles. For heavy-duty trucks, China again leads with 6500 vehicles.

One of the most critical elements used for transportation in hydrogen infrastructure is hydrogen refueling stations (HRS). These stations can be classified depending on the production technology of the hydrogen and whether the fuel is produced at the station (on-site production) or supplied from a centralized production facility (off-site production). Despite the differences in production and delivery methods, most retail stations require the following components: a hydrogen production unit, a purification unit to provide hydrogen of suitable purity for fuel cells, a hydrogen compressor, storage tanks, gas pressure regulator, refrigeration unit, safety equipment and dispensers (Qin et al. 2014). Off-site hydrogen refueling stations include all stations where hydrogen is transported from a centralized production facility by road transport or via specific pipelines. Transportation of hydrogen by road is carried out by heavy-duty trucks, which store hydrogen as compressed gas (GH_2) in tube trailers under pressure above 180 bar or as liquid (LH_2) in special tanks at cryogenic temperatures of -253 °C for long distances (Apostolou and Xydis 2019; Qin et al. 2014). However, due to the high amount of energy required to liquefy hydrogen, this method is much more costly than transporting gaseous hydrogen and is therefore not currently widely used to supply hydrogen refueling stations. On the other hand, gaseous hydrogen is first produced in large industrial plants at relatively low pressures of around 30 bar and then compressed to around 180–200 bar for transportation by truck or 70 bar for transportation by pipeline (Qin et al. 2014; Gillette and Kolpa 2008). Hydrogen transportation through the existing pipeline network is considered a low-cost solution, but the construction of new pipelines limits expansion due to high capital requirements. The potential for hydrogen to cause embrittlement of steel and welding materials used in pipelines is considered a serious safety concern. Furthermore, the high permeability and leakage problems of hydrogen are an important consideration for the safety of pipeline systems. To overcome these technical challenges, the use of fiber-reinforced polymer (FRP) pipelines, which are lower cost and easier to install compared to steel, is proposed (DOE 2024a).

The second type of HRSs is operationally similar to the first, but hydrogen is produced locally. The most common technologies for on-site production are steam methane reforming (SMR) and water electrolysis. On-site HRSs offer capacity limitations ranging from 100 kg/day to 1000 kg/day, depending on the output of the H_2 generators (Apostolou and Xydis 2019; Qin et al. 2014). However, the most common systems produce 100 kg H_2 per day. In steam-methane reforming, methane is reacted with steam under 3–25 bar pressure in the presence of a catalyst to produce hydrogen, carbon monoxide and a small amount of carbon dioxide. As this process is endothermic, external heat must be supplied to the process to sustain the reaction. Then, in the so-called "water–gas shift reaction", carbon monoxide and steam react with the help of a catalyst to produce carbon dioxide and more hydrogen. The final stage of the process, "pressure swing adsorption", removes carbon dioxide and other impurities from the gas stream, leaving largely pure hydrogen. Steam reforming can also be used to produce hydrogen from other fuels such as ethanol, propane and even gasoline (DOE 2024b). Water electrolysis is a process in which water is decomposed into hydrogen and oxygen by applying direct current voltage. The electrolysis technologies commonly used in HRSs are alkaline and proton exchange membrane (PEM) electrolysis (Apostolou and Xydis 2019; Gökçek and Kale 2018).

The main differences between the two main types of HRS stem from the hydrogen source used during refueling. In terms of the hydrogen storage layout of the HRS, off-site stations are simpler and in addition to the infrastructure required to receive only the delivered hydrogen, the system focuses more on the compression and refueling processes. The capacity of such stations can be adjusted according to demand, depending on storage possibilities. Today, most stations have capacities between 100 kgH_2/day and 520 kgH_2/day for compressed gaseous hydrogen (GH_2), while in the case of liquid hydrogen (LH_2) storage this capacity increases to over 1000 kgH_2/day (Apostolou and Xydis 2019). On-site hydrogen producing refueling stations include the necessary equipment for the production and processing of H_2 gas prior to its initial storage. This equipment consists of a purification system and a hydrogen generator (methane steam reformer or water electrolysis unit) to increase the purity of the hydrogen fuel to acceptable levels (i.e. fuel index higher than 99.97%) in accordance with ISO standards. While most of the existing on-site HRSs offer capacities of 100 kgH_2/day to 400 kgH_2/day, it is planned to increase the capacity up to 8000 kgH_2/day for future HRSs serving fuel cell trucks (Apostolou and Xydis 2019). In financial terms, on-site hydrogen production stations require a much higher capital cost due to additional hydrogen production components. In particular, for HRSs with on-site H_2 production by water electrolysis, the investment cost can be 1.5 times higher compared to an off-site station of the same capacity. Therefore, the cost of levelized hydrogen fuel is currently higher at stations with on-site production and in some cases can be up to double the cost at off-site stations (Apostolou and Xydis 2019).

Figure 13.3 shows the distribution of hydrogen refueling stations operating worldwide by region in 2023 (Grube and Rex 2024). Asia ranks first with 71.5% and 871 stations, reflecting investments in hydrogen infrastructure, particularly in Japan and South Korea. Europe comes second with 22.7% and 276 stations, making significant strides in the transition to a hydrogen economy. North America ranks third with 5.1% and 62 stations, but infrastructure is mostly concentrated in certain regions, such as California. In Latin America and Australia, on the other hand, there are very few hydrogen stations and hydrogen infrastructure in these regions is still in its infancy. Overall, it can be observed that hydrogen refueling stations show a large geographical variation and that hydrogen infrastructure is more widespread, especially in Asia and Europe. This distribution reflects the strategic approaches of countries and regions towards the hydrogen economy and their investments in renewable energy infrastructure.

13.3 Cost of Hydrogen Production

The cost of hydrogen production varies greatly depending on the production methods used and regional energy costs. The levelized costs of hydrogen production according to the technology applied in Europe in 2023 are shown in Fig. 13.4 (EHO 2024).

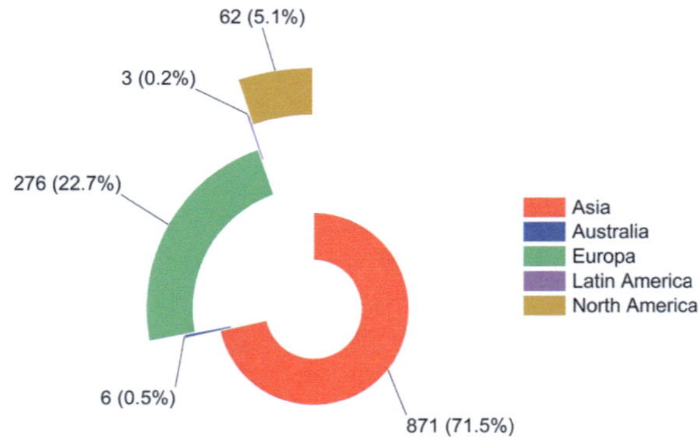

Fig. 13.3 Distribution of hydrogen fueling stations in operation by country

The technologies considered here are Steam Methane Reforming (SMR), SMR with carbon capture, grid connected electrolysis, water electrolysis with direct connection to a renewable energy source (renewable hydrogen). The levelized cost (LCOH) of hydrogen production by steam reforming (SMR) is 3.75 EUR/kg, of which 74.9% is due to natural gas, 15.7% to capital expenditure (CAPEX) and 6.9% to other operational expenditure (OPEX). With the integration of carbon capture technology, the cost of hydrogen produced by this method increased to 4.41 EUR/kg, with natural gas accounting for 68.7%, CAPEX 18.6% and CO_2 costs 9.8% of the cost breakdown. The electrolysis method powered by grid electricity has a higher cost structure with LCOH reaching 7.94 EUR/kg, where electricity costs account for 43.2% of the total cost, CAPEX for 34.8% and other OPEX for 13.1%. In the case of hydrogen production by electrolysis with direct connection to a renewable energy source, the cost is calculated at EUR 6.61/kg, with CAPEX accounting for 57.3% and electricity costs 39.6% of the total cost. This graph shows the different distribution of cost components in each production technology, providing an understanding of the determinant cost drivers for different technologies.

Figure 13.5 presents the levelized production costs (LCOH) of hydrogen in different countries in Europe, broken down by production technology (EHO 2024). The SMR method stands out as the lowest cost hydrogen production technology in many countries. Especially in countries where natural gas prices are relatively low, SMR offers an advantage over other methods. However, the addition of carbon capture technology increases the production cost, which in many countries is higher than the SMR method. Carbon capture reforming is a preferred option in some countries to reduce carbon emissions, but it comes at an additional cost. Grid-connected electrolysis is characterized by higher production costs across Europe. Since the cost of this method is highly dependent on electricity prices, LCOH rises significantly in countries with high electricity prices. This leads to significant cost differences between countries in hydrogen production dependent on grid electricity. The electrolysis method, which is directly linked to renewable energy sources, is considered a relatively more advantageous option in countries with high renewable energy potential.

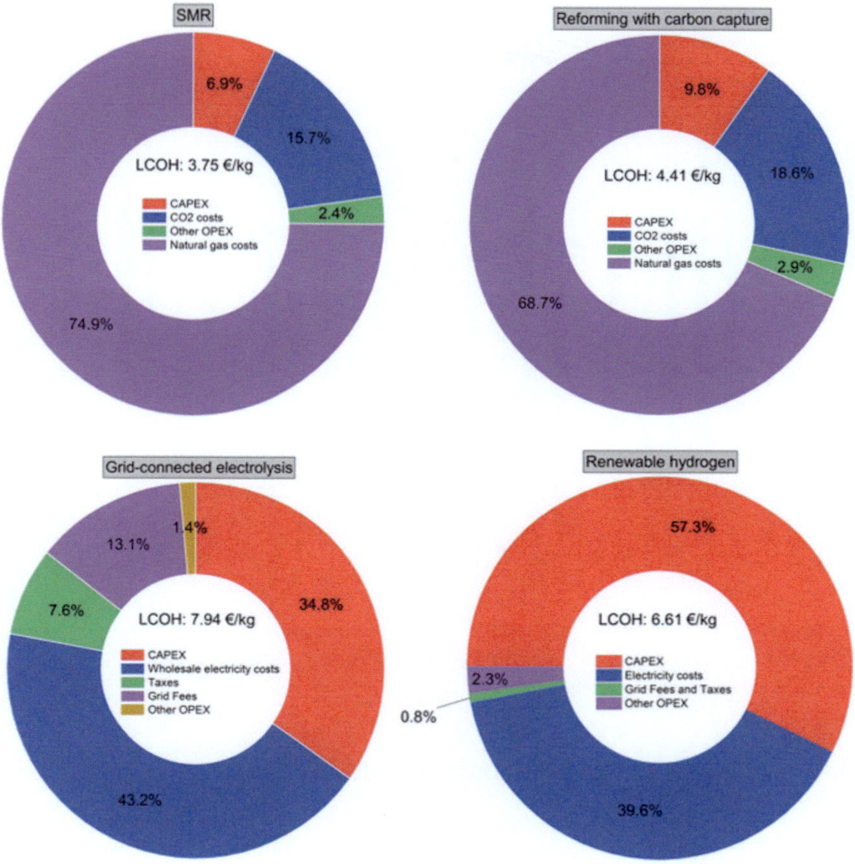

Fig. 13.4 Hydrogen production costs by technology used in Europe

This method offers a competitive cost structure, especially in countries where the cost of electricity from sources such as solar or wind is low. However, the cost of this method also varies from country to country and depends on the efficiency and availability of renewable resources. In countries where electricity from renewable sources is more abundant and cheaper, it can be more cost-effective, while in regions where renewable resources are limited, the cost increases. Overall, the graph shows that hydrogen production costs vary considerably across European countries and that these differences are shaped by the production method, energy sources and energy prices. This emphasizes the need for each country to choose the most appropriate hydrogen production technology according to its energy resources and price structure.

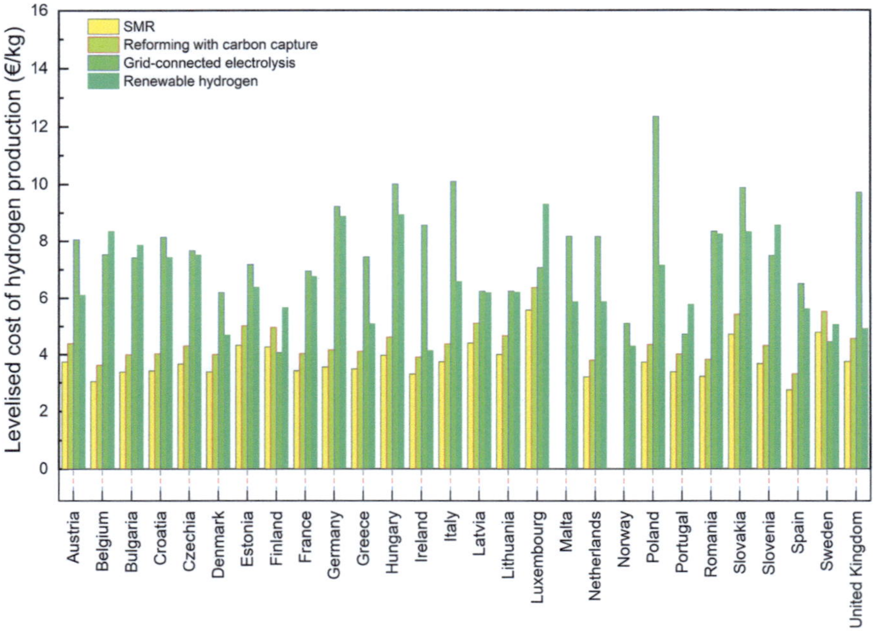

Fig. 13.5 Levelized production costs (LCOH) of hydrogen in different countries in Europe by production technology

13.4 Retail Prices at Hydrogen Refueling Stations

Retail prices at hydrogen fueling stations vary from country to country, depending on energy costs, local production methods and government incentives. Hydrogen prices are usually calculated per kilogram (kg). Figure 13.6 shows the retail prices of hydrogen at hydrogen refueling stations in different countries in Europe (H2.LIVE 2024). In some countries, such as Poland, Norway, Lithuania, Lithuania, Switzerland, the Czech Republic and Sweden, there is a single set price value for hydrogen. For example, in Norway the price is the highest at €25.35/kg, while in Sweden it is the lowest at €7.87/kg. On the other hand, prices vary widely in countries such as the Netherlands, France, Belgium and Germany. In Germany in particular, the price of hydrogen ranges from €9.50/kg to €24.90/kg (H2.LIVE 2024). The average price per kilogram of hydrogen fuel at public refueling stations in California-USA is $32.94, more than four times the average price of H_2 fuel in South Korea and Japan, at $7.18/kg and $7.71/kg respectively (Hydrogen Insight 2024). The largest integrated green hydrogen production and refueling complex in China can supply hydrogen at $4.86 per kilogram, almost on par with diesel. These price differences can be attributed to factors such as hydrogen production methods, energy sources, infrastructure costs and government incentives.

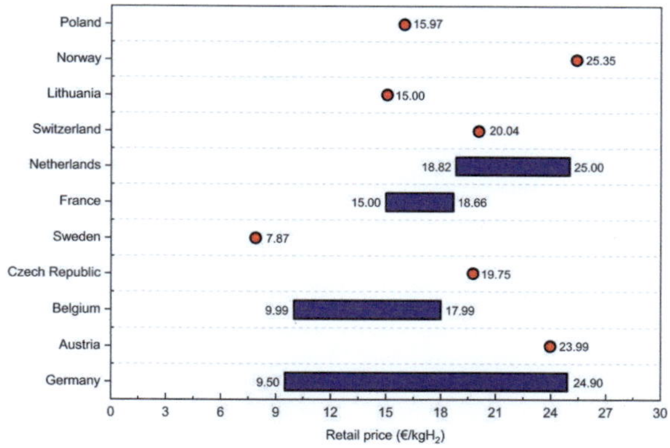

Fig. 13.6 Retail prices at hydrogen fueling stations by different countries in Europe

13.5 Case Study: LCOH Calculations for On-Site Hydrogen Refueling Stations

This case study presents a techno-economic analysis that calculates the LCOH and assesses the feasibility of stand-alone hydrogen refueling stations powered by hybrid renewable energy generation systems at selected locations in Türkiye, Norway, Poland, Spain and Japan, with the goal of ending the use of fossil fuels in the transportation sector and promoting the proliferation of fuel cell vehicles. The study considers the cost of hydrogen brought to a value by considering renewable power systems and hydrogen refueling stations together.

13.5.1 Locations and Renewable Resource Data Assessments

Location information about the locations considered in this case study is given in Table 13.1. Istanbul in Türkiye, Ørve Ervik nord in Norway, Warsaw in Poland, Shizuoka in Japan and Almeria in Spain were selected as potential locations for on-site hydrogen refueling stations that meet energy needs from renewable power systems.

Table 13.1 Data on the selected locations

Location	Coordinate values		Altitude (m)
İstanbul-Türkiye	41°7.2'N	28°3.6'E	217
Ørve Ervik nord-Norway	60°26.7'N	5°19.5.1'E	42
Warsaw, Poland	52°11.51'N	21°04.26'E	100
Shizuoka-Japan	34°39.1'N	138°10.6'E	30
Almeria-Spain	36°52.3'N	2°19.4'W	150

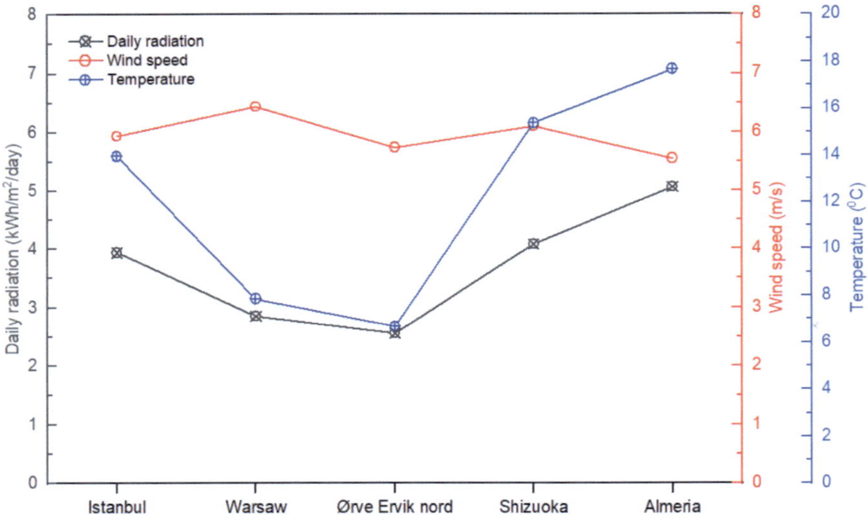

Fig. 13.7 Annual average meteorological data at the investigated locations

In this study, renewable energy data for the locations where the hydrogen station is considered to be established are taken from the NASA Prediction of Worldwide Energy Sources database (HOMER Energy 2024). Figure 13.7 shows a numerical comparison of annual average meteorological data for five different locations.

In terms of daily radiation, Almeria has the highest value with 5.06 kWh/m^2/day, while Ørve Ervik Nord shows the lowest value, 2.56 kWh/m^2/day. When comparing wind speeds, the highest value is observed in Warsaw with 6.44 m/s and the lowest in Almeria with 5.55 m/s. In terms of temperature, Almeria has the highest value of 17.68 °C, while Ørve Ervik Nord has the lowest value of 6.68 °C. These data show that each location exhibits significant differences in daily radiation, wind speed and temperature parameters, providing important information for the potential assessment of renewable energy systems.

13.5.2 Proposed Refueling Station and Hydrogen Load

The main elements of on-site hydrogen refueling station driven by renewable power system considered in this study are presented in Fig. 13.8. The station consists of wind turbines generating electricity from wind power, solar panels generating electricity from solar energy, electrolyzers for water separation, power converters, hydrogen compressor, hydrogen storage tanks, hydrogen cooling unit and a hydrogen dispenser.

Conventional fuel cell vehicles usually have a hydrogen storage capacity of about 5 kg, which allows them to travel about 500 km. Therefore, this study considers the design of a hydrogen refueling station equipped with renewable energy sources

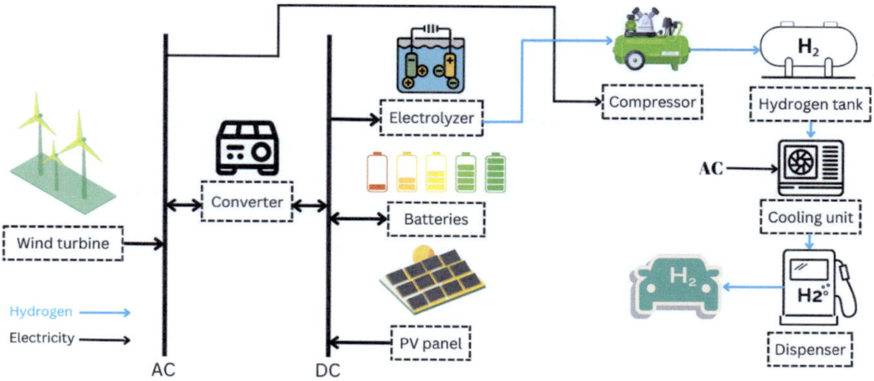

Fig. 13.8 Main elements of on-site hydrogen refueling station driven by renewable power system

to meet a daily demand of 120 kg of hydrogen. With this production capacity, the hydrogen station has the potential to meet the hydrogen demand of 24 fuel cell vehicles per day. The hydrogen load at the hydrogen station is shown in Fig. 13.9. The load estimation is based on an hourly demand assumption.

In this study, hydrogen production costs at hydrogen refueling stations supported by renewable energy sources are calculated using the methodology and equipment given in reference by Gökçek et al. (2024). In the considered on-site production hydrogen refueling stations, the cost of hydrogen brought to a value (LCOH) can be calculated using Eq. (13.1).

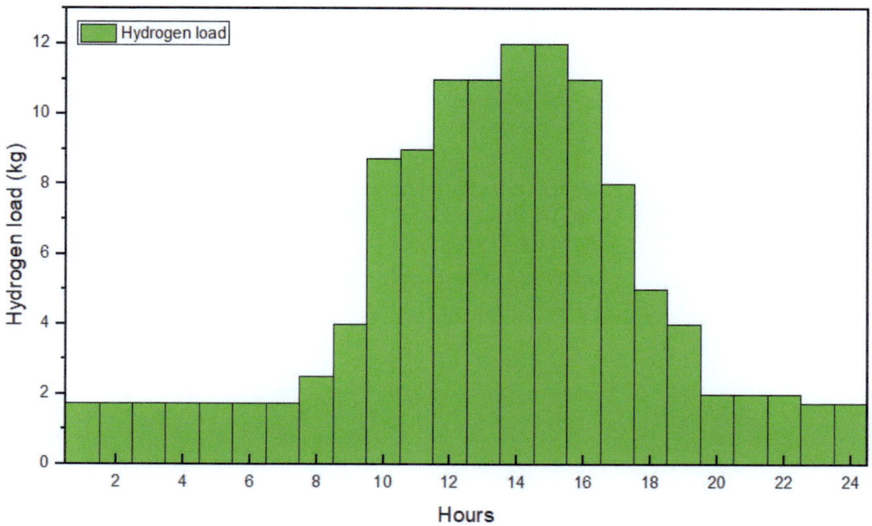

Fig. 13.9 Hydrogen load at the hydrogen refueling station

$$LCOH = \frac{C_{ann}}{M_{hydrogen}} (\$/kg) \tag{13.1}$$

C_{ann} is the total annual cost ($/year) of the entire system (power generation unit, hydrogen production, compression, cooling and distribution units) including capital, replacement, operation and maintenance costs. $M_{hydrogen}$ is the annual hydrogen production amount (kg/year). Table 13.2 summarizes the technical and economic characteristics of the on-site hydrogen refueling station components (Gökçek et al. 2024).

Table 13.2 Technical and economical characteristics of proposed hydrogen refueling station

System component	Parameters	Value
PV panel	Rated power (kW)	1
	Lifetime (years)	30
	Initial investment cost ($/kW)	650
	Operation and maintenance cost ($/year)	10
Wind turbine	Rated power (kW)	800
	Lifetime (years)	30
	Capital cost ($/kW)	2000
	Operation and maintenance cost ($/year)	32,000
Batteries	Nominal capacity (kWh)	1
	Capital cost ($/kWh)	110
	Operation and maintenance cost ($/year)	10
	Replacement cost ($/kWh)	100
Electrolyzer	Lifetime (years)	15
	Capital cost ($/kW)	1200
	Operation and maintenance cost ($/year)	200
	Replacement cost ($/kW)	1000
Converter	Lifetime (years)	15
	Capital cost ($/kW)	700
	Operation and maintenance cost ($/year)	10
	Replacement cost ($/kW)	630
Compressor	Nominal capacity (kW)	50
	Capital and replacement cost ($/kW)	2500
	Lifetime (years)	10
Cooling unit	Capital and replacement cost ($/unit)	12,000
	Lifetime (years)	10
Tank	Capital cost ($/kg)	1500
	Replacement cost ($/kg)	1000
	Operation and maintenance cost ($/year)	3
	Lifetime (years)	15
Dispenser	Capital cost ($/unit)	54,000
	Replacement cost ($/unit)	45,000
	Lifetime (years)	10

The results of the optimal sizing and techno-economic analysis of the on-site hydrogen station supported by renewable energy sources (wind and solar) are presented in Table 13.3.

The highest PV (photovoltaic) panel capacity is in Ørve Ervik Nord (947 kW) and the lowest in Shizuoka (511 kW). According to the optimal system design, there is 1 wind turbine in the hybrid power system created in all locations. The battery capacities, which are determined by considering the energy storage needs of the systems, are the highest in Shizuoka (266 kWh) and the lowest in Almeria (144 kWh). This difference shows that energy storage requirements may vary according to the continuity and seasonal variability of solar and wind resources. Electrolyzer power was set at 500 kW at all locations. Hydrogen storage tank capacity was highest at Ørve Ervik Nord (600 kg), while tank capacities at other locations ranged between 300–500 kg. The annual electricity production of the PV systems was highest in Almeria, reaching 1,107,069 kWh. This shows that Almeria has a high solar energy potential and although the installed capacity of the PV system is smaller than in some other locations, the amount of electricity produced is higher. In Warsaw, the annual PV production was the lowest at 539,993 kWh, which is due to lower solar radiation values. The capacity factor of the PV system created in Almeria is 19.2%, while in Warsaw it is 11.4%. The annual electricity generation from wind turbines was highest in Warsaw (3,135,732 kWh), indicating that the wind potential of this region is higher than in other locations, based on the average annual wind speed. In Almeria, the annual production was the lowest at 2,489,981 kWh, indicating that wind speeds in Almeria are lower than in other regions. The capacity factor of the wind turbine in Warsaw was 44.7%, while in Almeria it was 35.5%. From the economic analysis, the net present cost was found to be the highest in Ørve Ervik Nord ($6,855,951) and the lowest in Almeria ($5,782,128). These values reflect the total installation and operating costs of the systems. The higher cost of Ørve Ervik Nord is due to

Table 13.3 Optimal results of the hybrid power system for considered locations

Technical and economical parameters	Istanbul	Ørve Ervik nord	Warsaw	Shizuoka	Almeria
PV panel (kW)	675	947	541	511	659
Wind turbine (kW)	1 × 800	1 × 800	1 × 800	1 × 800	1 × 800
Battery (1 kWh LA)	149	165	153	266	144
Converter (kW)	378	370	408	394	341
Electrolyzer (kW)	500	500	500	500	500
Hydrogen tank (kg)	300	600	300	500	300
PV system electricity (kWh)	855,881	857,596	539,993	697,734	1,107,069
Wind turbine Electricity (kWh)	2,778,227	2,604,404	3,135,732	2,881,139	2,489,981
Net present cost ($)	5,857,537	6,855,951	5,782,530	6,297,177	5,782,128

the larger PV panel and hydrogen tank capacity as well as higher maintenance and operational expenses.

Figure 13.10 shows the annual cost breakdown of the optimized hydrogen refueling station for the five different locations considered. For the on-site hydrogen refueling station, this breakdown includes the wind turbine, solar panels, batteries, converter, electrolyzer, hydrogen tank and additional equipment including compressor, cooling unit and dispenser. The annual cost of the wind turbine in the optimized system is calculated as 90,267.59$. The PV cost is lowest in Shizuoka ($17,208.85) and highest in Ørve Ervik nord ($31,898.96). Battery cost is highest in Shizuoka ($5,502.56) and lowest in Almeria ($2,978.83). Converter cost is highest in Warsaw ($23,084.94) and lowest in Almeria ($19,289.70). The electrolyzer cost was calculated at $48,544.36 in all locations. Hydrogen tank cost is highest in Ørve Ervik nord with $54,608.33 and lowest in Istanbul, Warsaw and Almeria with $27,304.17. The annual cost of additional equipment including compressor, refrigeration unit and dispenser is calculated as $21,418,221. This cost analysis is important to understand the regional variations of hydrogen production and storage processes based on local renewable resource potential and to develop cost-effective sustainable energy systems. The cost data provided includes cost breakdowns of various components and shows in which items regional differences are concentrated. Furthermore, this cost distribution shows the importance of strategic positioning in terms of the financial sustainability of hydrogen refueling stations.

Figure 13.11 shows the LCOH values for on-site hydrogen refueling stations supported by renewable power systems in different locations. The lowest LCOH value is calculated in Warsaw with $5.275/kg, while the highest value is calculated in Ørve Ervik Nord with $6.157/kg. This difference directly affects the cost of hydrogen production at each location, depending on the electricity generation costs and changes in the system configuration.

13.6 Concluding Remarks

Fuel cell vehicles contribute to environmental sustainability by offering zero-emission transportation solutions. These vehicles, which have the potential to reduce carbon emissions by reducing dependence on fossil fuels, stand out with their high energy density, versatile distribution, fast refueling and zero emission advantages. However, there are serious obstacles to the adoption and diffusion of this technology. Globally underdeveloped hydrogen refueling infrastructure and high hydrogen production costs are the biggest challenges to the rapid adoption of hydrogen-fueled transportation. At the end of 2023, 88,025 fuel cell vehicles were in operation worldwide, with most of these vehicles concentrated in Asia, particularly in South Korea and China. While North America has a significant share, Europe is developing in this area, led by Germany and France. In vehicle categories, passenger cars are dominant, with China leading in the bus and truck segments. Global deployment of the technology may accelerate due to infrastructure development and cost reductions.

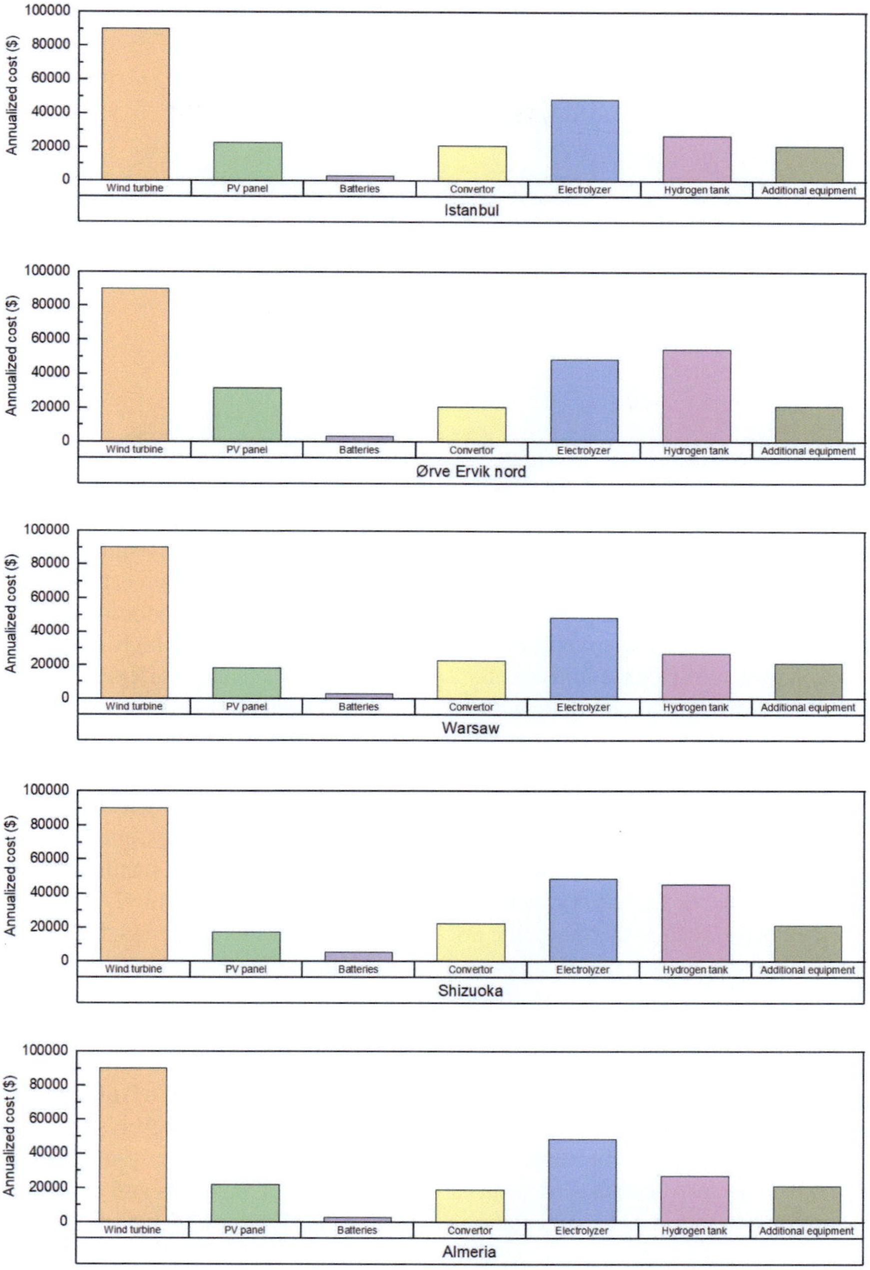

Fig. 13.10 The annualized cost distribution of the hydrogen fueling station for considered locations

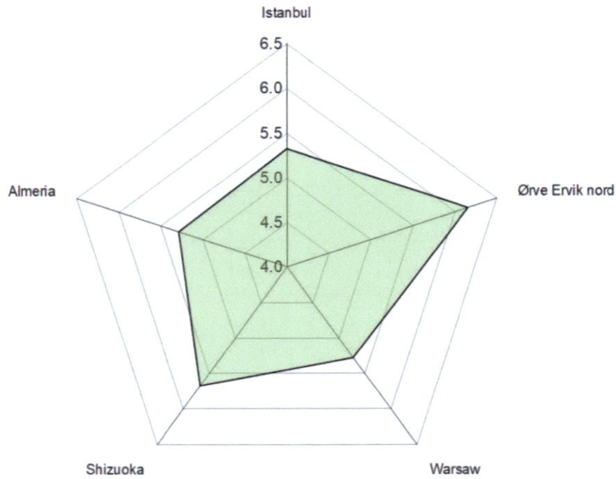

Fig. 13.11 LCOH values calculated for different locations

In Europe, hydrogen production costs vary widely depending on the technology used and the energy sources of the countries. While Steam Methane Reforming (SMR) is the most cost-effective method in most countries, SMR with carbon capture integrated is preferred for emission reduction purposes but comes with additional costs. Grid-connected electrolysis is more costly across Europe due to high electricity prices. The electrolysis method directly linked to renewable energy offers a competitive cost, especially in countries with low energy costs such as solar or wind. This diversity in production costs makes it necessary for countries to determine the most appropriate hydrogen production technology according to their energy resources and cost structures. Retail prices at hydrogen refueling stations vary widely from country to country, and these price differences are based on production methods, energy costs, infrastructure costs and government incentives. In Europe, Norway stands out with the highest price of €25.35/kg, while Sweden has the lowest price of €7.87/kg. In some countries, such as Germany, the price range is quite wide. In the US, California has a high average price of $32.94 per kilogram, while South Korea and Japan have lower prices. In China's green hydrogen production and supply complex, the price of hydrogen is $4.86/kg, close to the price of diesel fuel. This variation in prices is shaped by many factors such as regional energy costs and government support. A case study calculating the levelized cost of hydrogen production (LCOH) of independent hydrogen refueling stations powered by hybrid renewable energy systems was conducted for five different locations. The optimal renewable power system for the locations considered consists of wind turbine, solar panel, batteries and converter. The optimal system is based on hydrogen load demand. The hydrogen costs brought to a value are calculated as 5.331 $/kg for Istanbul, 6.157 $/kg for Ørve Ervik nord, 5.275 $/kg for Warsaw, 5.672 $/kg for Shizuoka and 5.287 $/kg for Almeria. Such regional cost differences reveal the importance of choosing the most suitable location for the installation of hydrogen fueling stations. Furthermore, such analyses can guide decision makers in identifying cost-cutting strategies for the deployment of hydrogen technologies. This study will be useful to better understand the impact of strategic decisions such as location selection and component-based cost optimization on the total cost structure of projects.

References

Acar C, Dincer I (2019) Review and evaluation of hydrogen production options for better environment. J Clean Prod 218:835–849. https://doi.org/10.1016/j.jclepro.2019.02.046

Apostolou D, Xydis G (2019) A literature review on hydrogen refueling stations and infrastructure. Current status and future prospects. Renew Sustain Energy Rev 113:109292. https://doi.org/10.1016/j.rser.2019.109292

DOE (2024a) Hydrogen pipelines. (n.d.). U.S. Department of energy. https://www.energy.gov/eere/fuelcells/hydrogen-pipelines

DOE (2024b) Hydrogen production: natural gas reforming. (n.d.). U.S. Department of energy. https://www.energy.gov/eere/fuelcells/hydrogen-production-natural-gas-reforming

EHO (2024), European Hydrogen Observatory (n.d.) European hydrogen observatory. https://observatory.clean-hydrogen.europa.eu/index.php/homepage

Genovese M, Cigolotti V, Jannelli E, Fragiacomo P (2023) Current standards and configurations for the permitting and operation of hydrogen refueling stations. Int J Hydrogen Energy 48(51):19357–19371. https://doi.org/10.1016/j.ijhydene.2023.01.324

Gillette JL, Kolpa RL (2008) Overview of interstate hydrogen pipeline systems. https://doi.org/10.2172/924391

Gökçek M, Kale C (2018) Optimal design of a hydrogen refueling station (HRFS) powered by hybrid power system. Energy Convers Manage 161:215–224. https://doi.org/10.1016/j.enconman.2018.02.007

Gökçek M, Paltrinieri N, Liu Y, Badia E, Dokuz AŞ, Erdoğmuş A, Urhan BB, Yoldaş Ö (2024) Optimum sizing of hybrid renewable power systems for on-site hydrogen refueling stations: case studies from Türkiye and Spain. Int J Hydrogen Energy 59:715–729. https://doi.org/10.1016/j.ijhydene.2024.02.068

Greene DL, Ogden JM, Lin Z (2020) Challenges in the designing, planning and deployment of hydrogen refueling infrastructure for fuel cell electric vehicles. eTransportation 6:100086. https://doi.org/10.1016/j.etran.2020.100086

Grube T, Rex M (2024) Deployment of fuel cell vehicles in road transport and the expansion of the hydrogen refueling station network: update

H2.LIVE: Hydrogen Stations in Germany and Europe (2024) H2.LIVE. https://h2.live/en/

HOMER Pro 3.15, HOMER energy (2024)

Hydrogen Insight (2024) | Hydrogen Insight (n.d.) https://www.hydrogeninsight.com/

Qin N, Brooker P, Srinivasan SS (2014) Hydrogen fueling stations infrastructure

Shi L, Hao Y, Lv S, Cipcigan L, Liang J (2020) A comprehensive charging network planning scheme for promoting EV charging infrastructure considering the Chicken-Eggs dilemma. Res Transp Econ 88:100837. https://doi.org/10.1016/j.retrec.2020.100837

Transport-Energy System—IEA (n.d.) IEA. https://www.iea.org/energy-system/transport

Chapter 14
Life Cycle Assessment for the Deployment of Green Hydrogen in Transport

Haiping Shen, Raquel Santos Jorge, and Anna Kalesnikava

Abstract Green hydrogen is expected to play a crucial role in decarbonizing the energy system. Hydrogen refuelling stations are one of the key elements in the large-scale deployment of green hydrogen. However, the decarbonization strategy is usually only evaluated in the usage phase of green hydrogen. To investigate the climate change impact and mitigation potential of the deployment of green hydrogen in the sector of transport, life cycle assessment (LCA) is introduced in this book. Firstly, this chapter illustrates the framework of LCA, covering its concept, targets, and categories of life cycle impact, as well as its methods (assessment procedure, databases, software, etc.). Secondly, the system of the hydrogen refuelling station is analysed and examples of the inventories of key components of the hydrogen refuelling station are displayed. Thirdly, a case study of LCA for the EU deployment of green hydrogen in transport towards 2050 is demonstrated. Process-based LCA is conducted within four designed scenarios (with green hydrogen, with blue hydrogen, without green hydrogen, and baseline). The analysis shows that green hydrogen wins on the climate impact compared to the other two decarbonization scenarios and baseline by 45–86% climate mitigation potential. Finally, a future framework for assessment of green hydrogen deployment is proposed, which includes LCA other assessments such as non-climate impacts, hydrogen leakage, risk analysis, and integrating with multi-dimension sustainability analysis.

14.1 Introduction

The energy transition requires solutions addressing multiple societal goals of energy security, economic implications, and environmental sustainability (Al-Breiki and Bicer 2021). Green hydrogen is expected to play an important role in the future energy system, as well as in decarbonizing end-use sectors such as transport (European

---

H. Shen (✉) · R. S. Jorge · A. Kalesnikava
Department of Industrial Economics and Technology Management, Norwegian University of Science and Technology (NTNU), Trondheim, Norway
e-mail: haiping.shen@ntnu.no

Y. Liu et al. (eds.), *Safety Measures for Hydrogen Fueling Stations*, Risk, Reliability and Safety Engineering, https://doi.org/10.1007/978-981-96-9412-9_14

Commission 2023). Hydrogen refuelling station is a key component for applying green hydrogen in transport. However, increasing attention on the climate change impact of hydrogen deployment has mainly focused on the use phase. This chapter explores the climate impacts associated with green hydrogen application in transport, from a life cycle perspective.

Life cycle assessment (LCA) is a tool for attributing environmental impacts to products and services. It is a holistic assessment of the environmental impacts which are considered over the life cycle including the production phase, use phase, and end-of-life phase, and covering the upstream in supply chains such as resource extraction and raw material production. Furthermore, LCA considers hundreds of the extracted resources in all inputs of energy and material flows, and the emitted substances in the outputs, that occur in all life cycle stages. It involves an assessment of a wide range of environmental impacts, such as climate change, human-toxicity, ecotoxicity, or natural resource use. There is a general need to support decision-making with complete, and detailed information about the environmental impacts of products, technologies, and organizations. This will enable the analysis of different technologies completely and comparably, achieving fair comparisons of alternatives and consistent assessments of burden shifting. Hence, LCA aims at true problem solving by avoiding burden shifting both from one life-cycle stage to another, as well as between different categories of environmental impacts.

The environmental life cycle assessment (LCA) can encompass several key tasks, such as:

- Evaluating the impact of climate change, and non-climate-impact such as resource depletion, land use, ecotoxicity, and human ecotoxicity.
- Comparing the climate change, and non-climate-impact between the research object and other scenarios/baseline in their complete lifetime.
- Estimating emissions associated with production, transmission and storage.
- Assessing the extraction of raw materials and resources needed in the production/service across various technical routes.
- Exploring the potential scarcity of critical resources used in the production/service to assess the risks in their supply chains.
- Identifying the main driver of environmental impacts and optimization potentials.
- Seeking opportunities for minimizing environmental footprint along the supply chain.
- Evaluating waste management strategies, focusing on minimizing waste generation, promoting recycling.

LCA is generally described as a procedure with four steps described in Fig. 14.1, which contains Goal and scope definition, Inventory analysis, Impact assessment, and Interpretation (International Organization for Standardization 2006).

- Step 1: Goal and scope definition.

The goal definition should unambiguously state the reasons for carrying out the study including the intended application and the intended audience. The scope definition should cover the product to be investigated including three important aspects:

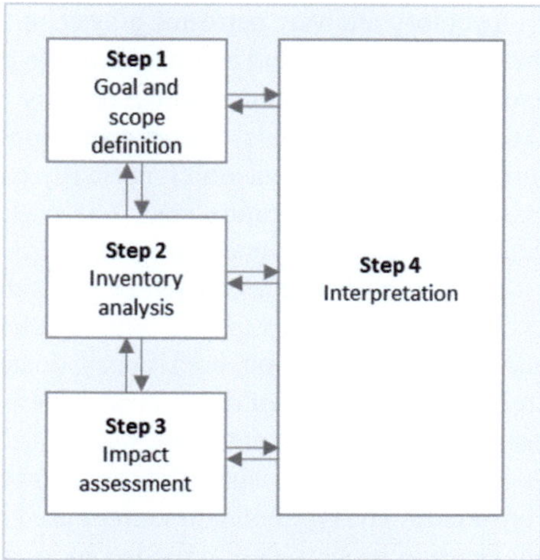

Fig. 14.1 Life cycle assessment framework (International Organization for Standardization 2006)

the 'functional unit', the environmental impact types to be investigated, and system boundaries (can be in relation to technical systems, geographical borders, time, and more). The functional unit is a reference unit for assessment and comparison between alternatives, which should capture the 'function' of the product(s) under study. This functional unit should present the scope of LCA. There are three typical LCA scopes shown in Fig. 14.2, such as Cradle-to-Gate, Cradle-to-Grave, and Cradle-to-Cradle. The first system boundary is only until the product is delivered at the gate before usage of the production. The second system boundary includes use stage and end-of-life. The third one investigates beyond the first life cycle of the product and focuses on circularity instead, representing the Ecological Loop (Parisi et al. 2020).

- Step 2: Inventory analysis

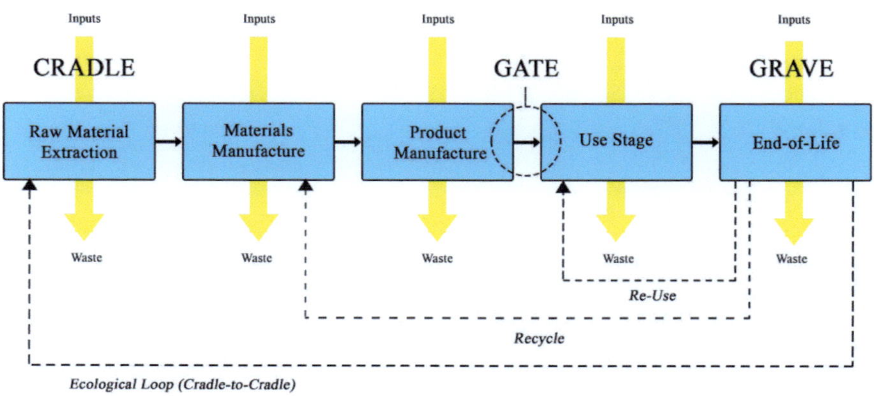

Fig. 14.2 Life cycle assessment scopes (Parisi et al. 2020)

Inventory analysis performs a systematic mapping of activities associated with the production, use, and end-of-life of the product. Usually, it constructs a flowchart and systematizes the quantitative data by use of matrix algebra, or LCA software. Data includes materials, energy, emissions, resources and waste flows. There are foreground system data which normally come from engineering process data from manufacturers or literature data. The background data source can apply the LCA database (for the supply chain) such as Ecoinvent (Wernet et al. 2016).

The data must be relevant, reliable, and accessible which includes the requirements of time coverage, geography coverage, technology coverage, completeness, representativeness, and precision, consistency, documentation, as well as reproducibility and credibility. The types of data can be involved in material, energy, emission, resource uses, and waste. Good data are often difficult to obtain or not accessible to the public, due to confidentiality issues. Data compilation can be challenging and generally time-consuming. The types of data sources used in LCA can include direct measurements, engineering process data provided by manufacturers, established LCA databases, findings from other LCA studies, literature sources, and expert judgments or estimates. In some cases, the use of proxy data—data from a similar technology or product that does not exactly match the one under study—may also be necessary. Several commercial and free databases are available on openLCA Nexus (GreenDelta (n.d.)), where Ecoinvent and GaBi are market-leading databases. Figure 14.3 presents a comprehensive list of LCA databases (GreenDelta (n.d.)).

- Step 3: Impact assessment

Impact assessment is the process to convert stressors to common impact units that are comparable and understandable. Stressors from inventory analysis include the emissions (to air, water and soil), resource (land, water, fuel) use, as well as mineral ore extracted, etc. Impact assessment can be contacted at 'midpoint' level or 'endpoint' level. The impacts at 'midpoint' level refer to more than ten categories such as Climate change, Particulate matter, Acidification, Ozone depletion,

Fig. 14.3 List of LCA databases (GreenDelta (n.d.))

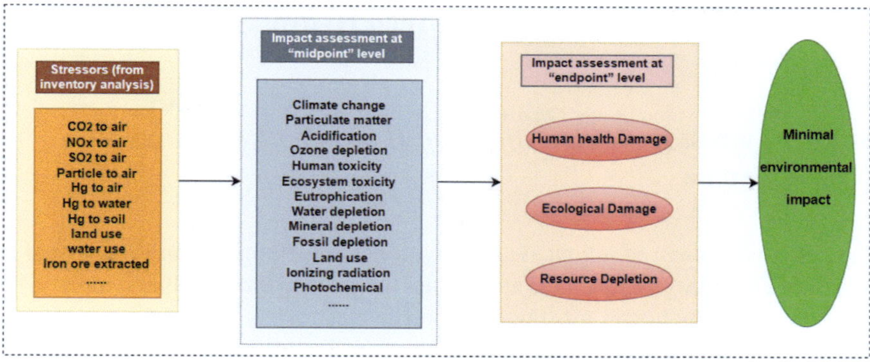

Fig. 14.4 Framework of life cycle impact assessment at midpoint and endpoint

Human toxicity, Ecosystem toxicity, Eutrophication, Water depletion, Mineral depletion, Fossil depletion, Land use, Ionizing radiation, Photochemical, etc. The impacts at 'endpoint' level are aggregated from the impacts at 'midpoint' level to three major categories: Human health, Ecosystem damage, Resources depletion. Life cycle impact assessment framework is displayed in Fig. 14.4.

There are different methods for conducting an LCA. Some methods only assess the life cycle impacts at midpoint level such as CML 2001, TRACI. Some conduct LCA both at Midpoint and Endpoint levels, such as ReCiPe, Eco-Indicator 99, Impact 2002+, Impact World+, LIME, LC-Impact (under development, currently only endpoints), which make full aggregation possible.

LCA software such as Simpro, openLCA, or Brightway can be applied to conduct the life cycle environmental impacts by inputting the data of the foreground system and connecting with database with defining the method. Python is getting more and more popular to establish models for LCA.

- Step 4: Interpretation

Interpretation includes both inventory analysis and impact results. This involves identifying critical assumptions and parameters, evaluating uncertainty, identifying important contributions from stressors (pollutants and/or resource), identifying important contributions from activities and finally drawing conclusions.

14.2 Life Cycle Assessment of Green Hydrogen Refuelling Station

14.2.1 System Definition of Green Hydrogen Refuelling Station

The system flowchart of a typical green hydrogen refuelling station is presented in Fig. 14.5. There are possibly three approaches for green hydrogen supply: trailer with hydrogen tanks, hydrogen pipeline, and on site electrolyser. These three hydrogen supply resources require different infrastructure. The mobile hydrogen tanks by trailer need the regular road transport; the hydrogen pipelines need hydrogen transmission and distribution pipes laying; the onsite hydrogen production needs renewable electricity and electrolyser. Inside of the hydrogen station, hydrogen is firstly collected and stored in the low pressure (0.6 MPa) tank from the mobile tank of the trailer, or from the pipeline, or from local production by the onsite electrolyser. The second process is to compress the hydrogen from low pressure to high pressure, which then is stored in high pressure (90 MPa) tank. Finally, the dispenser is ready with 35/75 MPa hydrogen after pre-cooling.

For the life cycle assessment, all the demand of the material and energy for the station construction and operation, as well as its supply chains are accounted for the system. The system boundary and scope can be adjusted based on the goal definition. The scope of LCA can be Cradle-to-Gate if the aim is to compare the three green hydrogen supply approaches for the station as the use phase and end-of-life are the same for the same fuel. However, if the aim is to study the mitigation potential of the life cycle environmental impacts from replacing fossil fuel or other fuels by

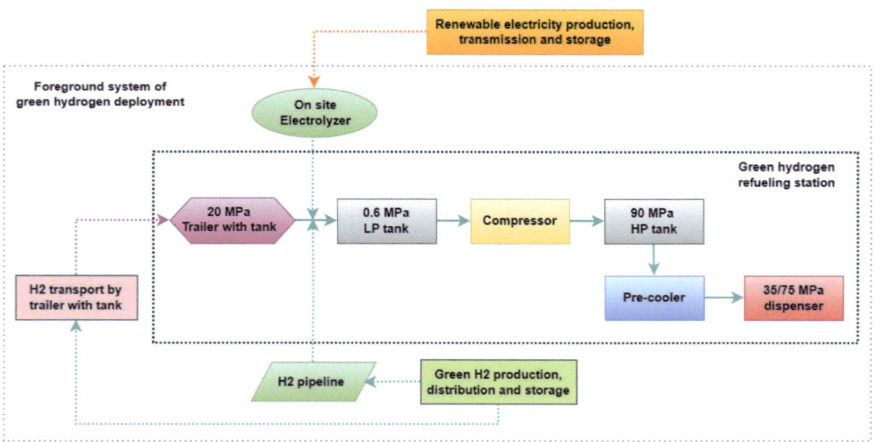

Fig. 14.5 Green hydrogen refuelling station with three approaches of the green hydrogen supply

deploying green hydrogen, the scope should include the use phase at least, as the impacts of different fuels are different during their usage phases.

14.2.2 Examples of Hydrogen Refuelling Station Inventories

Tables 14.1, 14.2, 14.3, and 14.4 provide examples of the inventories related to hydrogen refuelling station in the foreground system derived from literature data, including electrolyser, high pressure storage tank, basic hydrogen fuelling station, and key data for LCI calculation of the hydrogen stations. It is worthy to notice that the functional unit of the subsystem inventory may not be the same. Thus, it needs to be harmony with each other for LCA in the whole system.

The data of the supply chains in the background system can be collected from an LCA database such as Ecoinvent. It requires to define the name of the material/primary energy/product, region and key characters of the product or technology type to get the suitable data for the specific system.

14.3 Case Study of LCA for the EU Deployment of Green Hydrogen in Transport

The energy transition in a low-carbon path is generating challenges across diverse fields. In 2022, hydrogen accounted for less than 2% of Europe's total energy consumption, with 96% of it being produced using natural gas, resulting in massive CO_2 emissions (European Commission 2023). The European Union's hydrogen strategy prioritizes the development of renewable (green) hydrogen as a key solution (European Commission 2023). According to the International Energy Agency (IEA), nearly 90% of electricity is projected to be generated from renewable sources by 2050, with wind and solar PV contributing almost 70% of the total (Bouckaert et al. 2021). Within this framework, green hydrogen is anticipated to play a crucial

Table 14.1 LCI of basic hydrogen fuelling station for 850 kg/d (Burkhardt et al. 2016)

Low alloy steel	26.6	t
High alloy steel	5.21	t
Cast iron	2.94	t
Copper	1.18	t
Aluminium	499	kg
Polymer	372	kg
Electronics	64.0	kg
Concrete	210	t
Recycling concrete	210	t

Table 14.2 LCI for a high-pressure storage tank per 10 kg hydrogen (Boureima et al. 2011; Nath and Das 2011)

High pressure tank		
Rolling aluminium sheets	6.0	kg
Aluminium	6.0	kg
Rolling of chromium steel sheets	9.0	kg
Chromium steel	9.0	kg
Carbon fibre	71.4	kg
Epoxy resin	30.6	kg
Electricity	4.5	kWh
Rolling steel sheets	9.0	kg
Steel, low alloyed	9.0	kg
1 kg carbon fibre		
Acrylonitrile	2.25	kg
Vinyl acetate	0.20	kg
Resin	1.07	kg
Heat	175	kWh
Electricity	26.2	kWh
1 kg resin		
Epoxy resin	420	g
Styrene	400	g
Acrylic acid	180	g

role in decarbonizing end-use sectors such as industry, transport, and buildings. Its production from renewable energy sources (RES) is expected to offer a low greenhouse gas (GHG) emission profile to make it well-suited for the energy demand in these sectors.

This case study applies LCA to investigate various decarbonization pathways—including green hydrogen, blue hydrogen, and scenarios excluding green hydrogen—against a baseline scenario, evaluating the climate mitigation potential of deploying green hydrogen in transport. This session is derived from a previous study by NTNU on the LCA for a large-scale deployment of green hydrogen in Europe (Shen et al. 2024).

14.3.1 Scenarios Design

The deployment of green hydrogen encompasses its production, distribution, and application within a complete life cycle. Among the key sectors requiring decarbonization, transport stands out alongside industry and buildings due to its hard-to-abate emissions (Wang et al. 2021). Within the transport sector, heavy road vehicles and aviation present the most formidable obstacles to achieving decarbonization

Table 14.3 Detailed LCI for electrolyser (Haug et al. 2017)

Construction		
Balance of plant		
Liquid storage tank	88	m^3
Pump	9.8	kW
Inverter	500	kW
Steel	145	kg
Cell stacks		
Cell stack framework		
Copper	2.0	t
Unalloyed steel	200	t
Cells		
Nickel	30	t
Aluminium	450	kg
Calendered rigid plastic	780	kg
Tetrafluoroethylene	78	kg
Acrylonitrile butadiene styrene	160	kg
Polyphenylene sulphide	340	kg
Polysulphones	260	kg
N-Methyl-2-pyrrolidone	1.3	t
Aniline	49	kg
Acetic anhydride	54	kg
Terephthalic acid	88	kg
Nitric acid	33	kg
Hydrochloric acid	130	kg
Graphite	430	kg
Lubricating oil	0.48	kg
Zirconium oxide	1.1	t
Carbon monoxide	150	kg
Decarbonized water	11	t
Deionized water	86	t
Electricity	36	GJ
Heat	88	GJ
Steam	700	MJ
Industrial machine production	0.16	kg
Plaster mixing	780	kg
Operation		
Electricity	47.6	kWh/kg H_2
Water	18.8	l/kg H_2
Nitrogen	0.29	g/kg H_2

(continued)

Table 14.3 (continued)

Construction		
Potassium hydroxide	1.9	g/kg H_2
Steam	0.11	kg/kg H_2

Table 14.4 Data for LCI calculation of hydrogen fuelling stations (Reuß et al. 2017; Teichmann 2015; Wulf et al. 2018)

		LOHC	Pipeline	High pressure truck
Loss hydrogen	g/kg H_2	5	5	5
Loss dibenzyltoluene	g/kg H_2	16.1	–	–
Electricity compression	kWh/kg H_2	3.96	1.6	0.5
Electricity precooling	kWh/kg H_2	4.00	4.00	4.00
Liquid storage tank DBT, capacity	m^3	26.1	–	–
High pressure tank hydrogen, capacity	kg	125	125	125

(Wang et al. 2021). These sectors currently rely heavily on fossil-based fuels, such as diesel, kerosene, petroleum, and synthetic fuel blends, as their primary energy sources. This makes them critical decarbonization targets for the application of green hydrogen. To assess the mitigation potential for climate impact by the deployment of green hydrogen in transport sector, a baseline scenario and three alternative energy system scenarios have been developed. These scenarios explore the use of different fuels, both with and without green hydrogen. Their environmental impacts will be evaluated using LCA, focusing on emissions avoided across the entire life cycle of the systems rather than only during the operational phase (Shen et al. 2024).

(1) Baseline scenario: No change in energy mix.

The baseline scenario, depicted in Fig. 14.6, represents the continued reliance on fossil fuels in the transport sector, particularly within its hard-to-abate segments. This scenario is derived from the European Hydrogen Backbone report (Wang et al. 2021), which analyses the future potential of substituting fossil fuels with hydrogen in transport. However, in this scenario, no such transition occurs, and fossil fuels such as diesel, kerosene, petroleum, and natural gas continue to dominate energy consumption (Shen et al. 2024).

For the LCA, the analysis involves two systems, containing background system, which includes upstream activities such as the extraction of crude oil (both onshore and offshore), resource procurement, and the embodied energy required for these processes; and the foreground system, which covers the production, refinement, transportation, storage, distribution, and final use of fossil fuels.

Traditional fuel stations deliver fuel in this scenario (Shen et al. 2024).

(2) Scenario 1: Decarbonization with green hydrogen.

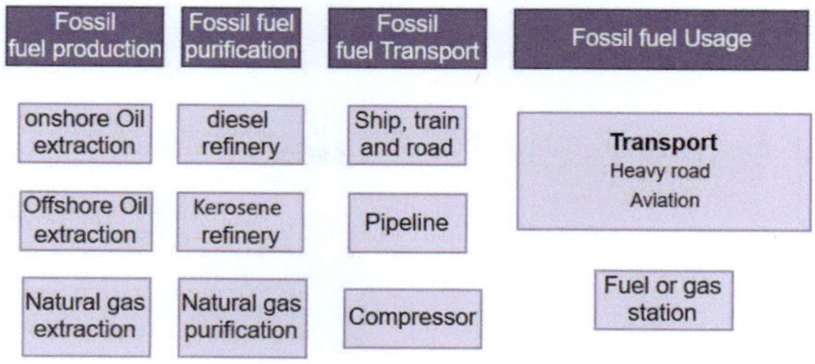

Fig. 14.6 Baseline: no change in energy mix

This scenario, illustrated in Fig. 14.7, explores the replacement of fossil fuels with green hydrogen in the hard-to-abate sectors of transport. It assumes a complete transition to green hydrogen.

The background system includes the production of renewable electricity, alongside the material and resource inputs required in the supply chain. Meanwhile, the foreground system covers the entire lifecycle of green hydrogen: its production through proton exchange membrane (PEM) electrolysis, transportation via pipelines, storage in tanks or salt caverns, distribution, and usage (Shen et al. 2024). PEM electrolysis is highlighted as the selected technology for green hydrogen production in this scenario. This method is particularly well-suited for integration with wind-powered systems, due to its adaptability and efficiency under fluctuating power conditions (Durakovic et al. 2022).

Green hydrogen refuelling stations are the infrastructure to deliver the fuel in this scenario.

(3) Scenario 2: Decarbonization with blue hydrogen.

This scenario, indicated in Fig. 14.8, examines blue hydrogen as a substitute for fossil fuels in the hard-to-abate sectors of transport. Similar in structure to Scenario 1 (green hydrogen), this scenario envisions the deployment of blue hydrogen as a transitional energy solution (Shen et al. 2024).

The background system encompasses upstream activities such as the extraction, purification, and transportation of natural gas, which serves as the primary feedstock for hydrogen production. The foreground system includes the processes for hydrogen production via steam methane reforming (SMR) integrated with carbon capture and storage (CCS), followed by the transportation of hydrogen through pipelines, storage

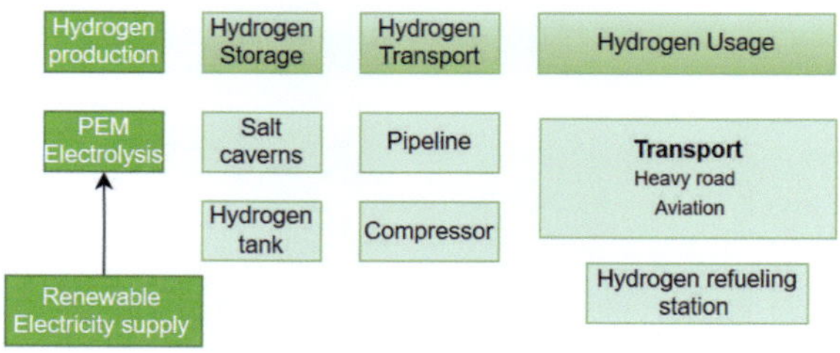

Fig. 14.7 Scenario 1: decarbonization with green hydrogen

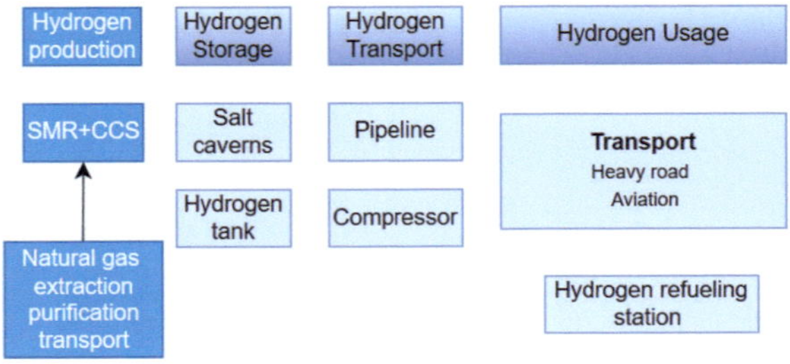

Fig. 14.8 Scenario 2: decarbonization with blue hydrogen

in tanks or salt caverns, distribution, and ultimate usage in the transport sector (Shen et al. 2024).

Same as scenario 1 Decarbonization with green hydrogen, blue hydrogen refuelling stations are the infrastructure to deliver the fuel in this scenario, which has the similar system in Fig. 14.3.

(4) Scenario 3: Decarbonization without green hydrogen

This scenario, displayed in Fig. 14.9, explores the replacement of fossil fuels in the transport sector with a mix of other clean energy sources, including biofuels, synfuels, and renewable electricity. It presents a vision of a diversified clean energy mix that excludes green hydrogen, offering a comparable alternative to Scenario 1, which focuses on green hydrogen deployment (Shen et al. 2024).

The background system includes upstream processes such as coal mining and the collection of solid biomass, which supply the raw materials and resources needed for clean energy production. The foreground system encompasses the production of clean energy through processes such as gasification and SMR with CCS. It also includes the transportation, storage, distribution, and end-use of these alternative fuels (Shen et al. 2024).

Biofuel/synfuel refuelling stations or electricity charging stations are the infrastructures to deliver the fuel or electricity in this scenario.

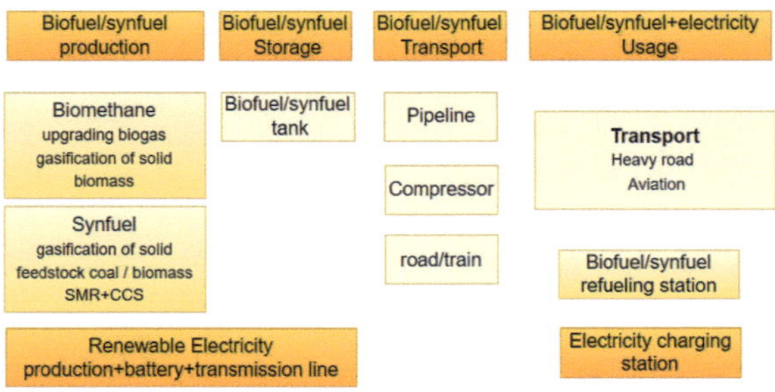

Fig. 14.9 Scenario 3: decarbonization without green hydrogen

14.3.2 *LCA Framework of Different Fuel Systems in Transport*

Process-based LCA is a widely used method for evaluating environmental impacts (Guinée et al. 2011). However, some studies apply other types of LCA such as the hybrid LCA integrating the input–output models (Gibon et al. 2017). This study employs a process-based LCA approach, applying 1 kWh of fuel usage as the functional unit. The energy system boundary contains two main components: the background system, which includes upstream processes such as resource extraction and energy supply, and the foreground system, which covers the flow of processes from energy production to distribution and final usage (Shen et al. 2024). A detailed representation of the system boundaries for the three decarbonization scenarios is demonstrated in Fig. 14.10, illustrating the energy flows and processes for LCA.

This case study derives all life cycle impact factors of various technologies from existing literature. Key data on hydrogen deployment in the transport sector, pipeline and storage infrastructure capacities, and assumptions regarding technical parameters are sourced from the European Hydrogen Backbone report (Wang et al. 2021). By integrating life cycle impact factors with the projections for hydrogen deployment in transport in Europe, a comprehensive assessment of life cycle climate impact across different scenarios towards 2050 is conducted.

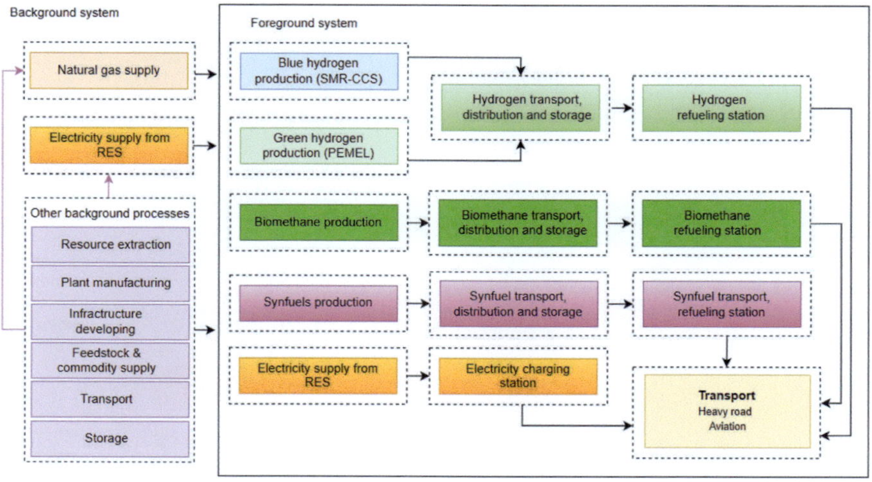

Fig. 14.10 Fuel system boundary for life cycle assessment

14.3.3 Decarbonization Options and Chosen Pathways

According to the European Hydrogen Backbone report, the total hydrogen demand across industries, transport, buildings, and power in Europe is projected to reach 2300 TWh by 2050 (Wang et al. 2021). This case study focuses on the transport sector, where hydrogen demand is expected at approximately 285 TWh by 2050 (Wang et al. 2021; Durakovic et al. 2022). The design of decarbonization pathways and options in the three decarbonization scenarios is derived from the projections in the European Hydrogen Backbone report (Wang et al. 2021).

In the transport sector, the estimated hydrogen demand includes 217 TWh for heavy road transport and 68 TWh for aviation (Wang et al. 2021). In scenarios 1 and 2, this demand is met entirely by green hydrogen and blue hydrogen, respectively. In Scenario 3, alternative clean energy solutions are applied. For heavy road transport, the demand is satisfied with 130 TWh of electricity and 87 TWh of bio-methane. For aviation, the energy mix includes 8 TWh of electricity, 30 TWh of bio-methane, and 30 TWh of synfuel (Shen et al. 2024; Wang et al. 2021).

Figure 14.11 provides the detailed decarbonization pathways and energy options for the transport sector in each scenario, emphasizing the strategies for transitioning toward a sustainable, low-carbon energy future (Shen et al. 2024).

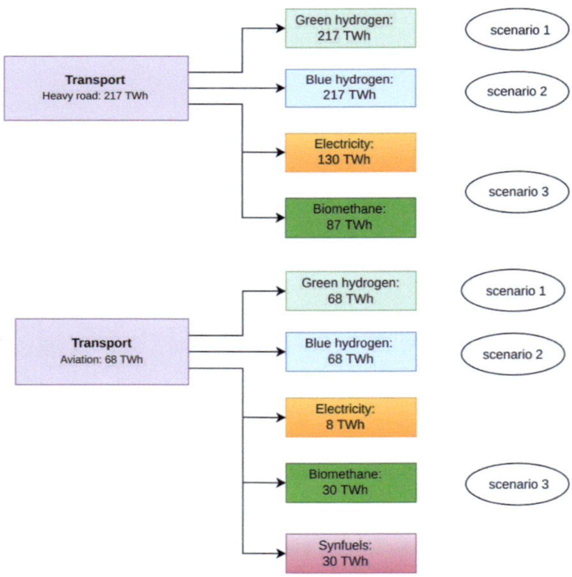

Fig. 14.11 Decarbonization options and paths in transport (Shen et al. 2024)

14.3.4 Mathematical Formulation and Performance Assumptions

The life cycle impact of different technologies is evaluated based on their impact intensity and the corresponding energy demand within each scenario projected for the transport sector in Europe towards 2050. This assessment is performed using Formula (14.1) (Shen et al. 2024).

$$I_{life,i} = \sum_{j} \left(Ii_{life,i,j} * E_j \right) \tag{14.1}$$

- $I_{life,i}$: total life cycle impact in category i, impact unit/year;
- $Ii_{life,i,j}$: impact intensity of technology j in impact category i, impact unit/kWh;
- E_j: energy demand of the technology j in each scenario in transport towards 2050, kWh/year.

In the field of hydrogen production, Delpierre et al. conducted an ex-ante LCA to evaluate the environmental impacts of wind-based hydrogen production in the Netherlands. Their study estimated the climate impact of green hydrogen produced via proton exchange membrane (PEM) electrolysis under future scenarios at 0.773 kg CO_2 eq/kg H_2 (Delpierre et al. 2021). Similarly, Hermesmann and Müller examined the life cycle environmental impacts of green, blue, and grey hydrogen production that analysed the effects of various electricity sources, such as the current grid mix and offshore wind power, across both present and future contexts (Hermesmann and Müller 2022). Their findings revealed that electricity supply has a minimal influence on the climate impact of blue hydrogen production but significantly affects on the impact of green hydrogen. The future green hydrogen production via PEM electrolysis powered by offshore wind in Germany has an environmental impact of 0.7 kg CO_2 eq/kg H_2, closely aligning with Delpierre's results (Hermesmann and Müller 2022).

The European Hydrogen Backbone report forecasts that by 2040, the total length of hydrogen pipelines across 21 countries will reach 39,700 km, with approximately 70% of the current natural gas pipelines (Wang et al. 2021). Furthermore, it speculates that hydrogen storage capacity will expand to 450 TWh by 2050, with salt caverns and tanks serving as the primary storage facilities (Wang et al. 2021). The report further assumes that 70% of this storage capacity will rely on retrofitted natural gas storage infrastructure (Wang et al. 2021).

Wulf and colleagues performed a comprehensive LCA of hydrogen supply chain architectures, encompassing transportation (pipeline), dispensing (hydrogen fuelling station), and storage (salt cavern and tank). Their findings estimated climate impacts of these processes to be 0.17, 0.21, and 0.03 kg CO_2 eq/kg H_2, respectively (Wulf et al. 2018).

The environmental impacts of hydrogen production, transportation, dispensing, and storage are determined by factors such as impact intensity, hydrogen demand,

and capacity factor. These variables are systematically represented in Formula (14.2), which outlines the equation for calculating life cycle impacts across various stages of the hydrogen supply chain (Shen et al. 2024).

$$I_{life,H_2,i,j} = Ii_{life,H_2,i,j} * E_j * f_j \tag{14.2}$$

- $I_{life,H_2,i,j}$: the life cycle impact of hydrogen technology j (production, transport, dispensing, and storage) in category i, impact unit/year;
- $Ii_{life,H_2,i,j}$: impact intensity of hydrogen technology j in category i, impact unit/kWh;
- E_j: energy demand of the hydrogen technology j in each scenario in Europe towards 2050, kWh/year;
- f_j: capacity factor of the hydrogen technology j, $f_{transport} = 0.3$ (30% new hydrogen pipeline and 70% existing natural gas pipeline); the rest of capacity factors are 1.

Table 14.5 presents the life cycle greenhouse gas (GHG) emission intensities of various technologies within the transport sector. These values are derived from literature data and are calculated based on performance assumptions. Notably, the biofuel included in this analysis is produced from waste feedstock (Florio et al. 2019), which results in a lower environmental impact compared to blue hydrogen.

Table 14.5 The life cycle GHG emission intensity of various technologies (Shen et al. 2024)

Technologies	GHG emission intensity	unit	Performance assumptions
Transport			
Green hydrogen delivered	0.021 (Wulf et al. 2018; Hermesmann and Müller 2022)	kg CO_2 eq/kWh H_2	Hydrogen energy density estimated at 33.33 kWh/kg H_2 (Yue et al. 2021)
Blue hydrogen delivered	0.115 (Wulf et al. 2018; Hermesmann and Müller 2022)	kg CO_2 eq/kWh H_2	Hydrogen energy density estimated at 33.33 kWh/kg H_2 (Yue et al. 2021)
Biofuel delivered	0.100 (Florio et al. 2019)	kg CO_2 eq/kWh biofuel	
Diesel (heavy road)	0.264 (Hannach et al. 2019)	kg CO_2 eq/kWh diesel	
Kerosine (aviation)	0.404 (Bicer and Dincer 2017)	kg CO_2 eq/kWh kerosine	
Synfuels (aviation)	0.115 (Hermesmann and Müller 2022)	kg CO_2 eq/kWh synfuels	
Electricity	0.007	kg CO_2 eq/kWh el	The estimated averaged climate impact of electricity mix in 2050

14.3.5 Comparisons of Climate Change Impact of Fuel Systems in Transport

In the transport sector, the life cycle climate change impact is lowest in the scenario 'decarbonization with green hydrogen' and highest in the baseline scenario 'no change in energy mix.' These impacts are estimated at 8.7 megatons CO_2eq (including 2.7 megatons CO_2eq from increased dedicated renewable capacity) and 62 megatons CO_2eq, respectively. The scenario 'decarbonization with blue hydrogen' results in a higher impact than the scenario 'decarbonization without green hydrogen,' with emissions of 33 megatons CO_2eq and 16 megatons CO_2eq, respectively.

The deployment of green hydrogen in transport leads to significant emissions reduction, with impacts avoided in the 'decarbonization with green hydrogen' scenario amounting to 85.9% compared to the baseline, 73.3% compared to the scenario with blue hydrogen, and 45.4% compared to the scenario without green hydrogen. These comparisons are visually represented in Fig. 14.12 (Shen et al. 2024).

It is noteworthy that the climate impact of heavy road transport employing green hydrogen is lower than that of using electricity and biomethane in the scenario without green hydrogen. However, the impact of electricity and biomethane in this scenario is significantly lower than that of blue hydrogen. In the aviation sector, utilizing green hydrogen results in a lower impact compared to biomethane and synfuel in the scenario without green hydrogen, which again exhibits lower impacts than blue hydrogen. Furthermore, the use of biomethane and synfuel in aviation results in a marginally lower climate impact compared to blue hydrogen (Shen et al. 2024).

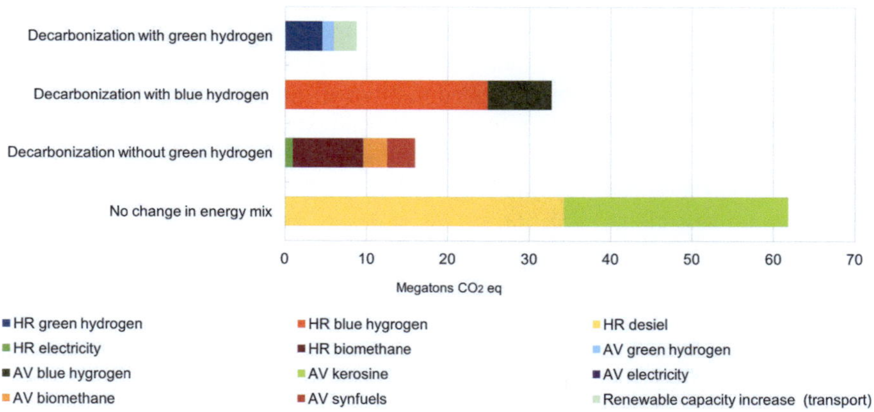

Fig. 14.12 Comparisons of life cycle climate change impact in transport (HR: heavy road, AV: aviation) within four scenarios in Europe in 2050 (Shen et al. 2024)

14.4 Life Cycle Climate Impact Analysis for Fuel Systems

Based on the projected energy demand in European transport sector by 2050, the decarbonization pathway of deploying green hydrogen demonstrates a clear advantage over blue hydrogen and other clean energy alternatives that exclude green hydrogen. Green hydrogen shows a significant climate mitigation potential compared to the baseline scenario of maintaining the current energy mix. Every fuel system has three main life cycle phases: the upstream supply chain, fuel production, and fuel usage. While all phases contribute to climate change impacts, the level of contribution varies significantly between those fuels (Eurostat 2023).

Clean energy carriers like renewable electricity and green hydrogen primarily derive their climate impacts from resource extraction and infrastructure construction (e.g., materials and their embodied energy), rather than from the energy usage phase. In contrast, fossil fuels such as natural gas, oil, diesel, and kerosene generate their climate impacts majorly during their usage phase, largely dependent on the effectiveness of greenhouse gas (GHG) reduction technologies like Carbon Capture and Storage (CCS). For fossil-derived clean energy carriers like blue hydrogen, biomethane, and synfuels, the impacts stem from both the fossil resource supply and the chemical processes involved in production, with CCS playing a critical role in mitigating emissions (Shen et al. 2024).

It is worth mentioning that renewable electricity, when directly applied in transport, has an even lower climate impact than green hydrogen because green hydrogen production involves additional energy conversion steps, such as electrolysis, which reduces overall efficiency. Furthermore, the biofuels in this case study are applied from organic waste (Florio et al. 2019), offering a lower climate impact compared to biofuels produced from crops or wood due to the benefits of energy recovery and reduced land-use impacts. Moreover, biofuel integrated with CCS can achieve negative climate impact. For this analysis, biomass power plants incorporate CCS technology, but CCS has not been applied to biomethane production due to limited data availability. To simplify the evaluation, the infrastructure impacts of transport modes are assumed to be consistent across all energy carriers (Shen et al. 2024).

As decarbonization pathways are projected toward 2050, certain resource-related trends warrant attention. Depletion of mineral and metal resources is expected to intensify, with declining ore grades (e.g., copper) increasing the embodied energy required for extraction and refining. Similarly, fossil fuel resource depletion will lead to higher energy demands for extraction processes, further exacerbating climate impacts. However, advancements in technology are anticipated to improve efficiency and offset some of the negative effects of resource depletion (Rötzer and Schmidt 2020). These opposing trends—resource depletion and technological improvement—will likely counterbalance each other to some extent. Optimistically, advancements in technology may outpace resource depletion, particularly as material recovery processes and fossil fuel independence gain momentum, offering a sustainable advantage (Shen et al. 2024).

In summary, the life cycle climate impacts of different technologies in decarbonization pathways primarily stem from energy consumption and land use. Energy consumption manifests in three distinct forms: (1) direct combustion of fossil fuels for heat generation in industrial applications or electricity production in thermal power plants, which constitutes a continuous source of emissions during the operational phase in lifetime; (2) embodied energy in the supply chain of the material and energy required for the construction of infrastructure for renewable energy technologies, such as wind and solar; and (3) resource utilization, including natural gas and biomass, for the production of clean energy carriers such as blue hydrogen, biofuels, and synfuels. The aggregate climate impact is determined by the total greenhouse gas (GHG) emissions accrued over the entire life cycle of the technology (Shen et al. 2024). In the baseline scenario, fossil fuel consumption represents the predominant driver of climate impacts. In contrast, the 'decarbonization with blue hydrogen' and 'decarbonization without green hydrogen' scenarios exhibit significant contributions from the fuel supply and utilization phases. The integration of CCS offers a pathway for mitigating GHG emissions by facilitating the conversion of fossil fuels into clean energy. Meanwhile, in the 'decarbonization with green hydrogen' scenario, the primary source of climate impact arises from renewable electricity. Specifically, the embodied energy of materials required for infrastructure development, such as wind turbines and solar panels, constitutes the major contributor to lifecycle emissions (Shen et al. 2024). The climate impact is fundamentally originated from the exploitation of finite resources, including fossil fuels and minerals. Green hydrogen, produced by renewable electricity, necessitate significant material inputs for infrastructure development. Conversely, blue hydrogen depends on continuous natural gas supply, albeit with a lower material intensity. The deterioration of resource depletion over the coming decades is a main driving force for rising climate impacts. Moreover, land use and land changes from the infrastructure further amplify climate impacts by degrading ecosystems and altering the Earth's surface albedo, thereby disrupting the planetary energy balance (Shen et al. 2024).

14.5 Future LCA for Green Hydrogen Deployment

14.5.1 Comparison of Other Environmental (Non-climate) Impacts

LCA enables to evaluate a wide range of environmental impacts, including both climate-related and non-climate-related impacts. The primary categories influencing the life cycle of green hydrogen production are climate change, water use, human toxicity, eco-toxicity, resource depletion, and land use. For wind and solar technologies, the LCA places particular emphasis on mineral resource depletion and land use as critical environmental considerations (Shen 2023).

Water use is a critical consideration, as green hydrogen production via electrolysis requires significant water input. In water-scarce regions, this demand may intensify competition with other uses, such as agriculture and drinking water. Human toxicity and eco-toxicity arise from the extraction of materials consumed by renewable infrastructure, and the disposal of spent components, potentially harming both human health and ecosystems (Pranjal et al. 2024).

Resource depletion is another key concern, as green hydrogen infrastructure relies on finite materials like lithium, nickel, and rare earth metals. Sustainable sourcing, recycling, and substitution are essential to mitigate these impacts. Similarly, land use affects biodiversity and ecosystems, particularly in the development of hydrogen production facilities and renewable infrastructure like wind and solar (Shen et al. 2024; Fragiacomo et al. 2024).

Green hydrogen's reliance on renewable energy, facing mineral resource depletion and end-of-life waste challenges. Balancing these trade-offs with climate mitigation benefits is vital to ensuring net-positive outcomes. Therefore, by addressing these non-climate impacts through innovative technologies and sustainable practices to avoid the environmental impact shift from on category to another, and from one life-cycle stage to another, green hydrogen can align with broader environmental goals (Shen et al. 2024; Hassan et al. 2024). Therefore, further LCA for fuel systems both in climate and non-climate impacts are necessary.

14.5.2 LCA Integrating with Hydrogen Leakage

The leakage of hydrogen is not considered in the LCA presented here. However, the leakage of hydrogen can occur during the whole life cycle including hydrogen production in electrolysers, transport in pipelines, and storage in tanks as well as dispenser. Hydrogen, as an indirect greenhouse gas (GHG), has a short lifespan but is prone to escaping into the atmosphere due to its small molecular size (Ocko and Hamburg 2022). Ocko and colleagues emphasizes that the climate impact of hydrogen deployment significantly depends on the rate of leakage and the time frame considered (Ocko and Hamburg 2022). For instance, if hydrogen were to replace all fossil fuels completely and the leakage reached 10%, estimations suggest it could contribute up to 0.25 °C of global warming by 2050. Similarly, if hydrogen were to supply half of the world's final energy supply, the contribution could still be at least 0.1 °C by 2050 (Ocko and Hamburg 2022). The climate mitigation benefits of both green and blue hydrogen would be diminished when hydrogen leakage is accounted for. Therefore, reducing hydrogen leakage and understanding the effects of varying leakage rates is essential to ensure that hydrogen is effectively integrated into the decarbonization pathways of energy transition (Ocko and Hamburg 2022).

14.5.3 LCA Incorporating Risk Analysis

Traditionally, LCA is performed in a deterministic approach. However, green hydrogen systems including refuelling stations have an associated risk of accidents. LCA of green hydrogen deployment can be extended by incorporating risk analysis results. The disturbances and unexpected events during lifecycle can be further explored. This approach is expected to yield more meaningful managerial insights, contributing to the overall improvement of sustainability practices.

14.5.4 LCA Integrating with Multi-dimension Sustainable Analysis

Besides environmental and safety considerations, understanding the social implications of hydrogen technologies deployment will be crucial for assessing how these technologies can be scaled up in a sustainable manner. Aspects such as social acceptance and socio-economic impacts (e.g., in terms of job creation) need to be better understood, using approaches which involve all stakeholders (citizens, industrial representatives, government, etc.).

Life cycle environmental impact assessment can be integrated with multi-dimension sustainable analysis to establish a comprehensive sustainable assessment of multi-dimension with the techno-economic-enviro-social-safety. This multi-dimensional evaluation provides the decision-makers a holistic solution for the green hydrogen deployment in the energy transition. Additionally, this integrated framework enables proactive identification of potential barriers to implementation and fosters innovation by aligning technical advancements with societal needs and environmental priorities.

References

Al-Breiki M, Bicer Y (2021) Comparative life cycle assessment of sustainable energy carriers including production, storage, overseas transport, and utilization. J Clean Prod 279:123481. https://doi.org/10.1016/j.jclepro.2020.123481

Bicer Y, Dincer I (2017) Life cycle evaluation of hydrogen and other potential fuels for aircrafts. Int J Hydrogen Energy 42(16):10722–10738. https://doi.org/10.1016/j.ijhydene.2016.12.119

Bouckaert S, Pales AF, McGlade C, Remme U, Wanner B, Varro L, D'Ambrosio D, Spencer T (2021) Net zero by 2050: a roadmap for the global energy sector. Retrieved from https://trid.trb.org/view/1856381

Boureima F, Van Mierlo J, Messagie M, Sergeant N, Govaerts L, Denys T, Michiels H, Vernaillen S, Schrooten L, Beckx C, Macharis C, Turcksin L, Bernardini A, Hecq W, Klopfert F, Englert M, De Caevel B, De Vos M (2011). Clean vehicle research: LCA and policy measures (CLEVER)—final Report. Belgian Science Policy, Brussels. Retrieved from http://www.belspo.be

Burkhardt J, Patyk A, Tanguy P, Retzke C (2016) Hydrogen mobility from wind energy—a life cycle assessment focusing on the fuel supply. Appl Energy 181:54–64. https://doi.org/10.1016/j.apenergy.2016.07.104

Delpierre M, Quist J, Mertens J, Prieur-Vernat A, Cucurachi S (2021) Assessing the environmental impacts of wind-based hydrogen production in the Netherlands using ex-ante LCA and scenario analysis. J Clean Prod 299:126866. https://doi.org/10.1016/j.jclepro.2021.126866

Durakovic G, Crespo del Granado P, Tomasgard A (2022) Powering Europe with north sea offshore wind: the impact of hydrogen investments on grid infrastructure and power prices. Energy 125654. https://doi.org/10.1016/j.energy.2022.125654

El Hannach M, Ahmadi P, Guzman L, Pickup S, Kjeang E (2019) Life cycle assessment of hydrogen and diesel dual-fuel class 8 heavy duty trucks. Int J Hydrogen Energy 44(16):8575–8584. https://doi.org/10.1016/j.ijhydene.2019.02.027

European Commission (2023) Hydrogen. Retrieved from https://energy.ec.europa.eu/topics/energy-systems-integration/hydrogen_en

Eurostat (2023) Shedding light on energy—2023 edition. Retrieved from https://ec.europa.eu/eurostat/web/interactive-publications/energy-2023

Florio C, Fiorentino G, Corcelli F, Ulgiati S, Dumontet S, Güsewell J, Eltrop L (2019) A life cycle assessment of biomethane production from waste feedstock through different upgrading technologies. Energies 12(4):718. https://doi.org/10.3390/en12040718

Fragiacomo P, Genovese M, Piraino F, Massari F, Boroomandnia M (2024) Analysis of a distributed green hydrogen infrastructure designed to support the sustainable mobility of a heavy-duty fleet. Int J Hydrogen Energy 51(1):576–594. https://doi.org/10.1016/j.ijhydene.2023.08.047

Gibon T, Arvesen A, Hertwich EG (2017) Life cycle assessment demonstrates environmental co-benefits and trade-offs of low-carbon electricity supply options. Renew Sustain Energy Rev 76:1283–1290

GreenDelta (n.d.) Nexus—openLCA databases. Nexus. Retrieved 16 Jan 2025, from https://nexus.openlca.org/databases

Guinée JB, Heijungs R, Huppes G, Zamagni A, Masoni P, Buonamici R, Ekvall T, Rydberg T (2011) Life cycle assessment: past, present, and future. Environ Sci Technol 45(1):90–96. https://doi.org/10.1021/es101316v

Hassan Q, Algburi S, Sameen AZ, Salman HM, Jaszczur M (2024) Green hydrogen: a pathway to a sustainable energy future. Int J Hydrogen Energy 50(1):310–333. https://doi.org/10.1016/j.ijhydene.2023.08.321

Haug P, Kreitz B, Koj M, Turek T (2017) Process modelling of an alkaline water electrolyzer. Int J Hydrogen Energy 42(15):15689–15707

Hermesmann M, Müller T (2022) Green, turquoise, blue, or grey? Environmentally friendly hydrogen production in transforming energy systems. Prog Energy Combust Sci 90:100996. https://doi.org/10.1016/j.pecs.2022.100996

International Organization for Standardization (2006) ISO 14040:2006—environmental management—life cycle assessment—principles and framework, 2nd edn. ISO. https://www.iso.org/standard/37456.html

Nath K, Das D (2011) Modeling and optimization of fermentative hydrogen production. Biores Technol 102(16):8569–8581. https://doi.org/10.1016/j.biortech.2011.03.108

Ocko IB, Hamburg SP (2022) Climate consequences of hydrogen leakage. Atmos Chem Phys Discuss 22:9349–9368. https://pdfs.semanticscholar.org/2901/7f35d70295af32860db77ae41de57823e48a.pdf

Parisi ML, Maranghi S, Vesce L, Sinicropi A, Di Carlo A, Basosi R (2020) Prospective life cycle assessment of third-generation photovoltaics at the pre-industrial scale: a long-term scenario approach. Renew Sustain Energy Rev 121:109703

Pranjal K, Abhijit D, Nasir M, Ratan KD (2024) Freshwater supply for hydrogen production: an underestimated challenge. Int J Hydrogen Energy 78(2024):202–217. https://doi.org/10.1016/j.ijhydene.2024.06.257

Reuß M, Grube T, Robinius M, Preuster P, Wasserscheid P, Stolten D (2017) Seasonal storage and alternative carriers: a flexible hydrogen supply chain model. Appl Energy 200:290–302. https://doi.org/10.1016/j.apenergy.2017.05.050

Rötzer N, Schmidt M (2020) Historical, current, and future energy demand from global copper production and its impact on climate change. Resources 9(4):44. https://doi.org/10.3390/resources9040044

Shen H (2023) Holistic economic and environmental optimization of future energy systems. Retrieved from https://ntnuopen.ntnu.no/ntnu-xmlui/handle/11250/3092852

Shen H, Crespo del Granado P, Santos Jorge R, Löffler K (2024) Environmental and climate impacts of a large-scale deployment of green hydrogen in Europe. Energy Clim Change 5:100133. https://doi.org/10.1016/j.egycc.2024.100133

Teichmann D (2015) Konzeption und Bewertung einer nachhaltigen Energieversorgung auf Basis flüssiger wasserstoffträger (LOHC) (PhD thesis). Technische Fakultät, Friedrich-Alexander-Universität, Erlangen-Nürnberg

Wang A, Jens J, Mavins D, Moultak M, Schimmel M, van der Leun K, Peters D, Buseman M (2021) Analysing future demand, supply, and transport of hydrogen. Hydrogen Knowl Centre. Retrieved from https://www.h2knowledgecentre.com/content/policypaper2256?crawler=redirect&mimetype=application/pdf

Wernet G, Bauer C, Steubing B, Reinhard J, Moreno-Ruiz E, Weidema B (2016) The ecoinvent database version 3 (part I): overview and methodology. Int J Life Cycle Assess 21(9):1218–1230. https://doi.org/10.1007/s11367-016-1087-8

Wulf C, Reuß M, Grube T, Zapp P, Robinius M, Hake J-F, Stolten D (2018) Life cycle assessment of hydrogen transport and distribution options. J Clean Prod 199:431–443. https://doi.org/10.1016/j.jclepro.2018.07.180

Yue M, Lambert H, Pahon E, Roche R, Jemei S, Hissel D (2021) Hydrogen energy systems: a critical review of technologies, applications, trends and challenges. Renew Sustain Energy Rev 146:111180. https://doi.org/10.1016/j.rser.2021.111180

MIX
Papier aus verantwortungsvollen Quellen
Paper from responsible sources
FSC® C105338

If you have any concerns about our products,
you can contact us on
ProductSafety@springernature.com

In case Publisher is established outside the EU,
the EU authorized representative is:
Springer Nature Customer Service Center GmbH
Europaplatz 3, 69115 Heidelberg, Germany

Printed by Libri Plureos GmbH
in Hamburg, Germany